Published in the United States of America by
IGI Global
Engineering Science Reference (an imprint of IGI Global)
701 E. Chocolate Avenue
Hershey PA, USA 17033
Tel: 717-533-8845
Fax: 717-533-8661
E-mail: cust@igi-global.com
Web site: http://www.igi-global.com

Library of Congress Cataloging-in-Publication Data

Names: Mellal, Mohamed Arezki, editor.
Title: Trends, paradigms, and advances in mechatronics engineering / Mohamed Arezki Mellal, editor.
Description: Hershey, PA : Engineering Science Reference, [2023] | Includes bibliographical references and index. | Summary: "This book investigates numerous issues of mechatronics to include mechanics, electronics, automation and informatics in the design of new more efficient systems"-- Provided by publisher.
Identifiers: LCCN 2022039958 (print) | LCCN 2022039959 (ebook) | ISBN 9781668458877 (h/c) | ISBN 9781668458884 (s/c) | ISBN 9781668458891 (eISBN)
Subjects: LCSH: Mechatronics.
Classification: LCC TJ163.12 .T74 2023 (print) | LCC TJ163.12 (ebook) | DDC 621--dc23/eng/20221012
LC record available at https://lccn.loc.gov/2022039958
LC ebook record available at https://lccn.loc.gov/2022039959

This book is published in the IGI Global book series Advances in Mechatronics and Mechanical Engineering (AMME) (ISSN: 2328-8205; eISSN: 2328-823X)

British Cataloguing in Publication Data
A Cataloguing in Publication record for this book is available from the British Library.

Trends, Paradigms, and Advances in Mechatronics Engineering

Mohamed Arezki Mellal
M'Hamed Bougara University, Algeria

A volume in the Advances in Mechatronics and Mechanical Engineering (AMME) Book Series

IGI Global
PUBLISHER of TIMELY KNOWLEDGE

Advances in Mechatronics and Mechanical Engineering (AMME) Book Series

J. Paulo Davim
University of Aveiro, Portugal

ISSN:2328-8205
EISSN:2328-823X

MISSION

With its aid in the creation of smartphones, cars, medical imaging devices, and manufacturing tools, the mechatronics engineering field is in high demand. Mechatronics aims to combine the principles of mechanical, computer, and electrical engineering together to bridge the gap of communication between the different disciplines.

The **Advances in Mechatronics and Mechanical Engineering (AMME) Book Series** provides innovative research and practical developments in the field of mechatronics and mechanical engineering. This series covers a wide variety of application areas in electrical engineering, mechanical engineering, computer and software engineering; essential for academics, practitioners, researchers, and industry leaders.

COVERAGE

- Control Methodologies
- Design and Manufacture
- Manufacturing Methodologies
- Autonomous Systems
- Computer-Based Manufacturing
- Tribology and surface engineering
- Biologically Inspired Robotics
- Medical Robotics
- Sustainable and green manufacturing
- Vibration and acoustics

The Advances in Mechatronics and Mechanical Engineering (AMME) Book Series (ISSN 2328-8205) is published by IGI Global, 701 E. Chocolate Avenue, Hershey, PA 17033-1240, USA, www.igi-global.com. This series is composed of titles available for purchase individually; each title is edited to be contextually exclusive from any other title within the series. For pricing and ordering information please visit http://www.igi-global.com/book-series/advances-mechatronics-mechanical-engineering/73808. Postmaster: Send all address changes to above address.

Titles in this Series

For a list of additional titles in this series, please visit: http://www.igi-global.com/book-series/

Handbook of Research on Aspects and Applications of Incompressible and Compressible Aerodynamics
Sathish K. Kumar (Nehru Institute of Engineering and Technology, India) and Naren Shankar Radhakrishnan (Vel Tech Rangarajan Dr. Sagunthala R&D Institute of Science and Technology, India)
Engineering Science Reference • © 2022 • 429pp • H/C (ISBN: 9781668442302) • US $335.00

Advances in Computational Approaches in Biomechanics
Pritam Pain (Haldia Institute of Technology, India) Sreerup Banerjee (Haldia Institute of Technology, India) and Goutam Kumar Bose (Haldia Institute of Technology, India)
Engineering Science Reference • © 2022 • 306pp • H/C (ISBN: 9781799890782) • US $260.00

Developing Charging Infrastructure and Technologies for Electric Vehicles
Mohammad Saad Alam (Aligarh Muslim University, Aligarh, India) Reji Kumar Pillai (India Smart Grid Forum, India) and N. Murugesan (India Smart Grid Forum, India)
Engineering Science Reference • © 2022 • 343pp • H/C (ISBN: 9781799868583) • US $240.00

Multidisciplinary Perspectives on Green Electromobility and Charging Stations
Yu Miao (East China University of Sciencer and Technology, China)
Engineering Science Reference • © 2022 • 305pp • H/C (ISBN: 9781799839972) • US $225.00

Electric Vehicles and the Future of Energy Efficient Transportation
Umashankar Subramaniam (Prince Sultan University, Saudi Arabia) Sheldon S. Williamson (Ontario Tech University, Canada) Mohan Krishna S. (Alliance University, India) and Febin Daya J. L. (Vellore Institute of Technology, India)
Engineering Science Reference • © 2021 • 293pp • H/C (ISBN: 9781799876267) • US $240.00

IGI Global
PUBLISHER of TIMELY KNOWLEDGE

701 East Chocolate Avenue, Hershey, PA 17033, USA
Tel: 717-533-8845 x100 • Fax: 717-533-8661
E-Mail: cust@igi-global.com • www.igi-global.com

List of Reviewers

Table of Contents

Detailed Table of Contents

Tuning the PID (proportional-integral-derivative) controller is one of the most important tasks to achieve accurate control. Many methods have been developed to tune the values of P, I, and D; however, classical methods showed limitations. During the last decade, many works addressed the problem as an optimization problem and showed that nature-inspired optimization techniques are prevalent, such as genetic algorithms and particle swarm optimization. This chapter provides a brief overview of applying nature-inspired optimization techniques to PID control optimization.

During the last decades, the word 'mechatronics' has been widely cited. A mechatronic system is a multidisciplinary system requiring metrics to be accounted at the design, and uses stages in order to improve its dependability and achieve high performance. The latter involves several elements that should be considered, such as reliability, availability, maintainability, safety, and cost (RAMS&C). This chapter provides an introduction to various elements of dependability and reliability calculation based on system configuration.

We live in the common Earth where the sustainable development depends on everybody's habit. To set up awareness, it is necessary to integrate this matter right from school teaching until the high education. An engineering education, concerning mechatronics in general and robotics in detail, will help to strengthen the skill and knowledge of students to find reasonable solutions for this problem. This chapter will contribute some experiences in the engineering education technology in university to show how the students can learn the robotics and apply their knowledge to design the machines with the software and simulation tools. All these activities are under the CDIO method - the modern approach in the engineering education. The research results illustrated in this chapter will clarify this concept.

Mechatronics engineering is a branch of engineering made by combining several other main branches as well as derived branches of engineering, which comprises mechanical engineering, electrical engineering, electronic engineering, computer science engineering, product engineering, robotics, and so on. In the initial development of mechatronics engineering, only mechanics as well as electronics were combined to form it, but as the time passed and with the help of technological advancements, it was expanded to combine other derived and subbranches of engineering. In this book chapter, mechatronics engineering is briefly explained, and how it originated as well as came to current level. It also highlights the role of mechatronics engineering in the construction sector, as well as how it transformed the current construction practices. It also briefly talks about building automation, CAD, and structural dynamics which are a form of applied mechatronics that help in the modern construction industry in its own ways.

Flexible manipulator real-time vibration control methods are effective, but finding the right control gain is difficult. The reason for this is that traditional approaches are not permitted by up-to-date and novel architectures. Current population-based

In this chapter, CFD analysis is performed to understand the aerodynamic characteristics of 3-D Entomopter wing. Entomopter is the MAV which is especially designed for Martian atmosphere exploration. It is a low Reynolds number flyer due to which it can reach to the places where human intervention is not at all possible. The aerodynamic simulations are carried out at different flap angles varying from -20° to 30° in Martian atmosphere. The Entomopter model considered for CFD analysis has a wingspan of 15 cm with varying chord length and Reynolds number ranging from Re=75000 Re=150000 i.e., at a velocity of 2 m/s to 30 m/s. The Entomopter 3-D printed model is designed in CATIA V5 and simulation is carried out using ICEM CFD and ANSYS FLUENT. Entomopters may undertake multimode operations and produce lift by continuously forming and shedding vortices from their flapping wings. So, this research is a viable aspect to understand the future development of MAVs for outer space exploration and can be promising for making impossible things possible.

This paper presents the fabrication and Stabilization of a low-cost in-house developed Rotary Inverted Pendulum Setup (STRIPS 1.0), commonly known as a Furuta Pendulum. The developed setup consists of a base frame on which an arm is mounted to allow rotary motion in the horizontal plane and a pendulum link subsequently mounted on the other end of the arm to allow rotation in the vertical plane. Essentially the setup possesses 2- DOF (Degrees of Freedom); however, since the actuation is provided at the arm axis only, the setup behaves as an under-actuated one. Owing to the inherently unstable nature of these under-actuated systems, full-state feedback and an LQR (Linear Quadratic Regulator) controller are designed and implemented, which maintain the pendulum link in the upright position even if disturbed about its linear region of control. In particular, a set of gains are chosen for a particular performance criterion and implemented using a MATLAB interface with the micro-controller on the developed setup, allowing a restricted arm movement.

Preface

In the last decades, most mechanical systems included electronic parts and the word mechatronics has emerged in most systems. The word Mechatronics was created by Tetsuro Mori, in 1969, in an official document in 1971, at the Yaskawa Electric Corporation company (Bishop, 2005; Mori, 1969). Mechatronics is composed of "mecha" referring to "mechanics" and "tronics" referring to "Electronics." In practice, mechatronics combines the fields of mechanics, electronics, control, and computers with their various subfields. A mechatronic system may be considered a complex system since multiple disciplines are involved. Searching for the word mechatronics in Google displays 54,200,000 results (November 2, 2022). Nowadays, many books, articles, journals, conferences, and research laboratories are devoted to mechatronics. Moreover, mechatronics has micro and nano sizes, i.e., Micro/Nano-mechatronics. The applications of mechatronics are very diverse, such as medicine (Dargahi, Arbatani, Sokhanvar, Xie, & Ramezanifard, 2014; Liang, Lin, Wu, & Chen, 2021; Rox et al., 2020), manufacturing (Ding et al., 2020; Hong et al., 2014; Perez, Landers, & Bristow, 2022), automotive (Galluzzi, Circosta, Amati, & Tonoli, 2021; Yu, Cheng, Evangelou, & Dini, 2022), aerospace (Keck, Schwarz, Meurer, Heckmann, & Grether, 2021), and unmanned aerial vehicle (Lee et al., 2012; Shao, Chen, Hou, & Zhao, 2021).

THE CHALLENGES

A mechatronic system includes actuators, such as motors, sensors, such as accelerometers, and software, such as C programming. Most often, an actuator itself is a mechatronic actuator. A mechatronic system facilitates life and industry; however, it should be designed, modeled, simulated, controlled, tested, and dependable. The mechanic part may involve materials, statics, kinematics, and dynamics of solids and fluids. The electronic part may involve components, signals, power, digital, and analog. The control part may involve mathematical models and controllers. The computer part may involve programming and artificial intelligence (AI). Measurement is very important in mechatronics. The accuracy of the measurement allows the mechatronic system to be accurate. Another point is related to the dependability of mechatronics systems due to many components and disciplines (Mellal, 2020). Therefore, the challenges are reliability, availability, maintainability, safety, and cost (RAMS+C). In addition, some metrics should be considered, such as confidentiality and resilience, notably in some domains. Ebert-Uphoff (2000) highlighted the importance of mechatronics in education for the present century. Turmanidze, Dašić, and Popkhadze (2020) discussed the role of mechatronics in Industry 4.0. Mechatronics has drastically evolved but it should be noted that mechatronic systems are highly impacted by obsolescence (Mellal, 2020).

SEARCHING FOR A SOLUTION

Several authors proposed solutions to mechatronic systems. Many modeling tools and optimization techniques were used for this purpose. Razi, Mokogwu, and Hashtrudi-Zaad (2022) used Möbius transformation to improve the stability. Liao et al. (2022) addressed the torque control of motor-driven exoskeletons. A mechatronic system involving electroactive polymer actuators has been investigated in Mutlu, Alici, and Li (2016). Jiang, Liu, Cui, and Jiang (2022) used rapidly exploring random tree for dynamic obstacle avoidance. Margetts and Ngwompo (2015) and Mellal (2011) modeled mechatronic systems with bond graphs. Mellal and Zio (2022); Mellal, Frik, and Boutiche (2021); and Mellal and Salhi (2021) improved the dependability by resorting to nature-inspired optimization techniques. These algorithms are particularly efficient to optimize such kinds of systems. They allow for finding optimal solutions in a reasonable practical time. Zhang et al. (2022) optimized the mass stabilization by a model predictive control. Fault-tolerant control is the most important element to be considered. Baimukashev, Rakhim, Rubagotti, and Varol (2022) addressed it using a new algorithm. Yury and Abramov (2011) used artificial neural networks to control a mechatronic machining center. Kulunk, Cakiroglu, Sahin, and Esen (2016) designed an actuator based on improved spring load. Song and Shen (2019) searched for a solution for smart home systems using mechatronics. Bazavan, Roibu, and George Bizdoaca (2022) and Lidia-Cristina, Horatiu, and Nicu-George (2021) introduced mechatronics to VirtuSphere for better design. A cloud E-learning management system for the teaching and learning process in mechatronics was investigated in Raheem and Ibrahim (2022). Shen et al. (2023) enhanced the vibration suppression performance using a fractional-order electrical network. An approved approach based on hybrid additive manufacturing/pick-n-place robotic was proposed in Kosmal et al. (2022) to improve the design. Rea and Ottaviano (2020) developed low-cost control systems using motor impairments.

ORGANIZATION OF THE BOOK

Trends, Paradigms, and Advances in Mechatronics Engineering presents the latest advances and applications of mechatronics. It highlights the recent challenges in the field and facilitates understanding of the subject. Covering topics such as the construction industry, design optimization, and low-cost fabrication, this premier reference source is a crucial resource for engineers, computer scientists, construction managers, students and educators of higher education, librarians, researchers, and academicians.

The book is organized into 13 chapters. A brief description of each of the chapters follows:

Chapter 1 provides a brief overview of PID (Proportional-Integral-Derivative) controllers. The meanings of control and different methods of PID tuning are illustrated.

Chapter 2 presents metrics of system dependability that are very important in mechatronics. The main elements of dependability are defined: reliability, availability, maintainability, safety, and cost (RAMS&C). It also provides an introduction to the basic calculation concepts of reliability depending on the system configuration.

Chapter 3 contributes some experiences in engineering education technology at university to show how the students can learn robotics and apply their knowledge to design the machines with the software and simulation tools. All these activities are under the CDIO method - the modern approach in engineering education. The own research results illustrated in this chapter will clarify this concept.

Chapter 4 explains mechatronics engineering and how it originated as well as came to the current level. It also highlights the role of mechatronics engineering in the construction sector, as well as how it transformed the current construction practices. It also briefly talks about building automation, CAD, and structural dynamics which are a form of applied mechatronics that help in the modern construction industry in its own ways.

Chapter 5 uses the Coronavirus Herd Immunity Optimization (CHIO) method, inspired by the herd immunity mechanism, which is a COVID-19 control method, for the optimization of flexible manipulator control gains. Gray-Wolf-Optimizer (GWO), another up-to-date population-based algorithm, and traditional Particle Swarm Optimizer (PSO) were used to compare the success of the method. The findings reveal that when it comes to optimizing the vibration controller gains of flexible manipulators, CHIO can outperform its contemporary and traditional competitors.

Chapter 6 performs the design optimization for a single-link flexible manipulator driven by a standard trapezoidal velocity input. The mass of the original manipulator is reduced by a quarter by design optimization without changing the payload. The finite element model, which is one of the non-rigid models, and topology optimization were utilized for the application. Dynamic stress and natural frequencies of the system are utilized as optimization objectives. The results were compared with the original manipulator in terms of both safety factor and vibration modes. In addition, the system parameters were compared with the tapered beam and beam with reduced width, which was prepared to have the same mass as the optimized designs. As a result, although the mass of the original manipulator was reduced by one-fourth with the proposed design, it was observed that the dynamic stresses decreased. The study is expected to have significant implications in terms of improving the benefits of flexible robots and providing a contribution to the attenuation of vibrations and stresses.

Chapter 7 develops the methodological basis for solving the problems of assessment, control, and formation of dynamic states of technical objects of transport and technological purposes in a state of vibration loading. Within the framework of structural mathematical modeling, mechanical oscillatory systems used as design schemes of technical objects are compared with schemes of dynamically equivalent automatic control systems. New results are presented in the field of technology for evaluating dynamic states and forms of interactions of elements of mechanical oscillatory systems with reduced elastic characteristics. For mechanical oscillatory systems, an interpretation of dynamic states and forms of interactions in the form of an oriented graph has been developed, creating prerequisites for the classification of a set of dynamic states in which a system can be located under the condition of various types of force excitation.

Chapter 8 introduces the design process of the wheeled climbing robot for welding inspection in the shipbuilding industry.

Chapter 9 uses the multi-criteria optimization technique TOPSIS (Technique for Order of Preference by Similarity to Ideal Solution) implementation procedures. The flowchart and result interpretation have been demonstrated to optimize or predict the best micromachining process parameters. A case study on the application of the TOPSIS technique in electrical discharge machining (EDM)) has been illustrated.

Chapter 10 performs CFD analysis to understand the aerodynamic characteristics of the 3-D Entomopter wing. Entomopter is the MAV that is specially designed for Martian atmosphere exploration. It is a low Reynolds number flyer due to which it can reach the places where human intervention is not at all possible. The aerodynamic simulations are carried out at different flap angles varying from -20° to 30° in the Martian atmosphere. The Entomopter model considered for CFD analysis has a wingspan of 15 cm with varying chord length and Reynolds number ranging from Re=75000 Re=150000 i.e., at a velocity of 2 m/s to 30 m/s. The Entomopter 3-D printed model is designed in CATIA V5 and simulation is carried out using ICEM CFD and ANSYS FLUENT. Entomopters may undertake multimode operations and produce lift by continuously forming and shedding vortices from their flapping wings. So, this research is a viable aspect to understand the future development of MAVs for outer space exploration and can be promising for making impossible things possible.

Chapter 11 presents the fabrication and Stabilization of a low-cost in-house developed Rotary Inverted Pendulum Setup (STRIPS 1.0), commonly known as a Furuta Pendulum. The developed setup consists of a base frame on which an arm is mounted to allow rotary motion in the horizontal plane and a pendulum link subsequently mounted on the other end of the arm to allow rotation in the vertical plane. Essentially the setup possesses 2- DOF (Degrees of Freedom); however, since the actuation is provided at the arm axis only, the setup behaves as an under-actuated one. Owing to the inherently unstable nature of these under-actuated systems, full-state feedback and an LQR (Linear Quadratic Regulator) controller are designed and implemented, which maintains the pendulum link in the upright position even if disturbed about its linear region of control. In particular, a set of gains are chosen for a particular performance criterion and implemented using a MATLAB interface with the microcontroller on the developed setup, allowing restricted arm movement.

Chapter 12 covers the implementation procedures of hydroponics, aeroponics, and aquaponics technologies in the real world; various inert and artificial mediums, possible types of plants; cultivating methods; technological advancements, and application of the Internet of Things (IoT); and pros and cons for implementations. The various cultivation methods are compared to conventional methods (soil-based cultivation) based on their technology advancements, environmental impacts, yields, and costs.

Chapter 13 highlights that the relationship between input and output parameters in wastewater treatment processes (WWTP) which is complicated, and it is difficult for designing models using statistics. Artificial Intelligence (AI) models are generally more flexible when compared with statistical models while modeling complex datasets with nonlinearity and missing data. Studies on WWTP of AI-based are increasing day by day. Therefore, it is crucial to review the AI techniques available which are implemented for WWTP. Such kind of review helps for classifying the techniques that are invented and helps to identify challenges as well as gaps for future studies. Lastly, can sort out the best AI technique to design predictive models for WWTPs.

Mohamed Arezki Mellal
M'Hamed Bougara University, Algeria

REFERENCES

Baimukashev, D., Rakhim, B., Rubagotti, M., & Varol, H. A. (2022). End-to-end deep fault-tolerant control. *IEEE/ASME Transactions on Mechatronics*, *27*(4), 2224–2234. doi:10.1109/TMECH.2021.3100150

Bazavan, L. C., Roibu, H., & George Bizdoaca, N. (2022). Strategy control of drive mechatronic system for Virtusphere with two actuators. *2022 11th International Conference on Modern Circuits and Systems Technologies, MOCAST 2022*. 10.1109/MOCAST54814.2022.9837781

Bishop, R. H. (2005). *Mechatronics: An Introduction*. Taylor & Francis Group. doi:10.1201/9781420037241.ch1

Dargahi, J., Arbatani, S., Sokhanvar, S., Xie, W. F., & Ramezanifard, R. (2014). A novel tactile softness display for minimally invasive surgery. *Mechatronics*, *24*(8), 1144–1156. doi:10.1016/j.mechatronics.2014.09.002

Ding, H., Gao, R. X., Isaksson, A. J., Landers, R. G., Parisini, T., & Yuan, Y. (2020). State of AI-based monitoring in smart manufacturing and introduction to focused section. *IEEE/ASME Transactions on Mechatronics*, *25*(5), 2143–2154. doi:10.1109/TMECH.2020.3022983

Ebert-Uphoff, I., Gardner, J. F., Murray, W. R., & Perez, R. (2000). Preparing for the next century: The state of mechatronics education. *IEEE/ASME Transactions on Mechatronics*, *5*(2), 226–227. doi:10.1109/3516.847096

Galluzzi, R., Circosta, S., Amati, N., & Tonoli, A. (2021). Rotary regenerative shock absorbers for automotive suspensions. *Mechatronics*, *77*, 102580. doi:10.1016/j.mechatronics.2021.102580

Hong, J., Shin, J. H., La, J., Koh, H. S., Lee, C., & Kim, S. W. (2014). *Design optimization for high manufacturability of optical fiber endcap for range detection LADAR*. doi:10.1080/15599612.2014.915602

Jiang, L., Liu, S., Cui, Y., & Jiang, H. (2022). Path planning for robotic manipulator in complex multi-obstacle environment based on improved_RRT. *IEEE/ASME Transactions on Mechatronics*, 1–12. Advance online publication. doi:10.1109/TMECH.2022.3165845

Keck, A., Schwarz, C., Meurer, T., Heckmann, A., & Grether, G. (2021). Estimating the wheel lateral position of a mechatronic railway running gear with nonlinear wheel–rail geometry. *Mechatronics*, *73*, 102457. doi:10.1016/j.mechatronics.2020.102457

Kosmal, T., Beaumont, K., Link, E., Phillips, D., Pulling, C., Wotton, H., Kudrna, C., Kubalak, J., & Williams, C. (2022). Hybrid additive robotic workcell for autonomous fabrication of mechatronic systems - A case study of drone fabrication. *Additive Manufacturing Letters, 3*, 100100. doi:10.1016/j.addlet.2022.100100

Kulunk, Z., Cakiroglu, S., Sahin, M., & Esen, Z. (2016). Designing spring load system for mechatronic actuator. *Proceedings - 2016 10th International Conference on Compatibility, Power Electronics and Power Engineering, CPE-POWERENG 2016*, 276–279. 10.1109/CPE.2016.7544199

Lee, J., Yoo, C., Park, Y. S., Park, B., Lee, S. J., Gweon, D. G., & Chang, P. H. (2012). An experimental study on time delay control of actuation system of tilt rotor unmanned aerial vehicle. *Mechatronics, 22*(2), 184–194. doi:10.1016/j.mechatronics.2012.01.005

Liang, J. B., Lin, W. P., Wu, R., & Chen, C. (2021). Correlation of accommodation and lens location with higher-order aberrations and axial length elongation during orthokeratology lens wear. *International Journal of Optomechatronics, 15*(1), 1–9. doi:10.1080/15599612.2020.1857889

Liao, H., Chan, H. H. T., Gao, F., Zhao, X., Liu, G., & Liao, W. H. (2022). Proxy-based torque control of motor-driven exoskeletons for safe and compliant human-exoskeleton interaction. *Mechatronics, 88*, 102906. doi:10.1016/j.mechatronics.2022.102906

Lidia-Cristina, B., Horatiu, R., & Nicu-George, B. (2021). Design of improved mechatronic system for Virtusphere. *2021 International Conference on Applied and Theoretical Electricity, ICATE 2021 - Proceedings*. 10.1109/ICATE49685.2021.9464938

Margetts, R., & Ngwompo, R. F. (2015). Mode switching in causally dynamic hybrid bond graphs. *Mechatronics, 28*, 96–102. doi:10.1016/j.mechatronics.2015.04.013

Mellal, M. A. (2020). *Soft Computing Methods for System Dependability*. IGI Global. doi:10.4018/978-1-7998-1718-5

Mellal, M. A., & Zio, E. (2022). Multi-objective availability and cost optimization by PSO and COA for series-parallel systems with subsystems failure dependencies. *Microprocessors and Microsystems, 89*, 104422.

Mellal, M. A. (2020). Obsolescence – A review of the literature. *Technology in Society, 101347*. Advance online publication. doi:10.1016/j.techsoc.2020.101347

Mellal, M. A., Adjerid, S., & Benazzouz, D. (2011). Modeling and Simulation of Mechatronic System to Integrated Design of Supervision: Using a Bond Graph Approach. *Applied Mechanics and Materials*, *86*, 467–470. . doi:10.4028/www.scientific.net/AMM.86.467

Mellal, M. A., Frik, A., & Boutiche, R. (2021). Reliability optimization of power plant safety system using grey wolf optimizer and shuffled frog-leaping algorithm. In Nature-Inspired Computing Paradigms in Systems: Reliability, Availability, Maintainability, Safety and Cost (RAMS+C) and Prognostics and Health Management (PHM) (pp. 1–13). doi:10.1016/B978-0-12-823749-6.00008-8

Mellal, M. A., & Salhi, A. (2021). Multi-objective System Design Optimization via PPA and a Fuzzy Method. *International Journal of Fuzzy Systems*, *23*(5), 1213–1221. doi:10.100740815-021-01068-z

Mori, T. (1969). Mechatronics. Academic Press.

Mutlu, R., Alici, G., & Li, W. (2016). A soft mechatronic microstage mechanism based on electroactive polymer actuators. *IEEE/ASME Transactions on Mechatronics*, *21*(3), 1467–1478. doi:10.1109/TMECH.2015.2502597

Perez, E., Landers, R., & Bristow, D. (2022). Optimal trajectory smoothing for direct energy deposition additive manufacturing processes. *Mechatronics*, *88*, 102910. doi:10.1016/j.mechatronics.2022.102910

Raheem, S. A., & Ibrahim, A. A. (2022). Effects of CLEM cloud based E-learning management system for teaching and learning process in mechatronics. *Optik (Stuttgart)*, *271*, 170048. doi:10.1016/j.ijleo.2022.170048

Razi, K., Mokogwu, C. N., & Hashtrudi-Zaad, K. (2022). Coupled stability analysis of cascaded leader–follower teleoperation networks using Möbius transformation. *Mechatronics*, *88*, 102897. doi:10.1016/j.mechatronics.2022.102897

Rea, P., & Ottaviano, E. (2020). Mechatronic design of low-cost control systems for rehabilitation and assisting devices. In Handbook of Research on Advanced Mechatronic Systems and Intelligent Robotics (pp. 82–97). doi:10.4018/978-1-7998-0137-5.ch004

Rox, M. F., Ropella, D. S., Hendrick, R. J., Blum, E., Naftel, R. P., Bow, H. C., Herrell, S. D., Weaver, K. D., Chambless, L. B., & Webster, R. J. (2020). Mechatronic design of a two-arm concentric tube robot system for rigid neuroendoscopy. *IEEE/ASME Transactions on Mechatronics*, *25*(3), 1432–1443. doi:10.1109/TMECH.2020.2976897 PMID:33746503

Shao, S., Chen, M., Hou, J., & Zhao, Q. (2021). Event-triggered-based discrete-time neural control for a quadrotor UAV using disturbance observer. *IEEE/ASME Transactions on Mechatronics*, *26*(2), 689–699. doi:10.1109/TMECH.2021.3051835

Shen, Y., Hua, J., Fan, W., Liu, Y., Yang, X., & Chen, L. (2023). Optimal design and dynamic performance analysis of a fractional-order electrical network-based vehicle mechatronic ISD suspension. *Mechanical Systems and Signal Processing*, *184*, 109718. doi:10.1016/j.ymssp.2022.109718

Song, Z., & Shen, S. (2019). The application of intelligent and modular development of mechatronics: A case of design of smart home system. *Proceedings - 2019 11th International Conference on Measuring Technology and Mechatronics Automation, ICMTMA 2019*, 216–219. 10.1109/ICMTMA.2019.00055

Turmanidze, R., Dašić, P. V., & Popkhadze, G. (2020). Investigation of operational characteristics of mechatronic systems in Industry 4.0. In Handbook of Research on Integrating Industry 4.0 in Business and Manufacturing (pp. 324–343). doi:10.4018/978-1-7998-2725-2.CH015

Yu, M., Cheng, C., Evangelou, S., & Dini, D. (2022). Series active variable geometry suspension: Full-car prototyping and road testing. *IEEE/ASME Transactions on Mechatronics*, *27*(3), 1332–1344. doi:10.1109/TMECH.2021.3097153

Yury, K., & Abramov, I. V. (2011). Building a knowledge base for intelligent control system of mechatronic machining center. *Proceedings of 14th International Conference on Mechatronics, MECHATRONIKA 2011*, 93–94. 10.1109/MECHATRON.2011.5961073

Zhang, R., Meng, L., Yu, Z., Chen, X., Liu, H., & Huang, Q. (2022). Stride length and stepping duration adjustments based on center of mass stabilization control. *IEEE/ASME Transactions on Mechatronics*, 1–11. Advance online publication. doi:10.1109/TMECH.2022.3159321

Chapter 1
Optimization of PID Controller:
A Brief Overview

Sara Benameur
M'Hamed Bougara University, Algeria

Sara Tadrist
M'Hamed Bougara University, Algeria

Mohamed Arezki Mellal
https://orcid.org/0000-0003-0667-8851
M'Hamed Bougara University, Algeria

Edward J. Williams
University of Michigan, Dearborn, USA

ABSTRACT

Tuning the PID (proportional-integral-derivative) controller is one of the most important tasks to achieve accurate control. Many methods have been developed to tune the values of P, I, and D; however, classical methods showed limitations. During the last decade, many works addressed the problem as an optimization problem and showed that nature-inspired optimization techniques are prevalent, such as genetic algorithms and particle swarm optimization. This chapter provides a brief overview of applying nature-inspired optimization techniques to PID control optimization.

DOI: 10.4018/978-1-6684-5887-7.ch001

INTRODUCTION

The process, a general term referring to a set of devices and intended to obtain a given product, is characterized by means of one or more measurable physical quantities to be mastered which will allow controlling the fixed objective. The objective of a regulation is to ensure the operation of a process according to criteria predefined by a specification that defines qualitative criteria to be imposed that are usually translated by quantitative criteria as stability, precision, and rapidity. The regulators are components intended for specific applications and relations between input and output whose coefficients are definitively fixed to define their regulation functions; however, an industrial regulator is a device whose parameters of its transfer function are adjustable and adaptable to the process to be controlled.

This chapter illustrates a brief literature review of PID controllers. The remainder of the document is organized as follows. Section 2 provides an overview of the control system. Section 3 discusses the performance of a servo system. Section 4 defines the PID controller, and the last section concludes the chapter.

CONTROL SYSTEM

Definition

A control system is a combination of equipment that manages, organizes, directs, or regulates the actions of other devices or systems in order to achieve a specific goal (Bongiorno & Park, 2020; Kondratenko & Chikrii, 2019). A control system does this by using control loops, which are procedures designed to keep a process variable at a specific set point. We can also define it as a system, which controls the output to produce the desired response. A control system involves sensors, actuators, reference input, and the system itself. The process or plant to be controlled is called

Figure 1. Block diagram of a control system

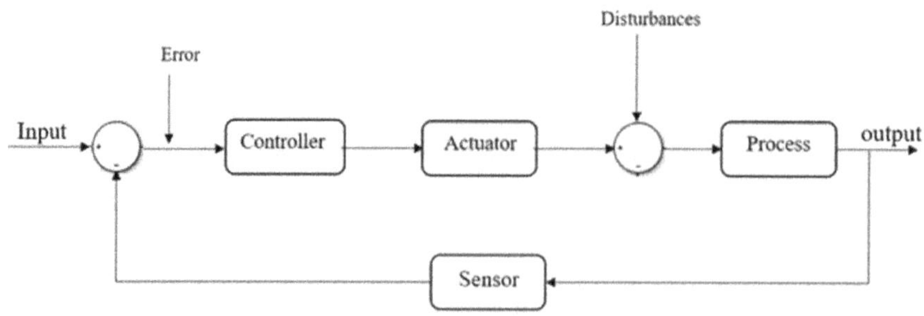

the system, whereas the sensor converts the physical signal to an electrical signal. The actuator converts the control signal to a power signal.

Control systems are divided into two categories:

- *Open Loop Control System:* an open loop control system is a type of control that ignores the reaction of the system. The system, on its own, is unable to produce the desired output and is unable to account for disturbances. Adjust changes in output case in these systems by manually updating the input. In principle, such control is to be used with caution if the system is naturally unstable.

Figure 2. Open-loop control block diagram

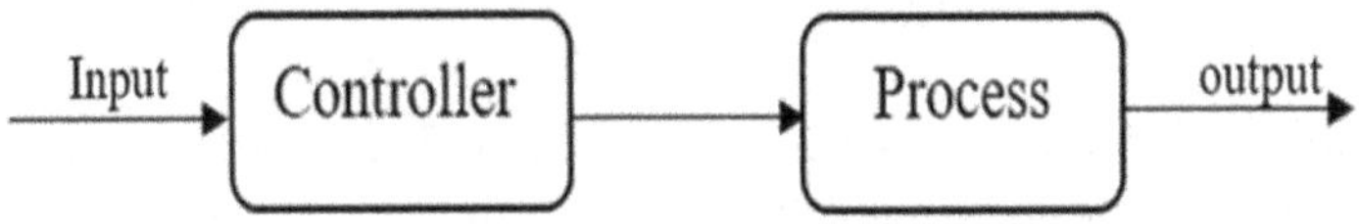

- ***Close Loop Control System:*** in this type, the input depends on the generated output of the system. In another meaning, the response of the system controls the input applied to the system. This control is obtained through the output generated by using a feedback path. The variation in inputs according to the output produces a more accurate system. Therefore, an open-loop control system can be converted into a close loop control system by providing feedback. This system makes suitable changes to keep control despite the disturbance.

Figure 3. Closed-loop control block diagram

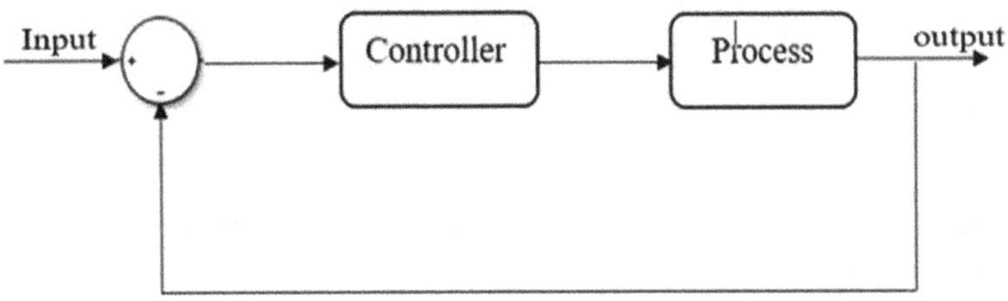

Hardware

Hardware is parts and electronic equipment that make the system operational; the term also refers to all equipment attached to computers (or other products) that require computer management. Hardware is fixed and cannot be altered without replacement or physical modification. It cannot work without the software which are the software part of the computer. The hardware found in a control system are:

- *Sensor:* is an organ for the avoidance of data, which creates from a physical quantity another physical quantity of a different nature (convert conditions of physical devices into signals appropriate for input of control system). This quantity representative of the quantity taken can be used for measurement or control purposes. The new generation sensors involve smart material sensors, micro sensors, and Nano sensors. Sensors can be classified as passive or active:
  - *Active Sensors*: they work as a generator. They consist of a set of power transducers (mechanical stopwatch, train gauge, gyro meter…). Active sensors are generally based on a physical size to be collected, thermal and mechanical or radiation energy. The classic ones are: thermoelectric effect, piezoelectric effect, electromagnetic induction effect, photo-electric effect, Hall effect and photovoltaic effect.
  - *Passive Sensors*: they don't need external energy to work, they generally rely upon impedance of which one of the determinant parameters is sensitive to the measure's quantity. The physical phenomena used influence the variation of this impedance.
- *Actuator:* is an organ designed to act on a process variable (usually a charge) under the control of control signal using auxiliary energy. In the meaning of automation, the actuator belongs to the operative part of an automated system, it's a part of a device or machine that helps it to achieve physical movement by converting energy often electrical, air or hydraulic into mechanical force.
- *Controller:* this part creates a stable closed loop by assessing the difference.

Software

This part includes all programing lines that describes the control system including mathematical models and hardware command.

Laplace Transform

The Laplace transform of $f(t)$, that is denoted by $L\{f(t)\}$ or $F(p)$, is defined by the Laplace transform formula:

$$F\left(p\right)=\int_{0}^{+\infty}f\left(t\right).e^{-pt}.dt \tag{1}$$

Transfer Function

The transfer function is a property of the system elements only. We obtain a transfer function by using the Laplace transform, transforming the differential equation or the mathematical model into an algebraic equation (time domain being transformed to the p –domain).

PERFORMANCE OF A SERVO SYSTEM

The system is generally subject to several basic criteria to prove its effectiveness according to the specification; we mention stability, precision and rapidity.

Stability

The essential quality for a regulated system is stability. A system is said to be stable as long as its response to a step approaches a finite value in steady state or when its response to an impulse which tends towards 0. For an unstable system, it is characterized by the presence of amplitude fluctuations and irreversible growth.

Figure 4. Stable and unstable system

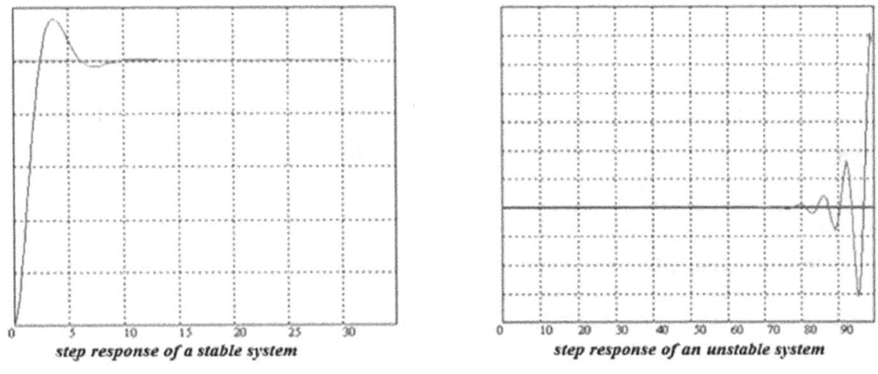

Rapidity

The rapidity in a control system is evaluated between $\pm 5\%$ of response time compared to its final value, this value can only be estimated in the vast majority of cases.

The figure below represents the step response of the system that allows us to see the characteristics of the speed by the response time at 5%:

- *Overshoot:* expressed as a percentage and defined by the maximum value of the output signal reduced to its final value.
- *Rise Time:* defined when the output signal first reaches its final value.
- *pic Time:* defined when the output signal reaches its maximum.
- *Settling Time* is the time it takes for the response to reach a steady state and stay within a specified tolerance around the final value.
- *Steady-State Error:* is the difference between the desired value and the actual value of a system when the response has reached steady state.

Figure 5. Tolerance, criteria of error in a step response

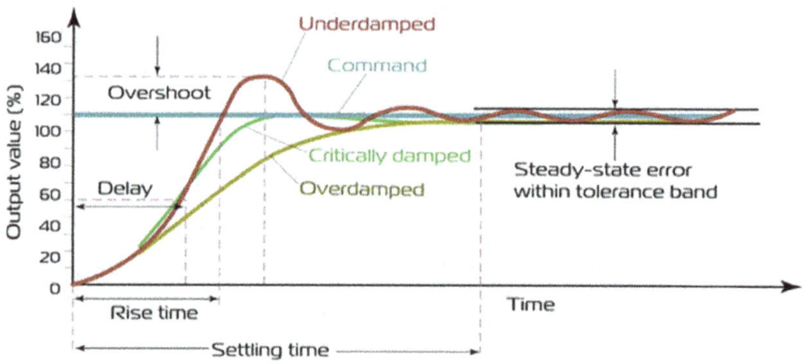

Precision

To have a good dynamic accuracy of a system with a step input, the transient regime must be characterized by a low exceedance and an optimal response time. For this, the parameters of a regulator are chosen in order to minimize the dynamic error, so one of the following criteria must be minimized:

- **Integral Absolute Error (IAE):** is given by:

$$IAE = \int_0^{\infty} \left|e(t)\right|.dt \tag{2}$$

This criterion expresses the area generated by the difference between the set point value and the actual value, and its role is to remove small errors.

- **Integral Square Error (ISE):** is given by:

$$ISE = \int_0^{\infty} e(t)^2.dt \tag{3}$$

The interest of this performance index is to correct systems whose transitional regime lasts too long, and takes much less account of exceedances below 1.

- **Integral Time Absolute Error (ITAE):** is given by:

$$ITAE = \int_0^{\infty} t.\left|e(t)\right|.dt \tag{4}$$

The introduction of the time parameter will correct highly oscillatory response systems.

- **Integral Time Square Error (ITSE):** is given by:

$$ITSE = \int_0^{\infty} t.e(t)^2.dt \tag{5}$$

This criterion places little emphasis on initial errors and heavily penalizes errors that occur towards the end of the transient response to a step entry.

PID CONTROLLER

Definition

The PID (proportional, integral, derivative) regulator is a control system; it improves system performances by adjusting the error between the output, setpoint and the control device based on its comparator. This is the most widely used regulator in industry and even in complex systems where its corrective qualities apply to multiple physical quantities.

Figure 6. System control with a PID controller

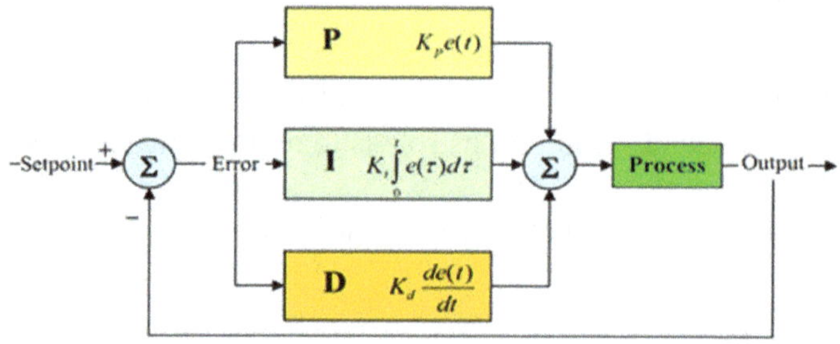

Proportional Action P

The proportional controller reduces the difference between the set point and the output as well as the rise time and the response time. Its disadvantage that it is unable to eliminate steady state errors and has degraded the stability of the system.

- Regulator control law:

$$u(t) = K_p e(t) \tag{6}$$

- Regulator transfer function:

$$G(p) = \frac{U(p)}{E(p)} \tag{7}$$

- Representation by block diagram: (see Figure 7)

Figure 7. Proportional corrector block diagram

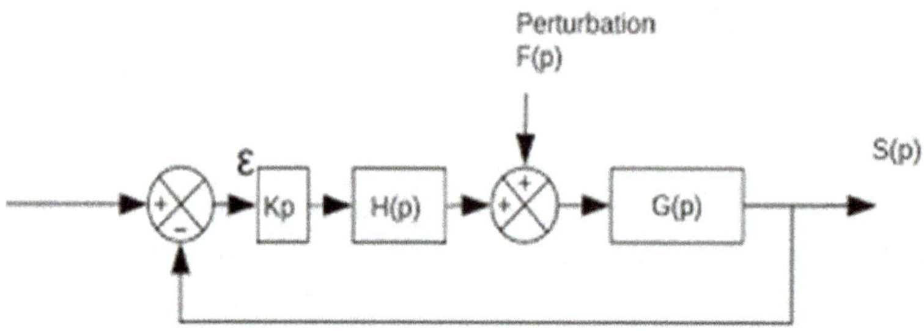

Integral Action I

It is the process of adding the sign of error at each instant and multiplying it by the gain K_i. It works to produce signals based on the accumulation of the error signal.

Integral action can cancel the (static) position error, but it can cause steady-state oscillations and overshoots and even decreases the stability of system.

- Regulator control law:

$$u(t) = \frac{1}{K_p} \int_0^t e(t)\,dt \tag{8}$$

- Regulator transfer function:

$$G(p) = \frac{U(p)}{E(p)} = \frac{1}{K_i p} \tag{9}$$

- Representation by block diagram: (see Figure 8)

Figure 8. Proportional corrector block diagram

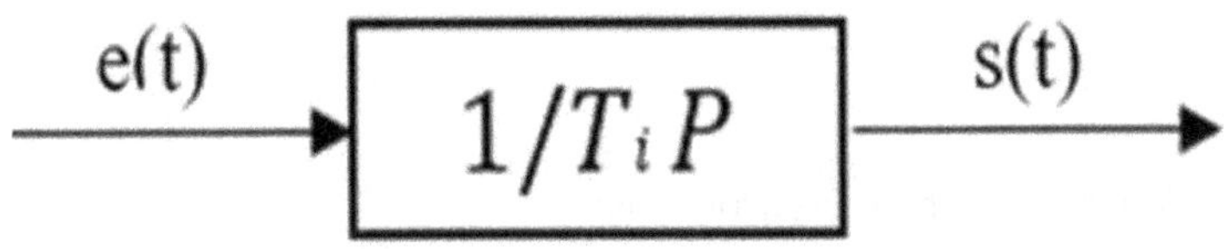

Derivative Action D

The controller D accelerates the response of the system and improves the stability by reducing the oscillations caused by other actions (P and I), it is applied to the variations of the quantity as regulated and not that of the error (set point-response).

- Regulator control law:

$$u(t) = K_d \cdot \frac{de(t)}{dt} \tag{10}$$

- Regulator transfer function:

$$G(p) = \frac{U(p)}{E(p)} = k_d p = T_d p \tag{11}$$

- Representation by block diagram: (see Figure 9)

Figure 9. Proportional corrector block diagram

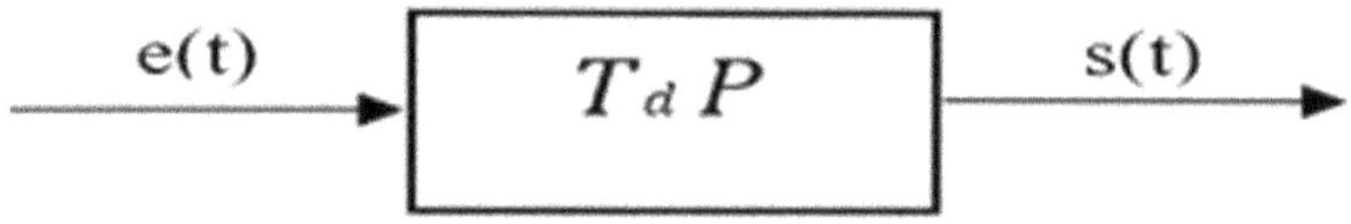

Types of Regulators

Proportional P regulator

The P regulator is the simplest type of regulator since it builds a $u(t)$ control proportional to the error $e(t)$. However, the instability inherent in the oscillations of the control loop increases simultaneously and the control loop can become unstable.

$$U(t) = K_p.e(t) \tag{12}$$

The transfer function of the P controller is:

$$G(p) = K_p \tag{13}$$

Proportional Integral (PI) Regulator

The integral term complements the proportional action as it compensates for static error, improves steady state accuracy; this works like a filter on an integrated signal, reducing the effects of interference (noise, disturbances) to attain a more stable system. Unfortunately, integral terms can also lead to overshooting and slower stabilization of the set point.

$$U(t) = K_p\left[e(t) + \frac{1}{T_i}\int_0^t e(t)dt\right] \tag{14}$$

PI controller transfer function:

$$G(p) = K_p\frac{1+T_ip}{T_iP} \tag{15}$$

Proportional Derivative (PD) Regulator

The PD regulator generally has a small static deviation compared to the pure P regulator. The addition of the derivative term reduced both the overshoot and the settling time and had a negligible effect on the rise time and the steady-state error.

$$U(t) = K_p\left[e(t) + T_d\frac{de(t)}{dt}\right] \tag{16}$$

PD controller transfer function is:

$$G(p) = K_p(1+T_dp) \tag{17}$$

Proportional Integral Derivative (PID) Controller

The PID (Proportional-Integral-Derivative) regulator is the combination of the three basic actions P, I and D. due to the terms P and I it allows the cancellation of a static error. Due to the D action, faster performance than a PI controller.

$$U(t) = K_p\ .e(t) + \frac{1}{T_i}\int_0^t e(t)dt + T_d\frac{de(t)}{dt} \tag{18}$$

The transfer function of a PID controller is:

$$G(p) = K_p + \frac{1}{T_ip} + T_dp \tag{19}$$

Structure

Here we will discuss the most commonly used binding structures to associate actions P, I, and D. These structures are functionally equivalent and it is easy to convert the coefficients used between them. The three most common types of configurations are:

- Series type
- Parallel type
- Mixed type

Table 1. Structure of proportional, integral, derivative (PID) controller

Structures	Equation	Diagram
Series	$u(t) = K_P \times \{\left(1 + \frac{T_d}{T_i}\right) e + \frac{1}{T_i}\int_0^t e \times dt + T_d \times \frac{de}{dt}\}$	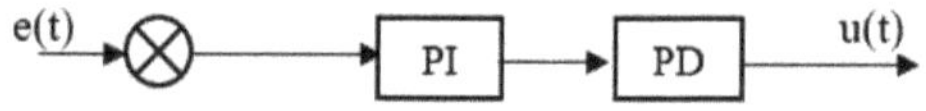
Parallel	$u(t) = K_p \times e(t) + \frac{1}{T_i}\int_0^t e(t)\,dt + T_d \times \frac{de}{dt} + U_0$	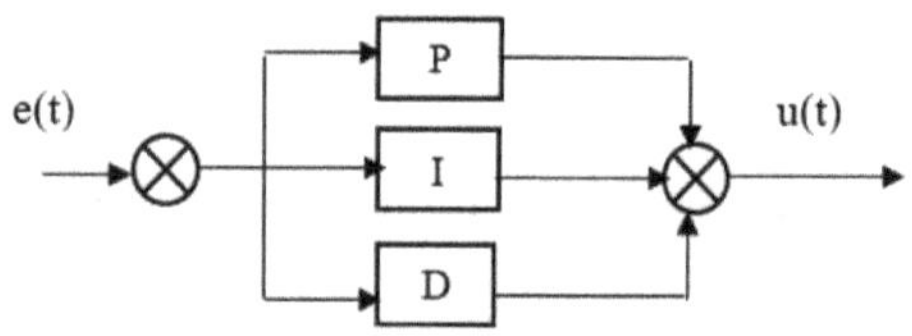
Mixed	$u(t) = K_p e(t) + \frac{K_p}{T_i}\int_0^t e(t)dt + K_p T_i \frac{de(t)}{dt} + u_0$	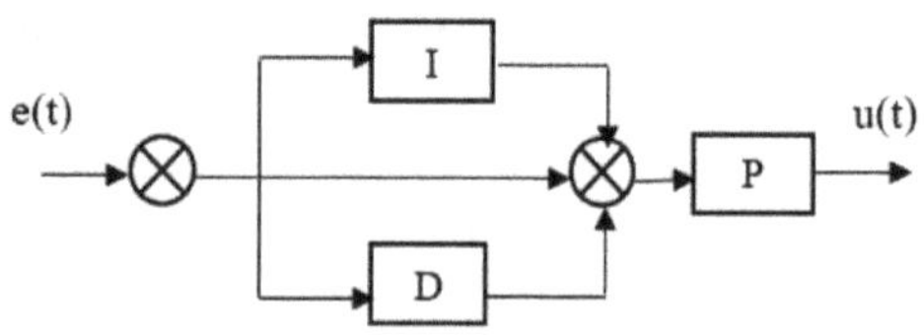

Tuning Methods of PID Controller

Manuel Tuning

Assuming that adequate data is accessible on the interaction to be observed, the ideal PID values can be determined. However, when assessing a complex process, with some data, such as the speed at which it responds to error corrections, it is possible to make fundamental level adjustments.

Manual change happens by changing the fundamental addition to its maximum value and the derivative gain to zero and increasing the proportional gain until the loop oscillates to a constant amplitude. Then, the proportional gain must be halved, and the integral gain adjusted to correct any deviation within an acceptable period. Finally, the derivative gain must be increased until the exceedance is minimized.

Heuristic and Metaheuristic Tuning Methods

The heuristic tuning method is one where general rules are followed to obtain approximate or qualitative results. The majority of PID loops in the world are tuned with such methods such as trial-and-error, Ziegler-Nichols, and optimizations methods (like PSO (Mellal & Zio, 2022)(Copot et al., 2018), GWO (Hamadache & Mellal, 2021; Mellal, Frik, & Boutiche, 2021), ACO (Mellal & Williams, 2018), and GA (Mellal & Williams, 2018)). A review of the literature on PID optimization can be found in Refs. (Borase, Maghade, Sondkar, & Pawar, 2020; George & Ganesan, 2022; Joseph, Dada, Abidemi, Oyewola, & Khammas, 2022). The most useful methods are ZN and optimization methods.

ZN Method

The ZN tuning method provides two different methods: the step response method and the frequency response method (Chidambaram & Saxena, 2018a, 2018b; Ziegler & Nichols, 1993).

- Step response method.
- Frequency response PID.

Optimization Methods

These are the integrated optimization methods use algorithms to develop an optimum solution to problems. The optimization problem is defined as the search for the minimum or maximum (the optimum) of a given function. These include fuzzy logic, neural networks, genetic algorithm and particle cluster optimization. The advantage of these approaches over conventional techniques is robustness and flexibility.

Auto-Tuning

The regulator assimilates how the process responds to a disruption or change in the set point and calculates the appropriate PID parameters. In the case of a temperature controller, when the automatic adjustment function is selected, the controller activates an output value by observing the timing and speed of the change. It then calculates the optimal P, I and D parameter which can later be adjusted manually if necessary (this regulator requires the set point to be at least 10° higher than the current process value of automatic adjustment to take place). Adaptive tuning can be found in Ref. (Åström, Hägglund, Hang, & Ho, 1993).

CONCLUSION

In this chapter, we have given a general idea about PID controllers and the control system. We knew the difference between hardware and software and that the hardware represents the sensor, the actuator and the controller. The action associated with the PID regulator allows optimal regulation by combining the advantages of each action: component "P" reacts to the appearance of a deviation of adjustment, component "D" counters variations in the set size and stabilizes the control loop and the "I" component eliminates the static error. There are many methods used to adjust and optimize the parameters of the PID controller, the ZN method and optimization methods have been mentioned.

REFERENCES

Åström, K. J., Hägglund, T., Hang, C. C., & Ho, W. K. (1993). Automatic tuning and adaptation for PID controllers - a survey. *Control Engineering Practice*, *1*(4), 699–714. doi:10.1016/0967-0661(93)91394-C

Bongiorno, J. J., & Park, K. (2020). Design of linear multivariable feedback control systems. In Design of Linear Multivariable Feedback Control Systems. doi:10.1007/978-3-030-44356-6

Borase, R. P., Maghade, D. K., Sondkar, S. Y., & Pawar, S. N. (2020). A review of PID control, tuning methods and applications. *International Journal of Dynamics and Control*, *9*(2), 818–827. doi:10.100740435-020-00665-4

Chidambaram, M., & Saxena, N. (2018a). Auto-tuning of decentralized unstable system with refined ZN method. In Advances in Industrial Control, (pp. 167–182). doi:10.1007/978-981-10-7727-2_9

Chidambaram, M., & Saxena, N. (2018b). Tuning rules for PID controllers for unstable systems. In Advances in Industrial Control, (pp. 151–165). doi:10.1007/978-981-10-7727-2_8

Copot, C., Mac Thi, T., & Ionescu, C. (2018). PID based particle swarm optimization in offices light control. *IFAC-PapersOnLine*, *51*(4), 382–387. doi:10.1016/j.ifacol.2018.06.096

George, T., & Ganesan, V. (2022). Optimal tuning of PID controller in time delay system: A review on various optimization techniques. *Chemical Product and Process Modeling*, *17*(1), 1–28. doi:10.1515/cppm-2020-2001

Hamadache, I., & Mellal, M. A. (2021). Design optimization of car side safety system by particle swarm optimization and grey wolf optimizer. In M. A. Mellal & G. M. Pecht (Eds.), *Nature-Inspired Computing Paradigms in Systems: Reliability, Availability, Maintainability, Safety and Cost (RAMS+C) and Prognostics and Health Management (PHM)*. Elsevier. doi:10.1016/B978-0-12-823749-6.00006-4

Joseph, S. B., Dada, E. G., Abidemi, A., Oyewola, D. O., & Khammas, B. M. (2022). Metaheuristic algorithms for PID controller parameters tuning: Review, approaches and open problems. *Heliyon*, *8*(5), e09399. doi:10.1016/j.heliyon.2022.e09399 PMID:35600459

Kondratenko, Y. P., & Chikrii, A. A. (2019). Advanced control techniques in complex engineering systems: Theory and applications. In *Studies in Systems*. Decision and Control. doi:10.1007/978-3-030-21927-7

Mellal, M. A. Frik, A., & Boutiche, R. (2021). Reliability optimization of power plant safety system using grey wolf optimizer and shuffled frog-leaping algorithm. In Mohamed Arezki Mellal & M. G. Pecht (Eds.), Nature-Inspired Computing Paradigms in Systems: Reliability, Availability, Maintainability, Safety and Cost (RAMS+C) and Prognostics and Health Management (PHM), (pp. 1–13). doi:10.1016/B978-0-12-823749-6.00008-8

Mellal, M. A., & Williams, E. J. (2018). A survey on ant colony optimization, particle swarm optimization, and cuckoo algorithms. In *Handbook of Research on Emergent Applications of Optimization Algorithms*. IGI Global. doi:10.4018/978-1-5225-2990-3.ch002

Mellal, M. A., & Zio, E. (2022). Multi-objective availability and cost optimization by PSO and COA for series-parallel systems with subsystems failure dependencies. *Microprocessors and Microsystems, 89*, 104422.

Ziegler, J. G., & Nichols, N. B. (1993). Optimum settings for automatic controllers. *Journal of Dynamic Systems, Measurement, and Control, 115*(2B), 220–222. doi:10.1115/1.2899060

Chapter 2
Metrics of System Dependability

Tafsouthe Saheb
M'Hamed Bougara University, Algeria

Mohamed Arezki Mellal
https://orcid.org/0000-0003-0667-8851
M'Hamed Bougara University, Algeria

ABSTRACT

During the last decades, the word 'mechatronics' has been widely cited. A mechatronic system is a multidisciplinary system requiring metrics to be accounted at the design, and uses stages in order to improve its dependability and achieve high performance. The latter involves several elements that should be considered, such as reliability, availability, maintainability, safety, and cost (RAMS&C). This chapter provides an introduction to various elements of dependability and reliability calculation based on system configuration.

INTRODUCTION

System dependability takes a very important place in the industrial design, which is a matter of concern for the last decade due to the growing complexity of modern engineering systems, and in the same way, their susceptibility to fail (Abdulgader, 2019).

Commonly defined as the science of failures proposing a solution to improve the reliability and safety of systems in a timely manner with taking to account numerous constraints, the definition of dependability may vary depending to the

DOI: 10.4018/978-1-6684-5887-7.ch002

field of application (nuclear, automobile, rail, etc.). Jean-Claude Laprie defined dependability as the ability to deliver a service that can justifiably be trusted (Avižienis, Laprie, Randell, & Landwehr, 2004). In addition, the International Electrotechnical Commission standards (IEC) determined dependability as the ability to perform when and as required. It is a domain that requires global knowledge of the system like the condition of use, external risks involving the system, its functional and material architectures, structure and fatigue of materials. The majority of advances are the result of feedback and accident analysis reports. It is crucial when it comes to high-risk systems whose failure leads to catastrophic consequences on the safety of the system itself or user(s) and the environment. It may also increase the repair time and cost, inducing a significant loss of production.

Likewise, dependability includes principal fundamentals of availability, maintainability, reliability, efficiency of maintenance logistics, and in some cases, durability, safety, and security.

METRICS TO ACHIEVE DEPENDABILITY

There are four approaches to attain dependability, and are the following (Laprie, 1995):

- **Fault Prevention** is an approach to enhancing development processes aiming to detect and reduce the number of faults in the system and eliminate the causes via process variations;
- **Fault Tolerance** is when the system still provides the service to satisfy the system function in presence of faults;
- **Fault Removal** is decreasing the presence of faults;
- **Fault Forecasting** is an estimation of the current number, and the possible consequences of the faults.

Metrics of dependability are all the proprieties related to the system to be a dependable one. Again, a system is not dependable in proportion to the level of lowliness of these attributes. Moreover, and in the majority of cases, dependability is not a binary event but grounded on successions and acceptable approaches. These approaches are specific to infrastructures as well as applications. Among the dependability attributes, some need to be enhanced over others in specific system applications (Aqel, 2021).

RAMS

Nowadays, engineers are confronting the ultimatum of high dependable system with the lowest cost of operations. This is where reliability, availability, maintainability and safety simultaneously come in. RAMS is an abbreviation for reliability, availability, maintainability, and safety which are defined as follows (Birolini, 2017):

- **Reliability** is the ability of a functional unit (system) to perform a required function under given conditions for a given time interval ("ISO/IEC 2382-14:1997 Information technology Vocabulary Part 14:," 1997). The reliability is expressed by the function $R(t)=P(E)$ that E is a non-faulty entity on $[0, t]$. Supposing that E is not faulty at time $t=0$:

$$R(t) = P(T>t) \tag{1}$$

where T the random variable designating the moment of failure of the device and P the probability;

- **Availability** is the ability of a functional unit to be in a state to perform a required function under given conditions at a given instant of time or over a given time interval ("ISO/IEC 2382-14:1997 Information technology Vocabulary Part 14:," 1997). Availability is a complementary relationship between reliability, maintainability and maintenance actions performed. Also, the associated probability $A(t)$ at time $t=0$ is expressed as follows:

$$A(t) = P(E) \text{ non faulty at the instant } t \tag{2}$$

- **Maintainability** is the ability of a functional unit, under given conditions of use, to be retained in, or restored to, a state in which it can perform a required function ("ISO/IEC 2382-14:1997 Information technology Vocabulary Part 14:," 1997). The maintenance is divided into preventive and corrective maintenance also into, among other things, scheduled and predictive maintenance. Preventive maintenance is the maintenance carried out before failures occur, and corrective maintenance is the maintenance carried out after failures occur. Preventive maintenance can further be divided into scheduled and condition-based maintenance. Scheduled maintenance is carried out in accordance with an established time schedule or established number of units in use. Condition-based maintenance is preventive maintenance that is based on performance and/or parameter monitoring and subsequent actions (Nilsson & Bertling, 2007). $M(t)$ corresponds to the probability that the restoration of a faulty entity E will be carried out within a time interval, which leads to the following expression (ZWINGELSTEIN, 2019):

$$M(t) = P(E)\ E \text{ is prepared on } [0,t] \quad (3)$$

- **Safety** is the ability of a system to not cause property damage or catastrophic consequences on the environment and humans(Laprie, 1995). Safety matters when the system is correctly working but also when it fails. Thus, it is necessary to take into consideration external influences such as human error or sabotage in order to prevent them.

$$S(t) = P[E \text{ avoids critical events on } [0,t]] \quad (4)$$

RAMS&C

RAMS&C is a shortcut for reliability, availability, maintainability, safety, and cost. The cost of a high level of RAMS is very expensive. The designer must make trade-offs between the dependability mechanisms needed and the costs. Users may reject systems that are unsafe, unreliable, or insecure as well as the cost of failure can be extremely high. combined to artificial intelligence, RAMS&C dealt with RAMS optimization problems with single or multi-objective function constraints in order to minimize system cost under availability constraints (Mellal & Zio, 2018; Mellal & Zio, 2020; Mellal, Zio, & Williams, 2020).

Other Metrics of System Dependability

- **Resilience:** there are different definitions and aspects relating to resilience, mainly depending on the fields of application such as psychology, economics or engineering. A general definition of resilience was raised by United Nations in 2009 as "The ability of a system, community or society which is exposed to hazards in order to resist, absorb, accommodate to and recover from the effects of a hazard in an efficient manner, through the preservation and restoration of its essential basic structures and functions" (Kahan, Allen, & George, 2009), (Mahzarnia, Moghaddam, Baboli, & Siano, 2020). Resilience is primarily concerned with the repercussions of disruptions, rather than the likelihood of their recurrence (Panteli & Mancarella, 2015). It is worth emphasizing that a system's dependability may be assessed without first identifying the risks, whereas resilience is linked to the capacity to deal with one or more specific hazard. To put it another way, a system that is resilient to certain types of events may be vulnerable to others (Chanda & Srivastava, 2016). Resilience has been characterized as a system's capacity to forecast a disaster, endure or absorb it, adjust to its repercussions, and swiftly restore its performance to an acceptable level after such an occurrence (Mahzarnia et al., 2020), (Chanda & Srivastava, 2015). As stressed in (Sansavini, 2017), a resilient system can recover as rapidly as feasible with the least amount of human involvement;

- **Integrity:** When an operating system is designed, implemented, and maintained to defend itself against unauthorized access to the point that security restrictions defined for that system cannot be compromised, it is said to have system integrity. System integrity is ensured via a layered secure trusted computing basis ("System integrity - IBM Documentation," n.d.). The trustworthy computing platform can defend itself against unwanted user access. In information security, the term "integrity" refers to the assurance that data has not been compromised with (Yee & Zolkipli, 2021);
- **Confidentiality:** Confidentiality refers to the activity of keeping a system's data private by limiting access to just those who are permitted. The most frequent security techniques used to maintain confidentiality include classifying data according to its value, ranging from public to top-secret, and calculating the impact of information exposure on organizations or individuals. Confidentiality is important in healthcare facilities, for example, when doctors and their patients exchange via emails, a breach of communication confidentiality might disclose patients' information and lead to serious implications (Crotty & Mostaghimi, 2014).

Figure 1. Metrics of system dependability

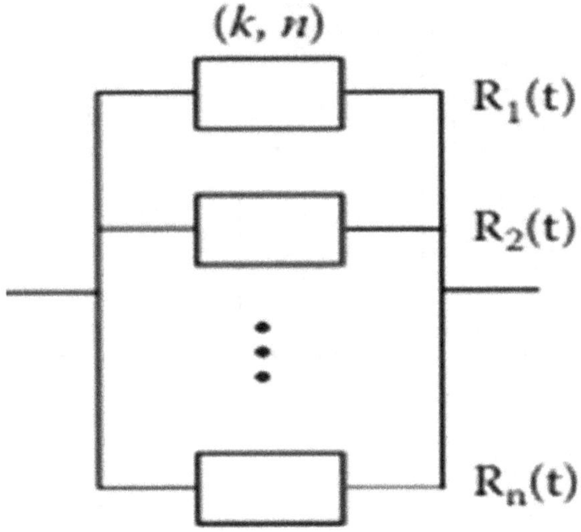

Figure 1 outlines the main principles linked to system dependability after we have gone through them. In this chapter, system reliability is presented.

SYSTEM RELIABILITY

This section describes how to design system reliability based on the configuration of the elements involved in the system. To represent the architecture and reliability modeling of the system, reliability block diagram (RBD) is considered. In general, the reliability block diagram is used to determine the overall system's reliability, if the fundamental component's reliability is available. However, there is a difference between the reliability modeling of simple systems and complex systems.

Reliability of Simple Systems

Series System

If a series system is to operate, all subsystems must be operational. This means that if any subsystem fails, the entire system will fail. In terms of system reliability, the system will be reliable if all of the components are. The components do not have to be physically linked in series to be considered a series system. The reliability of a series system decreases with the increase in the number of components in series. Figure 2 represents the RBD of a series system.

Figure 2. Series system

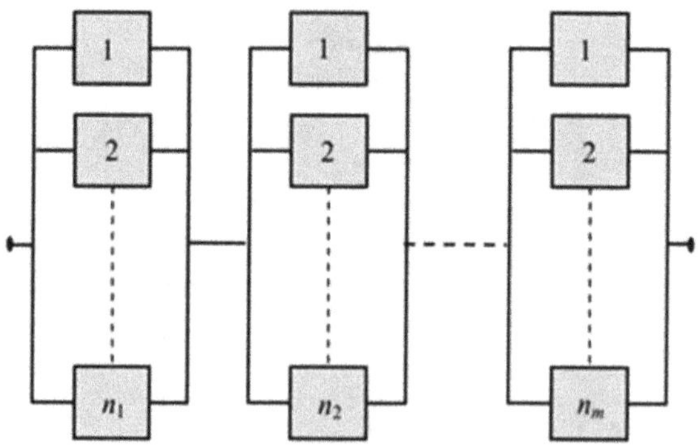

If the component failures are independent, the reliability of the series system is:

$$R_s(t) = R_1 \cdot R_2 \cdots R_n = \prod_{i=1}^{n} R_i(t) \tag{5}$$

Because the overall system's performance is determined by the success of each component, the reliability of each component's exponentially distributed failures i at time t is:

$$R_i(t) = e^{-\lambda_i t} \tag{6}$$

with λ the failure rate of the component at time t.

The failure rate of the series system is equal to the sum of the component failure. Thus:

$$R_s = \prod_{i=1}^{n} e^{-\lambda_i t} = \exp\left(-\sum_{i=1}^{n} \lambda_i t\right) \tag{7}$$

A personal computer, for example, may be made up of four core subsystems: the motherboard, the hard drive, the power supply, and the processor. These are connected in terms of reliability, and a failure of any of these subsystems will result in a system failure (Shi, Zhu, Xiang, & Feng, 2020).

Parallel System

In parallel systems, there are redundant pathways where all sub-systems (components) must fail before the entire system reliability is impacted. The parallel system can also be called *1*-out-of-*n* or (1, *n*) system which means that only one out of *n* subsystems is needed to operate for the system to be operational. A representation of a parallel system is shown in Figure 3.

For a parallel system where each component has its reliability at time *t,* we have the system reliability as follows:

$$R_s = 1 - q_s(t) = 1 - (Q_1 \bullet Q_2 \cdots Q_n) \tag{8}$$

where $Q_s(t)$ is the probability of system failure, for example:

$$Q_1 = 1 - R_1 = \text{failure of the component of 1} \tag{9}$$

$$\text{So } R_s = 1 - \prod_{i=1}^{n} Q_i = 1 - \prod_{i=1}^{n} (1 - R_i(t)) \tag{10}$$

Figure 3. Parallel system

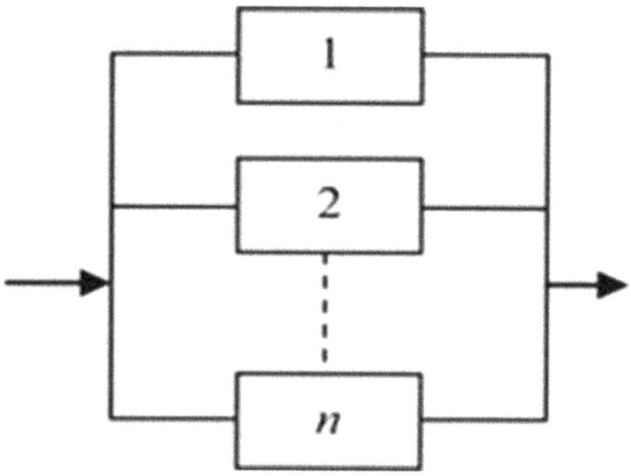

The parallel configuration is used for certain systems that need to have an extremely high reliability to avoid a failure that could cause catastrophic harm or loss of life such as a space shuttle, an airplane, or a lifesaving medical device. We may say that the larger the number of components in a system, the more reliable it will be (Barlow, 1998).

Series-Parallel System

The term "series-parallel" refers to a system that combines numerous series and parallel configurations (Figure 4). By dissolving the whole system configuration into similar subsystems and then calculating their reliability, we may determine the system's reliability.

Figure 4. Series-parallel system

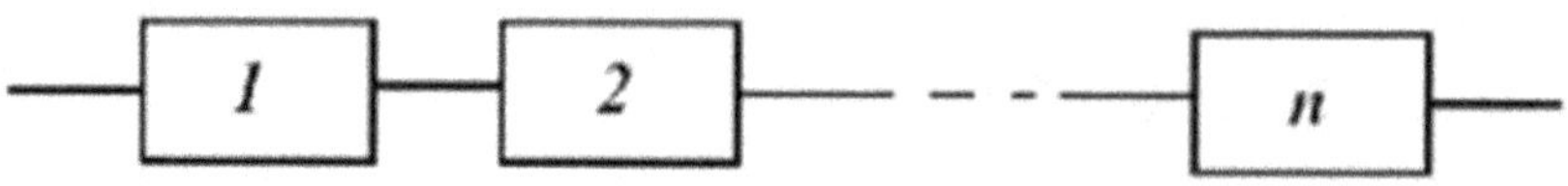

k -out-of- n System

A *k*-out-of-*n* system or (*k*, *n*) is an *n*-component system that functions or works if at least *k* of the *n* components work. Such a system can be divided into parallel and series systems, which are special cases. In a *k*-out-of-*n* system (Figure 5). The components can be arranged in a variety of ways, and several assumptions regarding component lifetimes can be made. The reliability function of the system is obtained by computing the probability that at least *k* components out of *n* components work. Assuming that the components have the same failure distribution $Q(t)$. The system reliability is obtained by using a binomial distribution; that is:

$$R_s(t) = \sum_{i=k}^{n} \binom{n}{i} [1 - Q(t)][Q(t)]^{n-1} \tag{11}$$

Figure 5. k-out of n system

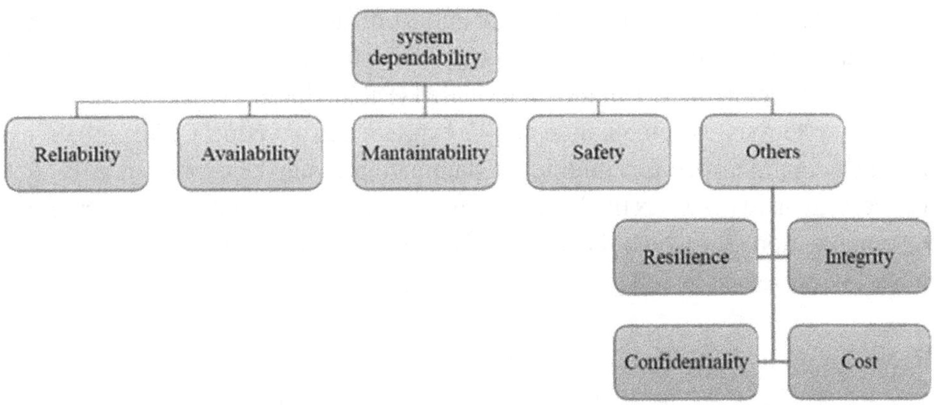

COMPLEX SYSTEMS

It is quite simple to obtain the mathematical or analytical formula that characterizes the system's dependability if it can be broken down into series/parallel configurations, which is not always the case for complex systems. In many circumstances, it is difficult to tell which components are connected in series and which are connected in parallel (Figure 6). Determining the system's dependability becomes more difficult. According to the book of M. Pecht (2009) (Pecht, 2009), three methods of complex system reliability analysis will be illustrated in the following subsection. Figure 6 represents an example of a complex system.

Figure 6. Example of a complex system

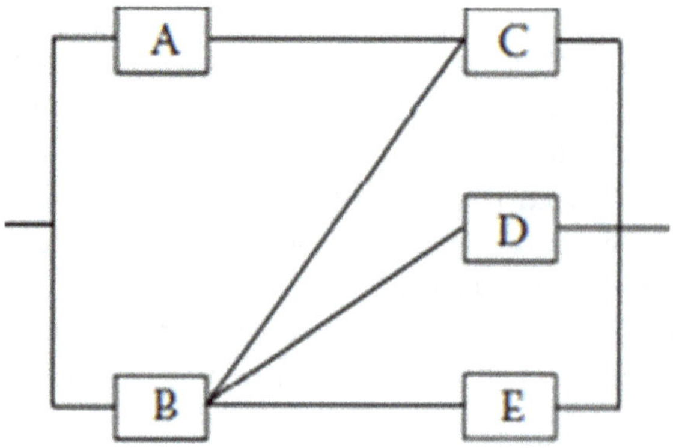

Complete Enumeration Method

The complete enumeration approach is based on a list of all conceivable component failure combinations. Table 1 lists all of the system's conceivable states (see Figure 6). The letters O and F stand for "system in the operational state" and "system in a failed state," respectively. Uppercase letters indicate an operational unit, whereas lowercase letters indicate a unit that has failed (Pecht, 2009).

Each system status combination may be expressed as a product of the probability of units being in a specific condition. Combination 2 may be expressed as $(1–R_A) R_BR_CR_DR_E$, where $(1–R_A)$ is the likelihood of unit A failing at time t. The total of all the combinations for which the system is in an operational state, O, is the system reliability expressed as follows:

$$\begin{aligned} R_s = {} & R_AR_BR_CR_DR_E + (1-R_A)R_BR_CR_DR_E + R_A(1-R_B)R_CR_DR_E \\ & + R_AR_B(1-R_C)R_DR_E + R_AR_BR_C(1-R_D)R_E + R_AR_BR_CR_D(1-R_E) \\ & + (1-R_A)R_B(1-R_C)R_DR_E + (1-R_A)R_BR_C(1-R_D)R_E \\ & + \ldots \\ & + (1-R_A)R_B(1-R_C)(1-R_D)R_E \end{aligned} \quad (12)$$

After simplification, the system reliability is given by:

$$\begin{aligned} R_s = {} & R_BR_CR_DR_E - R_AR_BR_C - R_BR_CR_D \\ & - R_BR_CR_E - R_BR_DR_E + R_AR_C + R_BR_C \\ & + R_BR_D + R_BR_E \end{aligned} \quad (13)$$

Table 1. Example of complete enumeration method

System description	System condition	System statuts
All components operable	ABCDE	O
One unit failed	aBCDE	O
	AbCDE	O
	ABcDE	O
	ABCdE	O
	ABCDe	O
Two units failed	abCDE	F
	aBcDE	O
	aBCdE	O
	aBCDe	O
	AbcDE	F
	AbCdE	O
	AbCDe	O
	ABcdE	O
	ABcDe	O
	ABCde	O
Three units failed	ABcde	F
	AbCde	O
	AbcDe	F
	AbcdE	F
	aBCde	O
	aBcDe	O
	aBcdE	O
	abCDe	F
	abCdE	F
	abcDE	F
four units failed	Abcde	F
	aBcde	F
	abCde	F
	abcDe	F
	abcdE	F
All units failed	abcd	F

Conditional Probability Method

The conditional probability technique is based on the law of total probability, which allows for the breakdown of a system into individual units and their states at time *t*. For instance, system reliability is equivalent to the reliability of the system multiplied by the reliability of unit A when it is in an operating state at time *t* (denoted by $R_s|A_G$). In addition to the reliability of the system times the unreliability of unit A when unit A is in a failed state at time *t* (denoted by $R_s|A_G$), which can be expressed as:

$$R_s = (R_s \mid A_G)R_A + (R_s \mid A_B)Q_A \tag{14}$$

This technique of segmentation repeats until each term is expressed in terms of the components' reliability and unreliability. Considering the complex system in Figure I.6 and its decomposition, using component C as an example of how this approach might be used, the system's reliability may then be written:

$$R_s = (R_s \mid C_G)R_C + (R_s \mid C_B)Q_C \quad (15)$$

The system reduces to the form depicted in Figure I.6 if the element C is operational at time *t*. As a result, the system reliability is equivalent to the series-parallel combination shown before, or:

$$R_s \mid C_G = [1-(1-R_A)(1-R_B)] \quad (16)$$

Figure 7. System reduction when unit C is operating

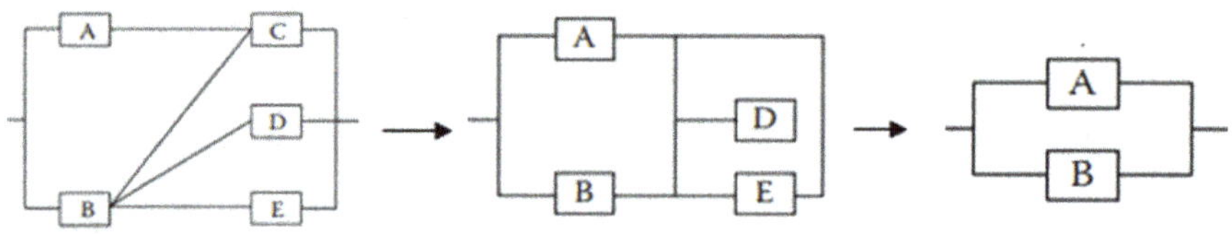

If the unit C fails at time *t*, the system configuration is reduced to the configuration illustrated below in Figure 8. Assuming the unit C fails at time *t*, the system reliability becomes:

$$R_s \mid C_B = R_B[1-(1-R_D)(1-R_E)] \quad (17)$$

Figure 8. System reduction when unit C fails

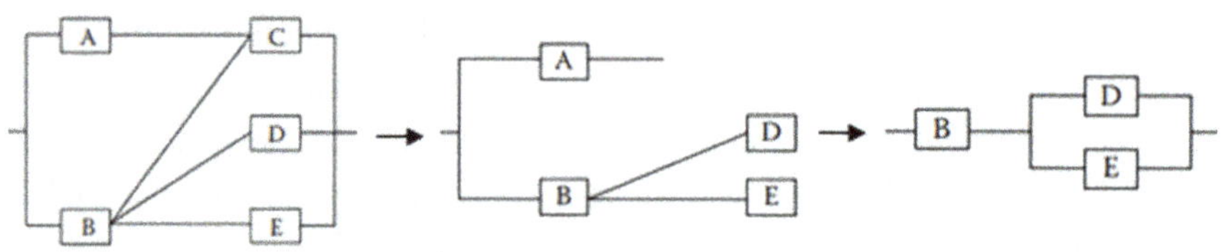

By substituting Equations 16 and 15 into Equation 13, the whole reliability of the system is as:

$$R_s = (R_s \mid R_G)R_C + (R_s \mid R_B)Q_C - [1-(1-R_A)(1-R_B)]R_C - R_B[1-(1-R_D)(1-R_E)](1-R_C) \quad (18)$$

Cut Set Method

The cut-set approach is adopted to compute system reliability, especially complex ones. The method is usually defined as a set of system elements which, when failed, lead to failure of the whole system. Therefore, if none of the set's components fails, the system will still operate.

To achieve system reliability using the minimal cut-set method, we need to follow these steps:

- Determine the minimum cut sets for a system;
- Model each cut set's components in a parallel arrangement;
- In a series setup, connect all of the minimal cut sets;
- System reliability is modeled and solved as a sequence of cut sets with parallel combinations of components in each cut set (Pecht, 2009).

Using the previous example, we can apply the first step of the cut-set method by representing the cut-sets of the system (Figure 9) and noted as C_1, C_2, and C_3:

$$C_1 = \{A, B\}$$
$$C_2 = \{B, C\} \quad (19)$$
$$C_3 = \{C, D, E\}$$

Figure 9. RBD of the system in term of minimal cut-sets

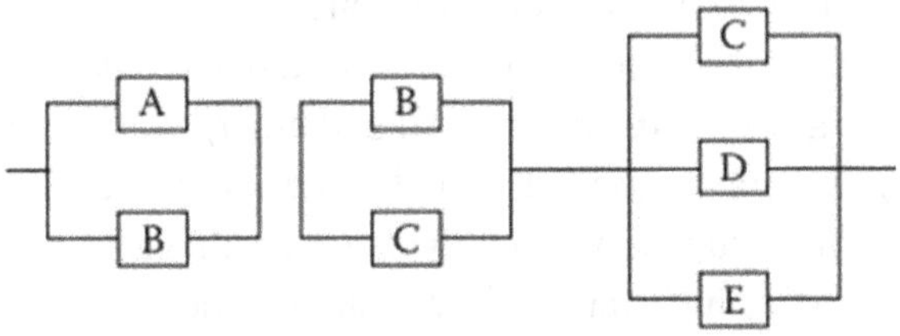

The system reliability is determined using techniques for series and parallel systems:

$$R_s = [1-(1-R_A)(1-R_B)][1-(1-R_B)(1-R_C)][1-(1-R_C)(1-R_D)(1-R_E)] \quad (20)$$

After simplification:

$$R_s = R_A R_C + R_B R_C + R_B R_D + R_B R_E - R_A R_B R_C - R_B R_D R_E - R_B R_C R_D - R_B R_C R_E + R_B R_C R_D R_E \quad (21)$$

Figure 10. Al-Khobar sub-distribution water network

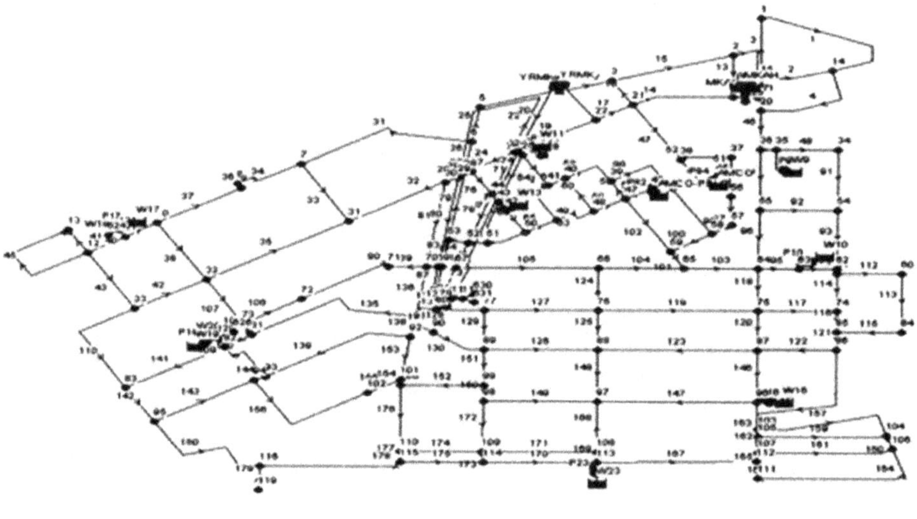

The minimum cut-set methodology can be very effective to solve very complex system reliability. For instance, the cut-set approach has been used for reliability analysis of the water distribution system in Al-Khobar, a city and a governorate in the Eastern province of the Kingdom of Saudi Arabia. The main concern was measuring the ability of the water distribution system to achieve consumers' needs in terms of quantity and quality, under usual and emergency circumstances.

After applying the cut-set method to calculate the hydraulic reliability of the water distribution, the result shows that the system reliability is 69.73%, which is very low regarding to the importance of the system in the daily basis of the citizens, knowing that the reliability value of Tucson, Arizona in the USA is 96%. Such low reliability in Al-Khobar is caused by the high failure rates of the water pipes resulting in higher cut-set failure probabilities. The Al-Khobar network's system reliability value might be enhanced by lowering yearly pipe failures through adequate maintenance and replacement of older pipes in the network, after identifying them (Al-Zahrani & Syed, 2005).

There are many methods for analyzing reliability. We have characterized in Table I.2 each approach or method studied according to three criteria:

- Inductive or deductive method;
- Quantitative or qualitative method;
- The intended objectives.

Table 2. Some methods and approaches for reliability

Method	Inductive/ Deductive	Quantitative/ Qualitative	Objectives
Preliminary Risk Analysis (PRA)	Inductive	Qualitative	Identify priori risks to be studied
Failure Modes, their Effects and their Criticalities Analysis (FMECA)	Inductive	Qualitative	Evaluate the consequences of failures
Fault Tree (FT)	Deductive	Quantitative	Evaluate the scenarios of a feared event
Reliability Diagram (RD)	Deductive	Quantitative	Represent a model of the system from the reliability of the components
State Space Method (SSM)	Inductive	Quantitative	Identify the passage through the failure states on the operation of the system
Petri Nets (PN)	Inductive	Quantitative	Identify the passage through the failure states on the operation of the system
Event Tree (ET)	**Inductive**	**Quantitative**	**Assess the possible consequences of an event**
Summarized Failure Combination Method (SFCM)	**Inductive**	**Qualitative**	**Determine combinations of failures**
Cause-Consequence Diagram Method (CCDM)	**Inductive/ Deductive**	**Quantitative**	**Analysis of an initiating event**
Truth Table (TT)	**Inductive**	**Quantitative**	**Identify all combinations of states**

Inductive approaches begin with the causes and end with the consequences, whereas deductive methods start with the results and end with the causes. We classified approaches as quantitative if they allow for significant quantification, and qualitative if they do not allow for it or if it is only a minor consideration. These approaches were tested on practical applications, the vehicle on-board electronic system (Anti-Lock Brake System) (ABS). By adjusting the braking torque, it improves braking efficiency. To do this, mechanical and/or hydraulic parts are controlled by the results of computations based on sensors data. They are associated with vehicle safety. One or more component failures can result in the entire system failing (Gissinger, Menard, & Constans, 2003).

CONCLUSION

In this chapter, we have gone through the concept and metrics of system dependability. Also, the different reliability analysis used to various system configuration, from the simplest systems' configuration to the complex ones.

REFERENCES

Abdulgader, E. A. S. (2019). *Optimization of Systems Reliability by Metaheuristic Approach*. École De Technologie Supérieure.

Al-Zahrani, M. A., & Syed, J. L. (2005). Evaluation of Municipal Water Distribution System Reliability Using Minimum Cut-set Method. *Journal of King Saud University -. Engineering and Science*, *18*(1), 67–81. doi:10.1016/S1018-3639(18)30822-5

Aqel, I. (2021). *Optimal heterogeneous components allocation in pharmaceutical production plant using AI techniques*. M'hamed Bougera University of Boumerdes.

Avižienis, A., Laprie, J. C., Randell, B., & Landwehr, C. (2004). Basic concepts and taxonomy of dependable and secure computing. *IEEE Transactions on Dependable and Secure Computing*, *1*(1), 11–33. doi:10.1109/TDSC.2004.2

Barlow, R. E. (1998). *Enineering Reliability*. American Statistical Association and the Society for Industrial and Applied Mathematics.

Birolini, A. (2017)... *Reliability Engineering*. doi:10.1007/978-3-662-54209-5

Chanda, S., & Srivastava, A. K. (2015). Quantifying resiliency of smart power distribution systems with distributed energy resources. *IEEE International Symposium on Industrial Electronics,* 766–771. 10.1109/ISIE.2015.7281565

Chanda, S., & Srivastava, A. K. (2016). Defining and Enabling Resiliency of Electric Distribution Systems with Multiple Microgrids. *IEEE Transactions on Smart Grid*, *7*(6), 2859–2868. doi:10.1109/TSG.2016.2561303

Crotty, B. H., & Mostaghimi, A. (2014). Confidentiality in the digital age. *BMJ (Clinical Research Ed.)*, *348*, g2943. doi:10.1136/bmj.g2943 PMID:24816731

Gissinger, G. L., Menard, C., & Constans, A. (2003). A mechatronic conception of a new intelligent braking system. Control Engineering Practice, 11(2), 163–170. doi: ISO/IEC 2382-14:1997 . doi:10.1016/S0967-0661(02)00041-2

IOS. (1997). Vocabulary, Part 14. In ISO/IEC 2382-14:1997 (2nd ed.). International Organization for Standardization

Kahan, J. H., Allen, A. C., & George, J. K. (2009). An Operational Framework for Resilience. *Journal of Homeland Security and Emergency Management*, *6*(1). doi:10.2202/1547-7355.1675

Laprie, J.-C. (1995). Dependability — Its Attributes, Impairments and Means. *Predictably Dependable Computing Systems*, 3–18. doi:10.1007/978-3-642-79789-7_1

Mahzarnia, M., Moghaddam, M. P., Baboli, P. T., & Siano, P. (2020). A Review of the Measures to Enhance Power Systems Resilience. *IEEE Systems Journal*, *14*(3), 4059–4070. doi:10.1109/JSYST.2020.2965993

Mellal, M. A., & Zio, E. (2018). Availability optimization of parallel-series system by evolutionary computation. *3rd International Conference on System Reliability and Safety*. Barcelona, Spain. 10.1109/ICSRS.2018.8688722

Mellal, M. A., & Zio, E. (2020). System reliability-redundancy optimization with cold-standby strategy by an enhanced nest cuckoo optimization algorithm. *Reliability Engineering & System Safety*, *201*, 106973. doi:10.1016/j.ress.2020.106973

Mellal, M. A., Zio, E., & Williams, E. J. (2020). Cost minimization of repairable systems subject to availability constraints using efficient cuckoo optimization algorithm. *Quality and Reliability Engineering International*, *36*(3), 1098–1110. doi:10.1002/qre.2617

Nilsson, J., & Bertling, L. (2007). Maintenance management of wind power systems using condition monitoring systems – Life cycle cost analysis for two case studies. *IEEE Transactions on Energy Conversion*, *22*(1), 223–229. doi:10.1109/TEC.2006.889623

Panteli, M., & Mancarella, P. (2015). The grid: Stronger, bigger, smarter?: Presenting a conceptual framework of power system resilience. *IEEE Power & Energy Magazine*, *13*(3), 58–66. doi:10.1109/MPE.2015.2397334

Pecht, M. (2009). *Product Reliability, Maintainability, and Supportability Handbook*, (2nd ed.). CRC Press. doi:10.1201/9781420009897

Sansavini, G. (2017). Engineering Resilience in Critical Infrastructures. *NATO Science for Peace and Security Series C: Environmental Security, PartF1*, 189–203. doi:10.1007/978-94-024-1123-2_6

Shi, Y., Zhu, W., Xiang, Y., & Feng, Q. (2020). Condition-based maintenance optimization for multi-component systems subject to a system reliability requirement. *Reliability Engineering & System Safety*, *202*, 107042. doi:10.1016/j.ress.2020.107042

Yee, C. K., & Zolkipli, M. F. (2021). Review on Confidentiality, Integrity and Availability in Information Security. *Journal of ICT in Education*, *8*(2), 34–42. doi:10.37134/jictie.vol8.2.4.2021

Zwingelstein, G. (2019). Sûreté de fonctionnement - Principaux concepts. *Automatique et Ingénierie Système*. doi:10.51257/A-V3-S8250

Chapter 3
Mechatronics in CDIO Projects for Green and Sustainable Environments

Duong Vu
Duy Tan University, Vietnam

ABSTRACT

We live in the common Earth where the sustainable development depends on everybody's habit. To set up awareness, it is necessary to integrate this matter right from school teaching until the high education. An engineering education, concerning mechatronics in general and robotics in detail, will help to strengthen the skill and knowledge of students to find reasonable solutions for this problem. This chapter will contribute some experiences in the engineering education technology in university to show how the students can learn the robotics and apply their knowledge to design the machines with the software and simulation tools. All these activities are under the CDIO method - the modern approach in the engineering education. The research results illustrated in this chapter will clarify this concept.

INTRODUCTION

Social activities enable all students to get in touch with companies and community, to form social obligation from the time they start taking their engineering courses. This encourages students to select and work in areas that closely match not only their areas of interest, but also impact community. This is the new way to integrate the "human education" to teach technical disciplines. As a result of these activities, students can determine the target area related to their study chosen by themselves to

DOI: 10.4018/978-1-6684-5887-7.ch003

learn better. This is the most valuable source of information that will help the team to set up their rough draft of entrepreneurship plan. (Duong et al., 2017). Interestingly enough, many of today's new technologies like Information Technology (IT), embedded system engineering and biotechnology no longer require vast amounts of upfront financial investment like before if we can find the niche market to get into. Given such opportunity, it will be the human capital that is more important than any other type of capital or resource (David et al., 2018). Engineering students were generally not equipped with knowledge and skills in social networking, money raising, financial dealing, Intellectual Property (IP), etc. to develop any motivation for entrepreneurship. Their curricula mostly focused on basic sciences and technical know-how. Worse, the big number of students in any one class made it hard to create an active and integrated learning environment, and their instructors are mostly of a past generation where "entrepreneurship" is still a strange word. The illustration for this solution will be found in case study of Mechatronics students at Duy Tan University who worked with their instructors on an entrepreneurship project of a robot which helps detect welding defects on ship and vessel hulls. As much as the technical aspect of the project is very complicated, we will look into what changes are made to our Learning Outcomes (CDIO Standard No. 2), Curriculum and Learning Experience to make them more active and integrated (CDIO Standard No. 3, 7, 8) in an effort to build students' entrepreneurial skills. Lessons learned from this case study, from miscellaneous issues to macro-level problems, will be of great benefit to universities and colleges which are looking for ways to improve their students' entrepreneurial skills and motivation. Finally, the green and sustainable environment set fort the effective treatment of plastic disposal. In this common struggle, mechatronics as the knowledge should be involved through the education to encourage the students make the innovation in reliable projects. As the practical and social impacted activity, this chapter described a new promotive environment friendly solution for plastic pollution by recycling bottles. The project has not only economic but also social value.

SHOWCASE ONE - AN INTEGRATED APPROACH FOR AN INTRODUCTION THE TECHNICAL SUBJECTS IN THE UNIVERSITY.

The integrated approach concerning the creative thinking, the innovation, the ability to adopt to changing environment, the team work skills and the active learning. This method encourages instructor and student to gain the increasing attention for the active learning for their outcome. It is noted that the project context is the main point of all inspirations and creativities. Therefor it is requested that instructor set fort not only simple presentation but also develop demand forecast for their products in their respective market areas through the CDIO project.

CDIO Standards 2 & 3- Refocus Conceive and Design Stages

Coming back to the CDIO Project, on stage conceive, all students actually discuss in teams the initial idea of the theme they wish to follow. At the beginning, these ideas may be unfeasible but if the proposal meets the basic plan and has practical sens based on CDIO syllabus (learning outcomes), the mentor encourage them follow. From the other aspect with the advanced CDIO Project, we require more personal skills, interpersonal skill - communication skills (especially oral and written English communication) and design ability, the guidance skill, systematic thinking, creative ability, selective criticizing, problem resolving, teambuilding (standard 2- CDIO Standards v 2.*0.*,2010) that they have to gain conceive stage through activities:

- As a first step the instructor ask the students to provide the information that they obtained in meetings with the companies or links they formed and identified practical problems related to their subject matter mainly in the area of Mechatronics. Each student describes and outlines the outcome they envision (mind map) (Johnson, 2001) – Fig .1. To finalize and improve this mind map, all team member needed the cognitive and affective development, creative and critical thinking. In this scenario, the important moment is pay attention on the professional ethics as the significant side of the "human education". Based on standard 2, this procedure facilitates learning outcomes.
- Help them to visualize what they have learnt through discussion with companies and associations.
- The instructor encouraged all member to be receptive to make changes in their initial ideas based on the information they obtain in meetings with partners, associations. Having this opportunity, the students can fine tune, be innovated and adjust their mind map themselves. This procedure helps students to improve their communication skills, promote creative thinking, become skilled at doing analysis and be open to critique theory in order to improve learning outcome and complete their mind map.
- This flexible model can be applied to catalyze discussions, reveal student's personal viewpoint on discerning information. Due to changing personal viewpoint, students can also reorganize their teams. It is focused on the problem which might be social in nature or demanded by the companies in the area of Mechatronics. The goals of the ensuing discussions may result in the requirement of delivering complete product according to one specific

operation in the production line, regulation of some product or some type of specific product which makes sense for disabled person.

The instructor then plays the role of a facilitator bridging and connecting all information, managing student's teams to conceive the idea, guide and trigger more the technological, engineering aspects to help them complete their engineering conception. Based on schedule of each student's team, the instructor is monitoring it so that all students during Project schedule must plan their workload and manage their time through all stages, because they have so many meetings, seminars, workshops, presentations with different partners, in limited time and placed.

Figure 1. Mind map on reasonable price electric wheelchair for disabled person

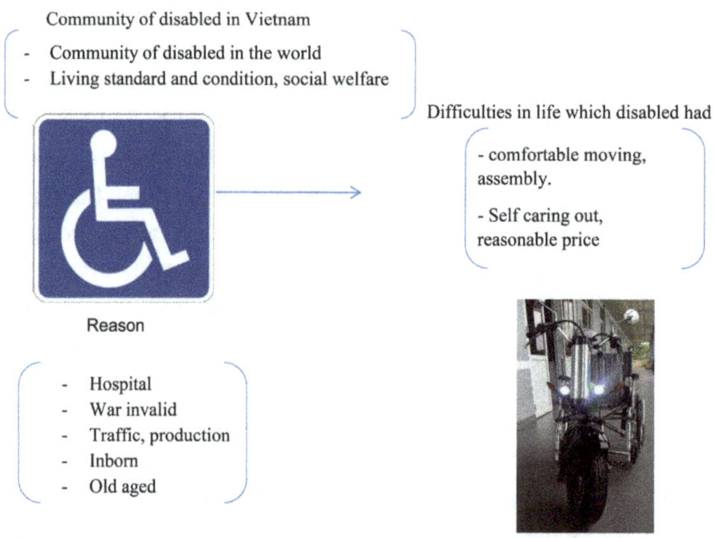

The idea for conception on wheelchair was created by our students using the abovementioned mind map. The concept was initiated from practical demand of disabled persons not only in Vietnam but also in the world. The statistics recently shown about 7% of total population in Vietnam is disabled. They need a wheelchair with the locomotive driver which could be disassembled very quickly and smoothly with the low price for moving mainly in indoor environment. All students worked hard on design the dismantle mechanism on driver, so that a disabled man could assembly and disassembly it in few seconds (Fig. 2).

Figure 2. Disassemble locomotive drive

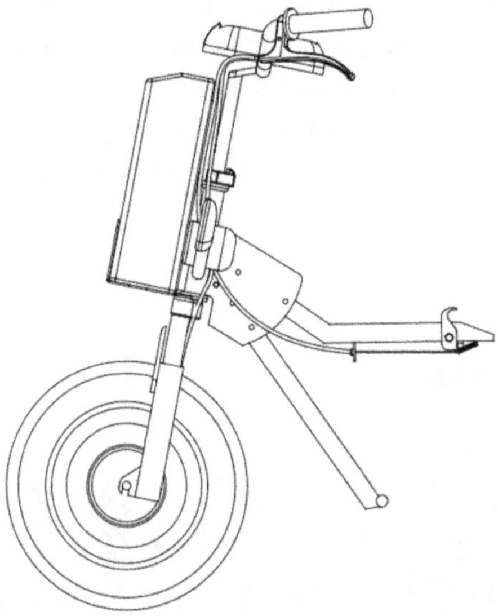

From this community many peoples lost moving ability therefore their mental and physical decreasing so strongly affected live quality. They face to nowadays heavy difficulties as moving, working, self- carrying, and their dream is some wheelchair to help them freely move and do homework. On Fig. 1 the mind map synthesizes all information, collected by one student's team (standard 1- CDIO Standards v 2*.0-2010*).

On the stage conceiving, this student's team was contacting with disabled Association and families for approaching the disabled people to learn the customer demand, consider the technology, develop the technical and maybe business plans. They also collected additional reference, specification type of wheelchair and start sketch designing of wheelchair (Martin, M. B., 2015). The instructor is responsible for arrangement contact with disabled association, their families, local authority to make comfortable conditions for student's investigation and survey.

Based on data collected in on site investigation, the student's team begin the second stage–design.

They analyze, and propose some draw and engineering of detail for the wheelchair (Fig. 3). On this step besides complete teamwork requirement, all students should be proficient in special knowledge and perfect skill with software as Microsoft Visio, AutoCAD (Sham Tickoo - AutoCAD, 2016), Autodesk Inventor (Sham Tickoo – Autodesk Inventor, 2016) to develop, simulate on mechanics. Result of this step is finding out complete thinking technical resolutions, which meet specification required and operation of wheelchair (Fig. 4). Meantime, the instructor guide and consult or answer student's questions in term of mechanical and dynamical calculations, method of data processing from sensors, to regulate limb movements. These activities are to support students to train on engineering prototype, processes, and system design.

In conclusion, through contact with disabled peoples help them simultaneously learn skills as communication, interviewing and data acquisition, observation, evaluation, technical report.

Figure 3. Student from DTU in onsite learning

Figure 4. Wheelchair design on Autodesk Inventor

Nevertheless, the instructor also encourages students to implement AutoCAD & Autodesk to model, then create 3D printer they use to make some details of prototype. This is very basic skill integrating in curriculum (Enelund, M., 2012) to design - implement the product (standard 2, 3 - CDIO Standards v 2.*0-2010*) and also could be necessary entrepreneurial skill. All students are stimulated in continuous improving design model together with disabled peoples in connection with how to cut down the cost by other technology instead of traditional one in the future.

In the third step – implementation, all team members from department of Mechatronics under the management of instructor, work hard in the University's laboratories, where they operate some machines as lase cutting machine, CNC milling and turning machines to manufacture the prototype of wheelchair. Since they are the junior student of Mechatronics, and having skills of manufacturing processes, all students can make the prototype. Through the practice they improved their implementation process. All member who joined this Project, performed many tests to reach the quality of wheelchair, especially the reliability in operation.

The last step – operation help the student to deliver their product to disable people. They instruct them how to operate the wheelchair in all conditions and strongly commit on repair service.

Figure 5. Wheelchair design on -site

Figure 6. Students finish the design of wheelchair in CME workshop

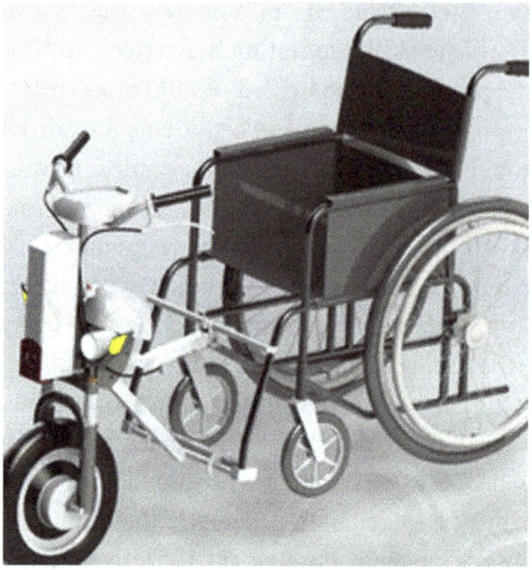

Refocus of Standards 6,7,8 (CDIO Standards v 2.0-2010) in Implement & Operate Stages

Following chain of training activities, all students are being willingly and actively suggested to implement models, fabrication proposal, product analyze in small team collaboration, discussion, cementation each other and feed back in practice. This stage provides students the experience; assignment in fabrication wheelchair for improving practical and professional skill; the problem- solving skill (standard 8), but the instructor only guides and suggest through connection with Center of Mechatronics Engineering – CME (one of research unit of the University, having facilities for manufacturing, programming, simulation…) so that some experts of this CME support them to realize their ideas. The students acquire a syntactic ability by themselves in one assignment including product analyze, engineering and especially be responsible for social obligations of designer (wheelchair). This is quite specific teaching and learning activity (standard 7) in CDIO Project which we apply in recent semester due to working practical space for students according (standard 6). All team members having acquired Center of Mechatronics Engineering take the initiative contact with technical staff, to show their proposal and then they are allowed to use technical tools, modern software, practical manufacturing processes right in CME to try fabrication of wheelchair (Fig 3,4,5,6). This methodology had some advantages as follow:

- The space in CDIO class was extended, the students are sitting not only in teaching separate rooms, but also moving to workshops, laboratories of Center of Mechatronics Engineering entirely close to families of disabled person. The learning in open space helped students apprehend affectively and more attractive in comparison with traditional classroom.
- Student communicated directly and worked with Instructor (Thomas, E. & Steven, S., 2006) so it improves professional experience. The fabrication under the monitoring and surveillance of experts eliminated a mistake, shortcoming of students due to lack of practical skills, increased the reliability of product, out coming device. Namely the wheelchair could be exactly and stably operated against design calculations.
- University explores maximum equipped technical facility, as research tool, to make business, also for training. Taking advantage available machinery, materials of Center of Mechatronics Engineering partly helped student save research expenses.

Figure 7. The testing of mechanical components and electronic automation for wheelchair

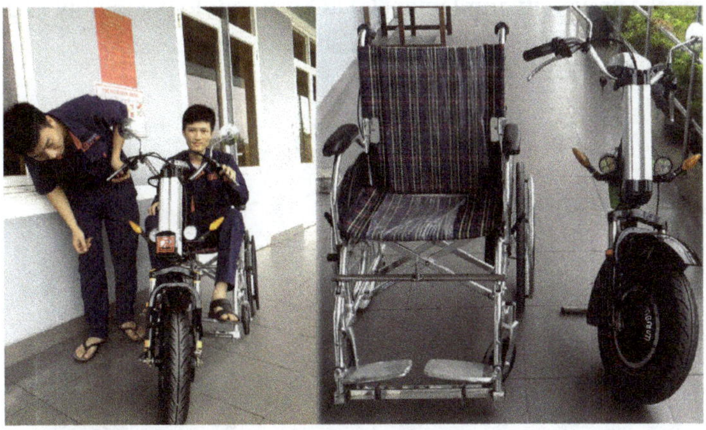

On stage - operation, the students are requested to test wheelchair under conditions right in company workshop (Fig. 7). In former Projects we do not request students do this because almost results of CDIO Project was just limited on grading completeness of volume theme based on target set forte and valuated knowledge content, gaining during its execution. In advanced CDIO Project, targeting on additional arming student some more experience and entrepreneurial skill, so the instructor got more criteria for evaluation practicability, fullness and flexibility in production scale.

Often one Project is evaluated by complex of criterions as relevance, fulfillment of objective, efficiency, impact value and sustainability.

INFERENCE

The Project engineering wheelchair for disabled person is a typical case motivating the perfect integrated and simultaneous implementation of some CDIO standards to develop, encourage professional skills for students in general, especially in Mechatronics and Mechanical Engineering also. This is one new approach to circumstance analyze in practice context to set the social obligation for students in market investigation and constructive implement, combining interdisciplinary knowledge in integrated education environment, extended space, taking advantage technology improving. It's aim for reasonable price practical product, paying attention on humanitarian scene. These are important knowledge and skills, arming students before joining labor market. In all steps of this typical CDIO project, the students of Mechatronics and instructors applied the CDIO Standars to make the very useful product in term of human education (Fig. 8) integrated with teaching the typical engineering technology subjects. This CDIO project once more show the evidence how the university implement CDIO methodology not only in preparation the curriculum, but also in all steps of the education process.

Figure 8. Handling ceremony in the University

SHOWCASE TWO - REFOCUS OF CDIO STANDARDS TO STRENGTHEN ENTREPRENEURSHIP SKILLS FOR STUDENTS.

Still, among the various skills of students that CDIO helped improve on, entrepreneurship skills have hardly been its focus. This is understandable given that successful entrepreneurism depends on many other variables besides the training quality and students' skillset such as the socio-economic environment or the legal policies or even the local cultures. As the example for initial success with this approach of the Faculty of Mechanical Engineering, a start-up project of students and staff at DTU for a robot which helps detect welding defects on vessel hulls automatically will be demonstrated stage by stage.

Entrepreneurship is an increasingly important aspect in engineering education today given the fact that innovative ideas on new technologies can bring about great success overnight to any individual. As a matter of fact, countries where there are the most entrepreneurs in new technologies are always more energetic and innovative than the rest of the world in reaching out for new standards of living and working conditions. It is hard to find where to start with given the fact that Vietnam is not among any of the developed countries around the world. However, the thirst for success in a country where the younger generations account for almost two thirds of the population is always high, especially when its economy is making great growth rates to catch up with other developed countries in Asia. Interestingly enough, many of today's new technologies like Information Technology (IT), embedded system engineering and biotechnology no longer require vast amounts of upfront financial investment like before if we can find the niche market to get into. Given such opportunity, it will be the human capital that is more important than any other type of capital or resource (David et al., 2018).

In Vietnam, for years, the college curriculum and teaching methodologies simply were designed to train graduates who will become employees in state-owned organizations or private companies. Engineering students were generally not equipped with knowledge and skills in social networking, money raising, financial dealing, Intellectual Property (IP), etc. to develop any motivation for entrepreneurship. Their curriculum mostly focused on basic sciences and technical know-how. Worse, the big number of students in any one class made it hard to create an active and integrated learning environment, and their instructors are mostly of a past generation where "entrepreneurship" is still a strange word.

In this show case, it is presented a study of Mechanical Engineering students at Duy Tan University who worked with their instructors on an entrepreneurship project of a robot which helps detect welding defects on ship and vessel hulls. As much as the technical aspect of the project is very complicated, it will look into what changes are made to our Learning Outcomes (CDIO Standard No. 2), Curriculum and Learning Experience to make them more active and integrated (CDIO Standard No. 3, 7, 8) in an effort to build students' entrepreneurial skills. Lessons learned from this case study, from miscellaneous issues to macro-level problems, will be of great benefit to universities and colleges which are looking for ways to improve their students' entrepreneurial skills and motivation.

CDIO Standards 2 & 3 Refocus for Design the Robot with Application in Welding Inspection

The robot for welding-defect inspection on ship and vessel hulls is a contracted project between the Center of Mechatronics Engineering (CME) of Duy Tan University with a major local military shipbuilder, Song Thu Shipyard. Involved are three different groups of students working under the direct supervision of staff members at CME. The teams are competing with each other for the best-designed model and the most cost-effective robot. With only 3 teams of students participating in the project, we processed most of the discussions and feedbacks in the project through short interviews and focus group discussion sessions based on model P-B-P (Dong et al., 2014).

Figure 9. Students investigating and discussing ship hulls on the dockyard

The original engineering requirement of the project by Song Thu Shipyard is to make some automatic mechanism which can help detect defects on ship and vessel hulls, which are still in their dockyard (Fig. 9). The project requires the development and delivery of a completely-functioning mechanism for that purpose, utilizing whatever technology needed to get the job done. For our students, even though most had worked on some complete CDIO projects before, this requirement posed certain uncertainty to their approach as they also needed to examine the use of other technologies which may not be within the scope of their specialty. This indeed later called for a major revision on the Learning Outcomes of engineering programs to better prepare students for situations under uncertainty in the real world. In addition, since all of the students involved are majored in Mechatronics Engineering, Electrical Automation or Telecommunication, they did not have very good knowledge of how to manage money for an entrepreneurship project. To add more to the problem, this project had a research and development component, and if the teams could not manage their budget properly, they might very well run out of funding before some effective design or prototype came to life. For this challenge, even though topics on financial planning and management were added into the Career Orientation courses of our students over the past three years of our CDIO adoption at Duy Tan University, they were not sufficient enough to prepare students for a real-world entrepreneurship project like this. As a result, we had to encourage our students to attend additional seminars on financial management and budgeting as well as to read more on the topics. Eventually, when the teams came up with some financial budgeting, we always asked them to do review with us and with a faculty member from our Faculty of Accountancy. Interestingly, one of the teams had managed to acquire a sponsorship for their project design from the Young Intellectuals Funding of the city. In all, we realized the need to add in additional Learning Outcomes accordingly, as followed:

- Students will have the ability to design and deliver complete project prototypes or products which require the integration of unfamiliar technologies and techniques (Figure 3).
- Students will have the ability to carry out financial work related to a typical entrepreneurship project amongst various uncertainty variables.

As the teams integrated various features into the design of their prototypes, it was amazing to realize how much the conceive and design stages were interwoven into each other in reality. For instance, at first, two teams came up with the concept of a crawling robot (Fig .10,11) for welding-defect inspection while another proposed an automatic crane or robotic arm for that same purpose. The problem with an automatic crane or robotic arm was that it could only handle certain sizes of ships and as the ships get bigger than its reach, we had to make a new automatic crane or robotic arm. This was not to mention of the expensive cost for the materials used to make such an automatic crane or robotic arm. As we agreed on the adoption of a crawling robot concept, the teams had tried out various types of grips so that the robot could stick itself to the vessel hulls. Unfortunately, that did not work out very well, and we had to switch to the use of magnetic wheels. As simple as that sound, the concept of our robot totally changed from the form of a spider-like robot to that of small car or train carriage. This turned out not to be a simple design modification but a total concept change as we had to go back and redo everything in the concept and structures of our robot.

Figure 10. Mechanical dynamics calculations for robot's movements

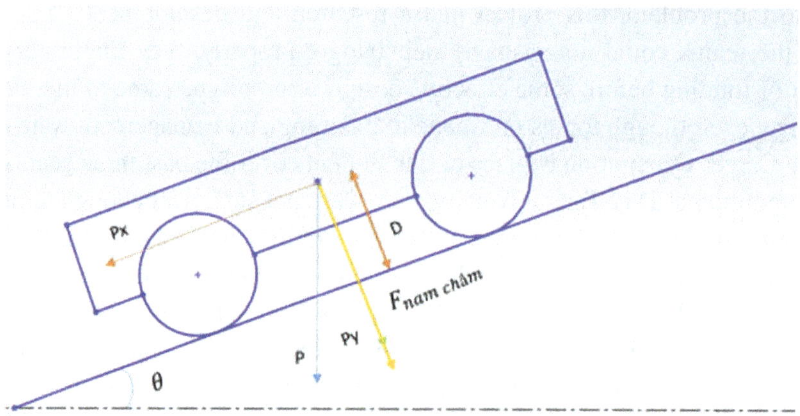

Figure 11. Various inspection positions of robot on weld seams in practice (1- position 1G, 2 and 3- position 3G, 4 and 5- position 4G).

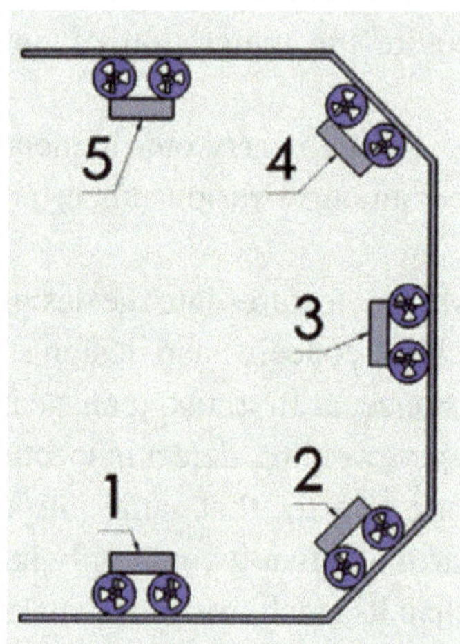

One lesson that emerged from the above conceive and design dilemma is that a strongly-Integrated Curriculum (i.e., CDIO Standard No. 3) would have helped make the conceive and design stages more efficient and effective. In fact, while all of our students were equipped with all the necessary skills and knowledge about mechanical dynamics (Fig. 12), material mechanics, sensor engineering, etc. (as these had been arranged before in our CDIO course mapping), they could hardly integrate knowledge of different fields into their design or it simply took them so long to make such an integration compared to the CME staff members (Fig. 13). There have been a lot of arguments inside our faculty as to whether that is a problem of the experience element or that was because we did not make our curriculum integrated enough. One thing all faculty members recognized was that while our students might not have lacked the knowledge, they lacked the confidence to quickly try out new concept and design ideas. This, however, is a very important element in the work of any entrepreneur in the engineering fields. As a result, an important lesson that we need to teach our students is that they should be more confident in what they are doing - those who strongly believe in themselves usually are successful, as a matter of fact, and that they should be able to deal with failure and learn from their mistakes when what they attempted to try out failed (Mark, D. C., & Amy, C.E., 2005).

Figure 12. Students working in CME labs and workshops with CME staff

Besides all the technical aspects, an integrated curriculum also includes the requirement that students need to actively integrate their personal and interpersonal skills into their project work. In particular, during the conceive and design stages, our students had to communicate frequently with managers and workers at Song Thu Shipyard as well as with the CME staff members and among themselves to arrive at the optimal design solution. Again, while all of these students have taken courses of Professional Speaking and Professional Writing or even Negotiation, they commented that it was much harder to discuss and exchange ideas with people who work at Song Thu Shipyard than with their team members or CME staff members. CME staff members, as a result, had found out that by preparing a communication plan, accommodating for various communication scenarios before meeting up with people at Song Thu Shipyard helped make the meeting much more productive for both sides. This tactic, as a result, is now widely recommended in all the project courses of our faculty.

Figure 13. Students assembling mechanical components and electronic control units of robot

CDIO Standards 7 & 8 Applied for Engineering and Testing the Robot

When it came to the Implement stage or putting together the prototype robot based on one final design chosen, we grouped the three teams together so as to have more individual labor taking care of various tasks. It is during this stage that our students had the most active interactions with managers and workers of Song Thu Shipyard (Fig. 14), in which they had to try their best to sell their ideas and results of their completed tasks. This is indeed an essential skill of every successful entrepreneur around the world. This stage would also become the most Integrated Learning Experience to most students if they were to acquire all the necessary skills and knowledge to carry out the project successfully. From our observations, students had to carry out mechanical shaping of the prototype robot, to use software tools like Microsoft Visio or Solid works to make frequent detailed updates to the design, to try out different sensors for the prototype, to test the functionality of the prototype in full or phase by phase, etc.

Surprisingly, the test cases and test procedures were where most of the Integrated Learning Experience (CDIO Standard No. 7) activities took place. To evaluate the efficiency of the robot, workers at Song Thu Shipyard simply made comparisons on certain performance outcomes between those of the traditional method and those of the new robot (Fig. 15). A revisit on the traditional welding defect inspection method here would provide an insight into how the testing evaluations were made for the robot: in practice, when the ship or vessel hulls are assembled, many big steel plates are welded together. Depending on the welding skills of the workers, certain weld seams may have some defects (Graham, J. A., 2016). In the past, after the welding was completed, a human inspector would use an ultrasound device to go through each weld seam inch by inch. This is very time-consuming and dangerous work as the human inspector had to use ladders or scaffolds or hanging ropes to reach out for each weld seam on the ship hulls. As for the robot, it was estimated that it only took one quarter of the time of a human inspector to complete the welding defect inspection for the whole ship. It also rarely misses out on any welding defect like a human inspector even though the quality of its inspection may not be much different from that of a human inspector as both have rely on the performance of the ultrasound device for the inspection. The robot, however, is not without its shortcomings: At times, if the magnetic field is not strong enough, it may fall off the ship hull and get broken (Fig. 15). It also requires continuous battery charging because its current battery capacity is not very high given the small size of the robot. Students had proposed a safety rope attaching to the robot to prevent it from falling off the hull in certain cases. In addition, proposals for new options of more long-lasting battery and more accurate device are being made.

Figure 14. Students taking initiative to discuss technical issues with managers and workers of Song Thu Shipyard

Figure 15. Shipbuilding engineers testing robotic manoeuvre and self-homing tracking device of inspection robot on a vertical dimension

In order to assess the total functionality as well as to identify performance weaknesses of the prototype or product, students need to have a good understanding of the complete design as well as various procedures and details in the testing phase. This is actually an essential capability of entrepreneurs in many IT and EE startups (Fig 16) as they usually know their products inside out and are quick to make any necessary improvement or modification to their products whenever needed (Eric, R., 2011). For that reason, our faculty has now put additional emphasis on design and testing courses in our curriculum as these will better prepare students for their future as an entrepreneur should they choose to be.

This welding defect inspection robot project also provides us with interesting clues and tricks in terms of Active Learning (CDIO Standard No. 8). Our students commented that they had actively learned a great deal through the project from the personnel at Song Thu Shipyard as well as from CME staff members and other students in their teams. In particular, in order to solve any technical problem, they simply had to dig as deep as they can into all the related technical aspects. For example, shipbuilding is a very special technical discipline, and there is no such discipline of study at Duy Tan University at the moment; hence, our students had to make the most out of the managers and workers at Song Thu Shipyard to learn as much as possible about shipbuilding for the design of their robot. Or by working with CME staff members right on-the-job when making and testing the robot, students gain the Design-Implement Experiences that they will never get from some lecture classes. One student even signified that the best Active Learning experiences came from working in adjacent workshops and controlling different machinery any time needed at CME. And yet, they also learned from each other to become more professional, to be on time, to be responsible for the work of others, and even to set an example for other team members when needed. All of these are good Active Learning practices that our faculty need to integrate back into many of our CDIO-based EE courses whether for the purpose of promoting entrepreneurism or not.

Figure 16. Welding defect inspection robot winning the First Prize in Startup Contest organized by Hoa Lac Hitech Park and the Ministry of Science & Technology

INFERENCE

In conclusion, the welding defect inspection project is a typical example about what aspects in training and learning we need to focus on to encourage students develop their entrepreneurial skills. In particular, it is important to reset the Learning Outcomes to make them more entrepreneurism-oriented first. Then, it is necessary to arm students with skills and knowledge in macro-level technical design, product testing procedures, financial management, Intellectual Property (IP), etc. And yet, by creating an Active and Integrated Learning environment, students will benefit from better exchange of ideas besides further integration of interdisciplinary knowledge. And while there is no certainty that these skills and knowledge alone or together may become the determinants for some successful entrepreneurships of our students in the future, they will make our students excel one way or another when they eventually join the labor market.

SHOWCASE THREE - THE INNOVATIVE SOLUTION FOR RECYCLING PLASTIC BOTTLES

Current levels of plastic usage and disposal is one of the biggest environmental challenges that we have to deal with (Pivnenko et al, 2015). Collecting and Recycling plastic is one of the most important actions currently used to meet this challenge, but it represents one of the most demanding areas in the plastics industry today. A major percentage of recycled plastic produced each year is used to manufacture disposable packaging items or other short-lived items that are discarded within a year or so. This clearly means that our current use of plastics is not sustainable. In addition, because of the durability of the polymers involved, large amounts of discarded end-of-life plastics are accumulating as debris in landfills and in natural habitats worldwide (Jefferson et al., 2009).

The high demand of plastic bottles specially in the food processing industry has brought with its severe environmental problems due to careless disposal of these bottles (Howell, 1992). Polyethylene terephthalate (PET) as packaging material accounted for 44.7% of single-serve beverage packaging in the US in 2021 and the plastic packaging accounts for 70% of the market of consumer products (Patnarin et al, 2022). The plastic waste is processed mainly by burning or sending to land fill. The constant buildup of plastic waste in the landfills has become a major source of concern for many sectors of sustainable society. It needs 450 years to initiate the decomposition of PET.

Design and Implementation

The process of recovering plastic waste and reprocessing it into useful products would be implemented in three stages:

1. Design and Manufacture of Production Line Components for Automatically Processing Plastic Bottles to Make Plastic Ribbon/Wire of Various Sizes.
2. *Artistic rendering of block diagram prototype of machine for cutting/connecting plastic ribbon.*
3. Setup and Organize Ecosystem for Collecting Plastic Used for Water and Soda Bottles

A successful plastics recycling operation requires a dependable source of plastic waste and an efficient collection and separation system. To encourage participation in the process, the consumers are employed and involved in the process. This also strengthen and enhances a sense of community ownership and involvement. It is suggested also the collection of plastic waste by changing on food.

The following set of machines have been designed that are available for converting the Polyethylene Terephthalate (PET) the clear plastic waste into plastic ribbons/wires. The ribbons/wires in turn are subsequently woven into scores of useful artifacts.

The Ribbon Cutting Machine presented in Fig.17 is the block diagram representation of the prototype machine designed and developed for the purpose of shredding the recycled plastic bottles into respective widths of plastic ribbons.

Figure 17. Block Diagram of the ribbon cutting machine

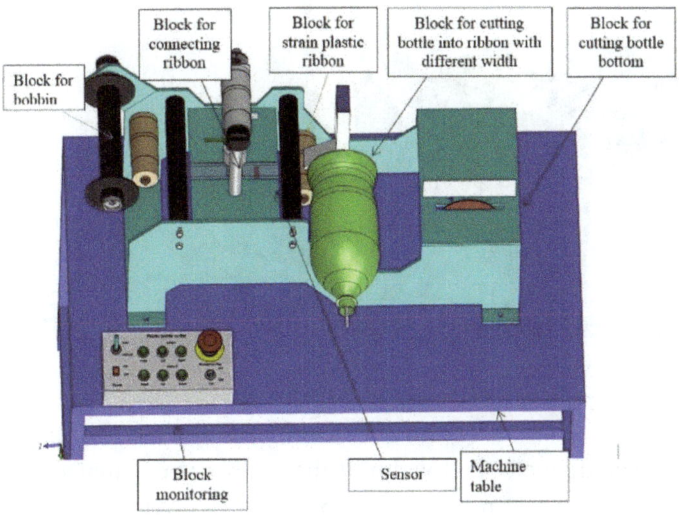

In Fig. 18, the ribbon is continuously feeding and is adjusted for some overlap. At this juncture the ultrasound heater is utilized for connecting individual pieces into a continuous plastic ribbon. The quality and appearance of the ribbon depends on welding parameters settings. The capacity of this machine is also a subject which is beyond the scope of this design.

Figure 18. Device for connecting plastic ribbon

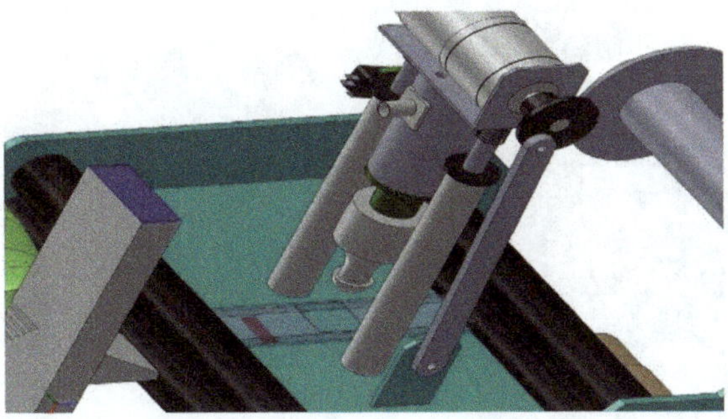

To adjust the different widths of ribbons for various end products, the machine (Fig. 19) must have the ability to cut plastics into various widths. The machine accomplishes this requirement by adjusting (sliding the blade-knife) the position of the knife which in turn regulate the width of cutting ribbons.

Figure 19. Cutting bottle into ribbon with different widths

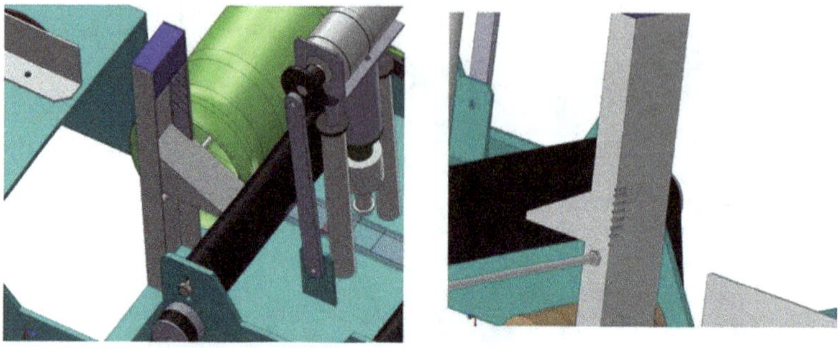

By using of this device (Fig. 20), the bobbin is freely rotating and the ribbon is being wound. The diameter and width of the bobbin is calculated so that we can process about 20-30 bottles per lot. It is expected to upgrade this device for the two scenarios:

- Automatic stopping the machine when the bobbin is full.
- Continuous recharging of the bobbin.

Figure 20. Device for bobbin ribbon

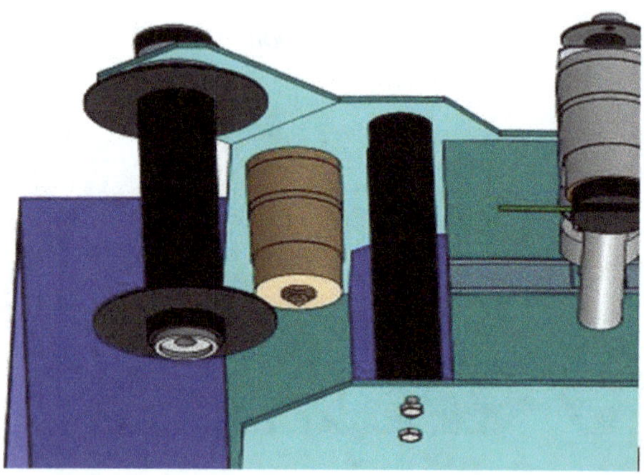

The general schema of machine controlling system for the relevant hardware functionality is shown in Fig. 21. Multiple inputs sought from the control panels and respective sensors are utilized as input to the processor. The respective output signals for controlling the two motors (cutting and bobbin/connecting) are being generated as per the algorithmic design for the machine.

Figure 21. Schema for machine controlling system

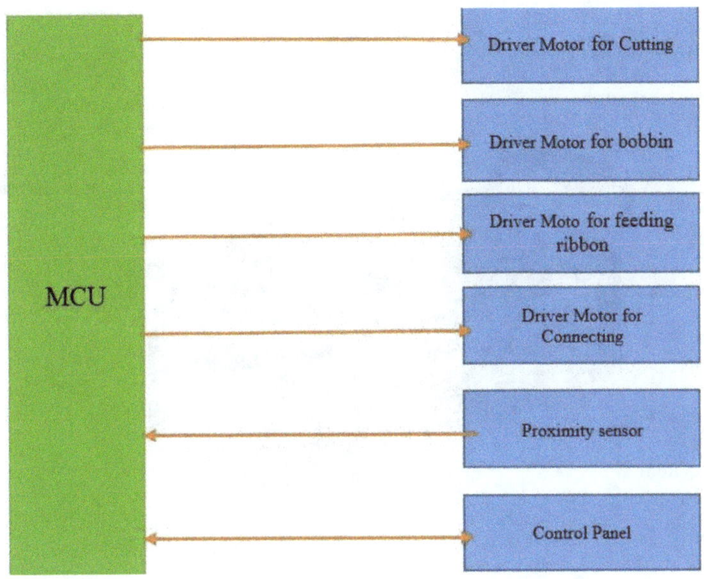

As the structure of the bottles is circular, the resulting plastic ribbon is spiral and not flat. As such the ribbon does not meet the requirements for product processing stages. This require further flattening of the ribbon without changing its characteristics by applying continuous modulating and regulating the temperature under the algorithmic control. A top view of the extricate machine is shown in Fig. 22 and Fig .23.

The Knitting Machine (Fig .24) is for knitting plastic Ribbon/wire. Depending on the specification of the end product the machine will regulates the knitting process.

The following steps discuss the concept in a simplified manner (Figure 25)

- Two wires are being fed from the two side continuously.
- One strip of plastic runs horizontally through the machine with the help of tension in the strip generated by the output spool (yellow strip below)

- One strip of plastic is fed with the help of clamping point it will stop when the end of the journey reaches 140 centimeters (green strip below)
- Using pressing mechanism to fold plastic strip in half. Then an ultrasonic welder is used to achieve the desired fusion of plastic strips.
- After fusion the scissors would cut the plastic strip 1, then plastic strip 2 would advance some to continue.

Figure 22. Extricate plastics machine

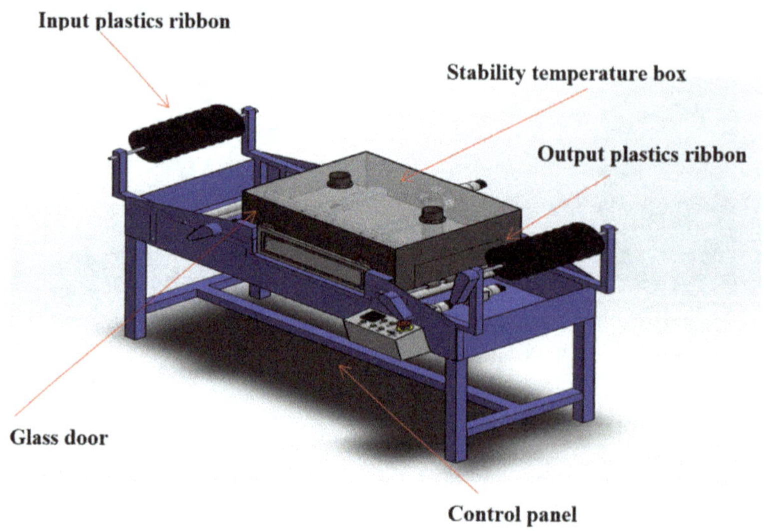

Figure 23. Top view of extricate machine

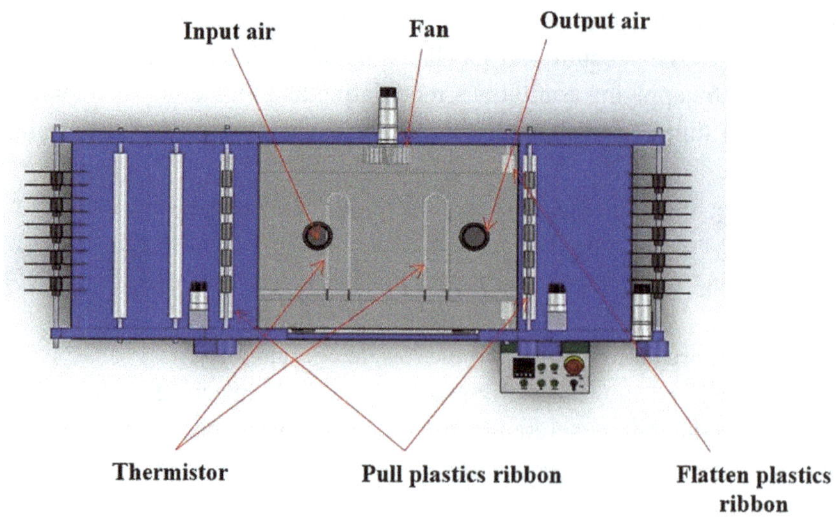

Figure 24. Plastic knitting machine

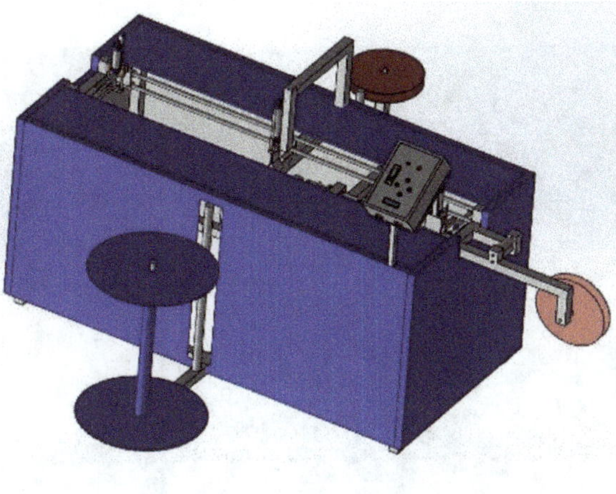

Figure 25. Top View of Knitting Machine

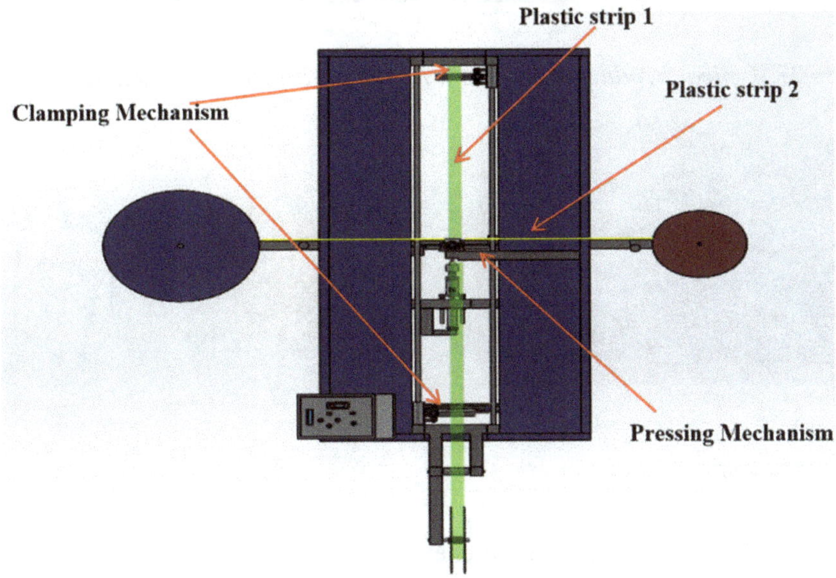

The physical prototype of this system shown in Figure 26.

The end product and artifacts shown above in Figure 27, have been tested as samples in the prototype machines and are ready for deployment for Recycling Plastic Bottles Entrepreneurship Partnership Program.

Figure 26. Physical Prototypes of Machines Designed and Manufactured for Green Environment Initiative for Recycling Plastic Bottles

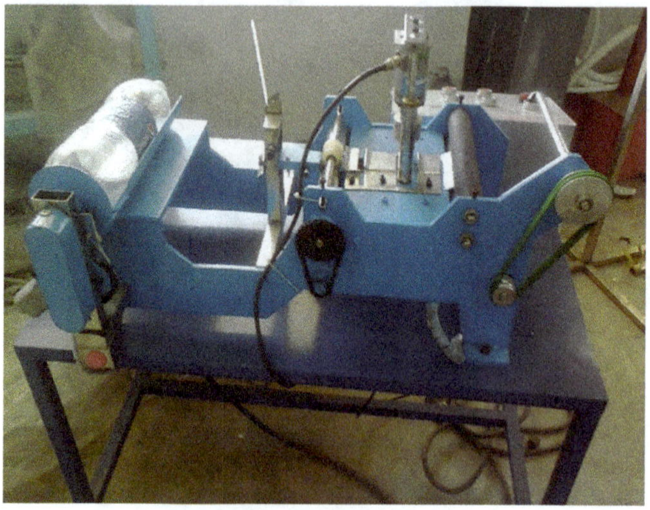

Figure 27. Typical Products Made from Recycling Plastic Ribbon/Wire

INFERENCE

The Relationship between Entrepreneurship and Economic Development for Developing Economies summarizes and updates the empirical evidence and presents the main lines of reasoning behind the relationship between economic development and entrepreneurship (Sander, W. et al., 2010). This is a big gamble with regard to scourging the hard-earned wealth and wellbeing of the families present and future.

The University, Government agencies and the private entrepreneur would form a three-prong partnership which would have a very high probability of success. The individual or joint entrepreneur would be responsible for a management of the entity. The University, the third party would provide the technical and scientific backbone to the entrepreneurship as described here.

CONCLUSION

- Mechatronics is the interdisciplinary field of the engineering technology and its successful development require the new engineering education technology in the university. There are many advanced engineering education technologies, but the CDIO method found the wide application and gained a valuable recognition in the world academic community.
- As a matter of fact, the entrepreneurship is an increasingly important aspect in engineering education today, it is suggested to focus CDIO standards in the program of education to strengthen the student's entrepreneurship skills right in university environment.
- Through the CDIO projects using the integrated approach for introduction the mechatronics engineering technology, future engineers are capable of innovative spirit and solution on how to help the disabled people, to protect the word from the pollution and to provide the safe conditions with the increasing productivity simultaneously.
- The important mission of the university education in mechatronics engineering besides the technical knowledge, must focus on various sciences to produce graduates, who possess a sense of community, self-confidence and capabilities to face to the global challenges.

ACKNOWLEDGMENT

The authors very strongly would like to express their indebtedness to the dedication and commitment of the faculty, staff and students of Duy Tan University, Vietnam for bringing this project to fruition. They also hope that this paper will contribute some idea in spirit of the conference.

REFERENCES

CDIO. (2010). The CDIO Standards v 2.0 . CDIO. http://www.cdio.org/files/standards/CDIOStds&Rubricsv2.0_2010Dec8.pdf

David, G. C., Geoffrey, T. W., & Paula, M. C. (2018). *The Routledge Companion to International Human Resource Management.* Routledge Publisher.

Duong, V. U., Dong, T. L., Tran., B., & Le, N. (2017). *Refocus of CDIO Standards to enhance students' entrepreneurship skills.* Proceedings of the 13th International CDIO Conference, University of Calgary, Calgary, Canada, June 18-22.

Enelund, M. (2012). Integration of Education for Sustainable Development in a Mechanical Engineering Programme. *Proceeding of the 8th International CDIO Conference*, Queensland University of Technology, Brisbane, July 1-4, 2012.

Eric, R. (2011). The Learn Startups: How Today's Entrepreneurs Use Continuous Innovation to Create Radically Successful Businesses. Crown Business.

Graham, J. A. (2016). Practical Arc Welding. Flairnet publisher, Australia.

Howell, S. G. (1992). Risk Reduction Engineering Laboratory Cincinnati, A Ten Year Review of the Plastics Recycling, Elsevier Science Publishers.

Jefferson, H., Robert, D., & Edward, K. (2009, July). Plastics recycling: Challenges and opportunities. *Philosophical Transactions of the Royal Society of London. Series B, Biological Sciences, 364*(1526), 27. doi:10.1098/rstb.2008.0311 PMID:19528059

Johnson, E. B. (2001). Contextual teaching and learning: What It Is and Why It's Here to Say. *Corwin Publisher., ISBN-13*, 978–0761978657.

Mark, D. C., & Amy, C. E. (2005). Failing to Learn and Learning to Fail (Intelligently): How Great Organizations Put Failure to Work to Innovate and Improve. *Long Range Planning, 38*(3), 299–319. doi:10.1016/j.lrp.2005.04.005

Martin, V.B., (2014) -Mobile Robotics R&D student team at Universidade Federal de Santa Catarina.

Patnarin, B., Pankai, K., Gregory, K., John, B., &Dharmendra, K. M. (2022). Polyethylene Terephthalate (PET) Bottle -to- Bottle Recycling for the Beverage Industry: A Review. *Polymers 14*(2366).

Pivnenko, K., Jabobsen, L. G., Eriksen, M. K., Damgaard, A., & Astrup, T. F. (2015). Challenges in Plastics Recycling. *Proceedings Sardinia 15th International Waste Management and Landfill Symposium.*

Sander, W., André, S., Martin, C. & Roy, T. (2010). The relationship between Entrepreneurship and economic development: is it U-shaped? *Foundations and Trends in Entrepreneurship, 6*(3), pp. 167-237. doi:. doi:10.1561/0300000023

Sham Tickoo. (2016). *AutoCAD 2016: A Problem- Solving Approach, Basic and Intermediate,* (22nd ed.). Purdue University Calumet.

Sham Tickoo. (2016). *Autodesk – Inventor 2016 for Designers,* (16th ed.). Purdue University Calumet.

Thomas, E., & Steven, S. (2006). Integrating the Study of Technology into the Curriculum: A Consulting Teacher Model. *Journal of Technology Education, 18*(1), 26–37.

Tran, T. L., Tung., D. M., & Nguyen, T. C. (2014). Improvement of Individual Learning and Instructor's Role for Better Team Learning. *Proceedings of the 10th Annual International CDIO Conference*, Barcelona, Spain, June 16-19.

Chapter 4
Mechatronics Engineering in the Modern World and Its Use in the Construction Industry

Aditya Singh
https://orcid.org/0000-0001-9347-5627
Lovely Professional University, India

ABSTRACT

Mechatronics engineering is a branch of engineering made by combining several other main branches as well as derived branches of engineering, which comprises mechanical engineering, electrical engineering, electronic engineering, computer science engineering, product engineering, robotics, and so on. In the initial development of mechatronics engineering, only mechanics as well as electronics were combined to form it, but as the time passed and with the help of technological advancements, it was expanded to combine other derived and subbranches of engineering. In this book chapter, mechatronics engineering is briefly explained, and how it originated as well as came to current level. It also highlights the role of mechatronics engineering in the construction sector, as well as how it transformed the current construction practices. It also briefly talks about building automation, CAD, and structural dynamics which are a form of applied mechatronics that help in the modern construction industry in its own ways.

DOI: 10.4018/978-1-6684-5887-7.ch004

INTRODUCTION

Mechatronics Engineering is also known by the name of Mechatronics, which is engineering's interdisciplinary branch, which can concentrate on the way to incorporate different main branches of engineering like Electrical Engineering, Mechanical Engineering as well as Electronics Engineering, in addition to other branches or subbranches of engineering like Robotics, systems, computer science, product engineering, control, as well as telecommunications. With the advancement of technology as the time passed, numerous engineering subfields were able to accomplish in growth or increase in number as well as in getting accustomed. Mechatronics engineering's main purpose has been to be able to produce a design sol which has the ability to combine all the numerous subfields as much as possible. Initially, mechatronics engineering or simply mechatronics was an area which was supposed to integrate only mechanics as well as electronics, due to which its name was made by combining the mecha from mechanics with tronics from electronics to get its name mechatronics. However, with the passage of time, the field of mechatronics started becoming more and more complex and its technical systems nonstop changed as well as advanced, the definition of mechatronics was broadened up, in order to comprise more technical areas in it.

What is the Importance of Mechatronics in the General Sense?

Mechatronics is an area which is significantly needed for the advancement and development of future engineering factories. It is also not an engineering area which people needed to know in their workplaces in the past. However, in the recent decades more and more devices are getting connected with the internet and there has been increased cooperation between mechanical engineering, electrical engineering, electronics engineering as well as computer science engineering to create new products, which are highly advanced and versatile than the existing products in the past. Thus, this trend has been spreading as the time passes and with the arrival of new such products, the working efficiency of people has reached considerable improvement. Due to this the earlier isolated fields and subfields of engineering are now becoming more and more interconnected. This further requires mechatronics engineers who can work as a specialist wherever necessary and needs to handle complex situations in this field. It is further improving the lives of general people with the arrival of interconnected devices and providing them comfort.

What is the Need and Significance of Mechatronics in the Construction Sector?

Originally, the construction sector was seen as a separate area and it wasn't connected with mechatronics even during the past decades when a number of developments were going on in the area of mechatronics. It is true that the construction sector wasn't much affected by mechatronics like mechanical engineering, electronics engineering and so on. However, by looking closely, it can be noticed that there are applications of mechatronics are present even in the construction sector, like CAD, structural dynamic systems, automation as well as robotics, BIM, building automation and home automation. These all are part of applied mechatronics in the construction sector, some of them helps in designing of structures, creation of 2D and 3D building drawings, whereas some of them helps in the construction process and execution of construction projects. Compared to the old days of the construction sector, with the assistance of applied mechatronics in construction projects, the manpower and labour requirements in projects has been decreased; efficiency, quality, time of completion of construction projects has also been increased considerably. It is expected that with the advancement of technologies and development of mechatronics engineering, the construction industry can have more significant developments in the future.

Objectives

There are some major objectives of this book chapter, which are as follows:

- The chapter focuses on the present scenario of mechatronics engineering and explains it briefly.
- This chapter highlights the importance of mechatronics engineering in the construction field.
- The chapter further describes the applications of applied mechatronics in the construction sector briefly.
- This book chapter further briefly explains the history and evolution of mechatronics engineering along with a systematic methodology of the way these applied mechatronics are applicable in different stages of construction projects.
- Further, there is a thorough review of literatures in the area of construction mechatronics and possible research gaps, mentioned in the chapter.

MOTIVATION

The study was performed by the author because of the tremendous growth of mechatronics sector in the recent decades. The author observed that in the past two decades, especially the last decade, the mechatronics engineering has become a major sub-branch of engineering, which has drawn the attention of the people on the global level towards itself. The interest of people is also keep on increasing in mechatronics engineering and its being applied in many other areas. The boundary of mechatronics engineering is also keep on increasing as the time passes. The author being a civil engineer, believed that it is important to highlight the utility of mechatronics engineering in the construction sector and allow the people to know that the mechatronics engineering is no longer only limited to mechanical engineering, electronics engineering, automation and robotics, but to other areas also as well as there is a possibility to further expand its application in other areas in the future. The author also focused the condition of mechatronics engineering in the present scenario.

ORIGIN, HISTORY AND EVOLUTION

The origin of the mechatronics word was through Japanese – English in addition to, Yaskawa Electric Corporation senior engineer, named Tetsuro Mori created it in the year 1969. The company in the country Japan, registered this word as its trademark and by the year 1971, it was given the registration number of "4632714". After some time, the company released the right to use the word mechatronics for the people to use it freely. That's how the word mechatronics started being used worldwide. At present, the translation of the word mechatronics in different languages started, as for advanced automated industry, the term mechatronics is an important one which can't be ignored. There are many people who consider the word mechatronics as a novel catch word identical to robotics, automation as well as electromechanical engineering. Based on French standard NFE 01-010 it is defined as a method with the goal of synergistic incorporation of mechanics, computer science, control theory as well as electronics within the manufacturing process in addition to include in the design of products, so that its functionality can be optimized or enhanced. As time passed, there were engineers created especially to take care of the field of mechatronics or mechatronics engineering. These engineers were having a specialization in mechatronics engineering and they were called as mechatronics engineer, whose role was to combine the principles of various areas, like mechanics, computing as well as electronics, in order to create a highly dependable, cost-effective as well as simple system in comparison to the traditional ones. The main instance of a mechatronics system is an industrial robot, which comprises features of mechanics, electronics as well as computing to perform its given tasks on a daily basis. The control engineering part of a mechatronic system's queries are solved with the assistance of engineering cybernetics, in order to govern or manage such systems. With cooperation, the mechatronic units accomplish the targets related with production as well as receive supple in addition to non-rigid properties related to manufacturing in the production system. New equipment associated to production, comprises of mechatronic units which are incorporated in accordance to a control architecture. The highly recognized architectures include heterarchy, hierarchy, hybrid as well as polyarchy. The ways for attaining technical outcomes are defined by control algorithms that is not necessarily use official approaches in their given designs. The hybrid systems that are significant to mechatronics involve planetary exploration rovers, production systems, synergy drives, automotive subsystems like spin assist as well as antilock braking systems, in addition to daily equipment like hard disks, autofocus cameras, phones, video, as well as CD players.

LITERATURE REVIEW

Klein and Reinhart (2016) studied the mechatronic systems and they worked on implementing Agile procedures in machinery as well as plant construction. According to them, earlier before their research work the Agile techniques were limited to IT projects and they were able to utilize it in mechatronic production systems, in order to enhance their development process. They stated that they classified as well as compared the Agile engineering and incorporated its techniques into mechatronic engineering process to get the desired results.

Habib and Komoto (2014) worked on comparatively analysing the design concepts of mechatronic systems with the assistance of computer aided design tool. They found out that designers are required to face complications resulting from the incorporation of subsystems with the numerous disciplines of engineering. They used function state modelling as well as CAD system to propose a method in order to develop product architecture with the assistance of comparative analysis of the functional overview and somatic breakdown of the current mechatronics systems to get the desired results.

Yudin et al. (2021) worked on the application of CDIO standards for cyber physical education in robotics as well as mechatronics and to solve the rising problem of the earlier mentioned application of the CDIO standards. They particularly worked in the area of nuclear physics as well as technologies, nuclear reactors as well as materials. They stated that their work will be able to content the novel necessities needed to modernize the existing education programs. They also developed the 3D modelling as well as design skills to get the desired outcome.

Zeitler et al. (2016) worked on CAD as well as simulation of spatial opto-mechatronic interconnected devices. They found that optical technologies became significant in the current sensor as well as communication applications but they weren't able to find manufacturing as well as design methods in the area of spatial opto-electronics so far. They worked on technology particular methods which were practical in order to design spatial opto-mechatronic applications with the emphasis on CAD as well as raytracing simulation of spatial waveguides to obtain the desired results.

Komoto and Tomiyama (2012) proposed a product modeling framework as well a computer aided design system intended for conceptual design of complex mechatronics products which emphasised on hierarchical system breakdown as well as constancy management of design info through distinct disciplines of engineering. They also proved the conceptual design case at a printer manufacturer in their study and they produced the desired study result.

Numan et al. (2016) worked in a computer simulated environment aimed at designing as well as analysing of outrunners brushless motors as well as its justification through the measurement of the chief parameters in actual motors. They worked on these special brushless motors intended for mechatronic mobile robotics systems and they used two models for their study, where the first one was a concentrated parameters model which they used for designing as well as analysing purpose. On the other hand, the second model was built on two dimensional electromagnetic finite element analysis on a computer aided design of the supposed motor.

Yakimov and Iovev (2019) worked to develop a novel leaning environment to improve education in the sectors like industrial automation in addition to mechatronics. To achieve their goal they worked on developing appropriate practice assignments as well as dependable electronic system in field of mechatronics, which comprised of actual parts like sensors, motors, PLCs as well as motor drives. They further focused on the establishment of the desired lab and the required essential software to get the desired study outcome.

Carvalho et al. (2021) worked to design advanced products related mechatronics. They incorporated Design Thinking technique in the initial stages of the products, Concurrent Engineering as well as Agilism during product development phase, and then Intellectual Property Management activities with it. This way they were able to incorporate the three most significant current philosophies regarding product design and at the same time, they showed the way to manage the intellectual assets without any problem.

Heuser and Bi (2021) performed a qualitative study on companies having mechatronic products and the way the technical debt effects them. To proceed with their study they initially performed a comprehensive survey which involved fifty experts out of which 21 of them were head of their respective departments and all of the experts were from twenty one famous companies in addition to ten sectors in the region influenced with German speakers. They classified the technical debt in 12 categories and found that two-third were still present, which can cause impending risk to the system.

Yuan et al. (2022) studied passive mechatronic vibration absorbers and worked on designing an appropriate methodology for them. They found that the current techniques have obvious precincts and the requirement to consider the parasitic effects of the device like transducer resistance, in order to assure the anticipated accuracy in the performance, which can be time consuming in some cases. They claimed that their proposed methodology was able greatly improve the performance and they verified it with the needed experiments. They also stated that their study can be implemented on other mechatronic absorber types as well as directly applied on vibration suppression of other engineering structures.

Sanchez et al. (2022) studied mechatronic systems and they worked on their state estimation problems, particularly underactuated mechanical systems. They also found one major challenge during their study, which was to evade the peaking phenomenon in case of nonlinear systems while performing state observation. They used Attractive Ellipsoid Method to obtain reliable estimation while checking the unmeasured state signals to get the desired study outcome.

Xia et al. (2021) worked on modular inexpensive atomic force microscope in order to use it for precision mechatronics teaching purpose. They found that the current atomic force microscope is not appropriate to be used for education purpose for mechatronics, as they were made originally for a different purpose. In their study they focussed on the design aspect of the new atomic force microscope as well as the way to implement them. They also stated that their proposed atomic force microscope was inexpensive, as its cost was less than or equal to four thousand US dollars, additionally it can be operated easily.

Geraldes et al. (2022) studied mechatronic architecture as well as they worked on designing a High Dynamic Double Crystal Monochromator. They stated that their proposed monochromator was intended for Sirius light source and it was capable of implementing an isolated mechatronic architecture having high closed loop performance. They also used dynamic error budgeting workflow in order to assist in developing it to provide positioning performance enhancements by factors of five as well as one hundred.

Wang et al. (2021) studied mechatronic systems and they performed primary failure mode and effects analysis with the assistance of function motion action hierarchy structure. They found that FMEA was initially utilized in aerospace systems after that in electronic systems in order to analyse failures, failure clearance and so on. They stated that the application of FMEA was found to be effective, so it was utilized in mechatronic systems to enhance the reliability but they found that it has certain limitations. They then proposed function motion action hierarchy structure particularly appropriate for mechatronic systems to obtain the desired outcome.

Afolalu et al. (2021) took a case study of mechatronics to proceed with their study and studied the sustainable 4th industrial revolution as well as desirable role of manufacturing processes in it. They highlighted the evolution of industrial revolution from the early 19th century till now and they stated that at present the novel digital technology, which consist of cyber physical systems as well as Internet of Things, is the important influencing factor behind the current 4th industrial revolution. In their research work, they mainly focused on mechatronics applications and how they are going to be affected during the fourth industrial revolution.

Braun et al. (2022) worked on brownfield systems and designed a methodology in order to find functional relations of mechatronic constituents as well as assemblies in them. They worked in Germany for their research study and they stated that their proposed methodology will be useful for automated reconstruction of functional relations as well as benefit the creation of Digital Twin, to produce the desired outcome.

Tokarev et al. (2021) worked on applying virtual reality instruments in pre-professional education, which includes schools and so on in Russian Federation. Their study area was in the direction of mechatronics as well as robotics and they performed their research after gaining experience from National Research Nuclear University MEPhI, in addition to their proposed work was in accordance to the standards of CDIO.

Oguzlnci et al. (2022) studied multilevel mechatronic systems and worked on physically adapting co-simulation time steps with the assistance of the ability to control the past co-simulation runs in their study. They found out that modeling as well as simulation of mechatronic systems needs an approach which should be multi-disciplinary one and they performed their research work in Belgium. They stated that their proposed method is faster than the existing method having similar accuracy in one dimensional in addition to multilevel mechatronic systems.

Bhowmick and Ganguly (2022) worked on vibration control of a group of mechatronic systems with the assistance of Negative Imaginary theory. They proposed a novel controller design method for the above mentioned group of mechatronic systems, which can please the stable as well as minimum phase of NI property. They performed their study in India and their methodology depends upon the principle of classical Internal Model Control. They used a lightweight cantilever beam and its vibration attenuation problem to show the effectiveness of their proposed methodology.

Wang et al. (2021) worked on an industrial grade parallel tool head application in aviation manufacturing in addition to physics based mechatronics modeling in China. They stated that their study results demonstrates that driving torque as well as tracking error are consistent with the real values, which further authenticated their proposed mechatronics modeling's efficacy. Their research work will help the technicians in addition to engineers to enhance their analysis as well as effectiveness of their design.

Baryshev et al. (2021) worked on applying CDIO standards in cyber-physical education in the area of robotics as well as mechatronics. They performed their study in a Russian research university – National Research Nuclear University MEPhI, and they found out that the current technical education is not having a particular standard for quality control in respect to the knowledge of a specialist. They stated that novel standards in the education programs are needed to be introduced and with the help of their proposed work, the quality of the qualification of trained specialists can be enhanced. They also took the instance of developing digital electronic skills in their research work.

Mehdi and Boudi (2021) worked in the direction of application of a sustainable conceptual design of mechatronic systems in order to assist regenerative braking system. They performed their study in Morocco and they found out that currently there is a need of a sustainable design methodology for mechatronic systems, so they proposed one accordingly and took regenerative braking system as an instance. They stated that through this, environmental aspects can be considered easily while developing them.

Dyboy and Stark (2021) worked on mechatronic products and they analyzed common functional models for them. They performed their study in Germany and they performed their research on functional flows of two mechatronic products which were complex in nature. They stated that their study result showed the hierarchy of functions as well as its relationships in regards to functional modeling. They also claimed that their work will be beneficial in finding the potential for function exchanging in the future.

Goodall et al. (2020) worked in the direction of railways finding the benefits of mechatronics as well as health monitoring, like other modes of transportation. They performed their study in United Kingdom and they explained the present research work going on as well as system wide benefits from low level like vehicle sub-systems to high level.

Gurtner et al. (2022) worked on creating effective oscillation detection in order to verify mechatronic closed loop systems with the assistance of search based testing. They found the existing established oscillation methods in the process industry cannot be applied directly to mechatronic systems, as frequent alterations of the operating points as well as tracking of shift trajectories have to be considered. To overcome the above mentioned challenges, Empirical Mode Decomposition joint with the decomposed mode's assessment are required.

Bula et al. (2021) used mechatronic scrap to present a cost efficient autonomous robot which can clean windows – MechCleanBot. They performed their study in Kosovo and they stated that the robot can be used in offices, skyscrapers, numerous objects, homes and even at dangerous places where it's unsafe for humans as well as it will save time and effort. They claimed that the robot can be controlled by Bluetooth RC Controller application through smartphone also.

Shi and Zhang (2021) studied autonomous intelligent mechatronic systems and they worked in the direction of advanced model predictive control framework for them. They wrote a review paper and covered an overall outline of their study as well as their perspective in their paper. They performed their study in Canada and they proposed for a unified model predictive control framework to overcome the existing challenges. They also analyzed the applications of model predictive control in mechatronic systems of different types like autonomous ground robots, autonomous marine vehicles as well as unmanned aerial vehicles.

Yang et al. (2021) studied new mechatronics – electro – hydraulic power coupling electric automobile and they researched on driving control strategy as well as optimization of fuzzy logic in them. They stated that their proposed model can decrease large peak torque as well as enhance the dynamic performance of an automobile during frequent starts as well as stops. They performed their study in China and they claimed that the consumption rate of the battery is also decreased up to a considerable extent.

Heimicke et al. (2021) worked in the direction realizing agile principles in the process of developing mechatronic systems. They performed their work in Germany and to complete their study, they used individual selection of appropriate practices, approaches as well as process models. They found out that the current agile methods are unable to meet the complete demands to develop mechatronic systems, like current organizational structure is not even thought of, and they claimed that their proposed study is able to overcome these challenges as well as produce desired outcome.

An et al. (2020) worked on mechatronic indicators which are dependable, intending for self-powered liquid sensing in the direction of safe transportation as well as smart manufacturing systems. Their work was based on Nano-generator's principle and they used block inserting mechatronic panel as well as an active electronic sensor in order to overcome the existing challenges in addition to support their study, to give the desired results. They also claimed that their study will be beneficial in encouraging the development of transportation systems, smart cities as well as industry 4.0.

RESEARCH GAPS

After thoroughly reviewing a number of scientific and research papers, particularly the recent ones, the author wasn't able to find much about role of mechatronics engineering in the construction sector. There weren't much direct studies done by researchers in order to relate the construction projects with mechatronics engineering. Very limited study was done by past researchers in this area.

MAIN FOCUS OF THE CHAPTER

This book chapter mainly focuses on the present scenario of mechatronics engineering in the current world. It briefly explains about mechatronics engineering and mentions the history as well as evolution of it as the time passed. Further, the chapter points out the major parts of applied mechatronics which are applicable in the construction sector. This chapter then describes those applied mechatronics parts which helps in the construction projects as well as their usages in different stages of the construction projects. Through this chapter, the author is highlighting the ways the construction industry is being modified and influenced at present. Thus, the growth of mechatronics engineering will increase further, especially in the construction sector.

ISSUES, CONTROVERSIES, PROBLEMS

The author has discussed some major issues in this book chapter briefly, which can't be ignored. Even though mechatronics engineering is capable of transforming the construction sector completely and allowing it to reach new heights, there are still limitations because of the existing technologies. Automation as well as robotics have the ability to improve the construction project's quality and decrease the execution time of projects. However, with the current technologies and initial cost of using robots or automated technologies, it is difficult to implement them on a large scale or throughout the world. Plus, sophisticated skills are needed for people handling, using, manufacturing, repairing and maintaining them. Similarly, in regards to home automation and building automation, which can modernise typical old styles homes and buildings, are hard to execute in all construction projects related to commercial as well as residential structures, mostly because of high initial cost of the project. With the advancement in the technologies in the future and possible reduction of initial cost in the future, these major issues can be solved with time and further improve the construction sector.

METHODOLOGY

In this section, phases of a construction project in order to construct any structure is explained briefly, along with the need of applied mechatronics to be used in different phases of a project: 1) Survey is to be done – which can be done with the help of automation such as drones for surveying purpose. 2) Planning of the project is required – with the help of CAD, 2D plans can be made for any structure easily, including residential and commercial buildings. 3) Design stage – Structure Dynamics Systems can help in assigning different types of loads applicable on the structure, like Dead Load, Live Load, Wind Load, Seismic Load and so on based on terrain condition, Building Codes, etc.; CAD and BIM software can be used for 3D modelling of the structures; in case of automation of home or building is required, then it needs to be considered in the designing stage. 4) Excavation of the soil – Robots or Driverless construction vehicles or excavators can help in the excavation stage. 5) Construction stage – Robots can help in the construction of structures in a project; in case automation of home or building is required then construction of residential or commercial building can be done accordingly. 6) Handover stage, where the construction projects are required to be handed over to the client by the main contractor. 7) Operation stage, where the client starts operating or using the structure based on its needs.

APPLICATIONS

In this section, some important applied mechatronics which are used and can be used in the construction projects are mentioned here, which are as follows:

- Computer Aided Design (CAD)
- Automation
- Robotics
- Drone
- 3D Printing
- Structural Dynamic Systems
- Home Automation
- Building Automation

Figure 1. Showing phases of a construction project on the left hand side and applied mechatronics on the right hand side of the reader

Computer Aided Design (CAD)

Computer Aided Design (CAD) is a software which engineers, construction managers, designers and architects uses to create designs or plans of a structure. It uses a computer to help them in creating, optimizing, analysing and altering the designs or plans of a structure. Productivity and quality of designs and plans greatly increased by it. There are different types of software which includes some which completely works in 2D, whereas some software completely works in 3D, while mostly translate between 2D and 3D. It is extremely useful in construction. The designers of a subcontractor are able to take the drawings from architects in a project, then they are able to add more details before proceeding with the construction. Plus, it is easy to compare with the actual plan and to recheck the details. Nowadays the designs or plans of structure are being stored in the cloud and at different locations, engineers and designers are able to obtain it easily.

Automation

Automation in the general sense means using wide range of technologies which can decrease the human interference as much as possible. The interference of humans are almost replaced by predetermined criteria, related actions as well as sub-process relationships embedded in machines. In the construction sector, methods, systems as well as processes which are combined in bigger machines making them almost independent while performing activities related to construction. They have the ability to decrease the injuries happen to humans at the site, reduction of completion time of activities, and help in improving quality assurance as well as quality control of the project, in addition to reducing the lack of skilled labours at the site.

Robotics

Robotics is an interdisciplinary branch which involves different branches of engineering like computer science engineering, mechanical engineering, electronics engineering, electrical engineering, bioengineering, and so on. Robots are used and can be used in the construction projects to help humans or to minimize the manual labour as much as possible. The robots can work at a fast speed, for long duration of hours without getting tired at a construction site unlike humans. They greatly increase the work output, accuracy in the work, replaces the lack of skilled labours as well as quality of the work. They are also suitable to work in dangerous environment at the construction site. The SAM100 robot made by a Construction Robotics company, which was New York State based one, has the ability to lay three hundred fifty bricks/hr. Then there are some driverless construction heavy vehicles which were seen in Japan in the recent years. However, at present either such advanced robots are either present in some parts of the world on a small scale or still in the testing period.

Drones

In construction projects, Drones are highly useful in inspecting the site, they help in the process surveying vast areas at the site. Traditionally, a team of surveyors are needed to survey the vast areas to be used for construction of big projects and this time consuming. However, with drones only one pilot is needed who can control the drone and survey the whole area within a short period of time. The drone pilot can be even present offsite and control the drone. It helps in performing the aerial surveys in a construction project.

3D Printing

3D Printing is famous by the name of additive manufacturing. They are used being used in construction projects to build walls. Either they help in fabricating walls at the site or they help in fabricating walls in separate sections in order to be transported to the give location of the construction site. Cazza system which is UAE based one and Building On Demand system from Copenhagen are some examples of 3D printing in the world. They have the ability to decrease the labour as well as time needed in a construction project considerably.

Structural Dynamic Systems

Structural Dynamic Systems is a type of system which helps in analysing the structures, like when structures are exposed to dynamic loading. These dynamic loads comprises of earthquakes, wind, blasts, people, traffic as well as waves. In order to find dynamic displacements, modal analysis as well as time history, Dynamic Analysis must be done. Through this engineers are able to understand how will the structure behave when it is exposed to certain types of loads, like wind load, dead load, live load, snow load and seismic load. This helps in designing a safe structure and software like Staad Pro is common software to be utilized for this purpose.

Building Automation

Building Automation is a general term which is used for commercial buildings which are having automatic centralized control of air conditioning, shading, heating, security systems, ventilation, electrical systems, access control as well as other interconnected systems by using Building Automation System or Building Management System. It is made in order to maintain and improve the comfort level of people residing in it. Basically, these buildings are also called as smart buildings or intelligent buildings.

Home Automation

Home Automation is similar to a building automation in nature, but it is used for residential buildings or homes. It is also called as Domotics, which will give the user to automatically control entertainment systems, electrical systems, electronic appliances, security systems and so on. It removes the manual process of handling all the things in a residential building and improves the comfort level of the inhabitants. The devices can be connected to IoT, which will become an important part of Internet of Things. Smartphones, tablets, desktop computers, and laptops can be used as user interface to control the whole system or there can web interface where internet will be used from a different location.

RESULTS AND DISCUSSION

In this section, number of graphs are made on the basis of data collected by different organizations to support the study. By analysing these graphs, the readers will be able to understand the current trends of mechatronics engineering in the present scenario.

Table 1. Showing percentage of male mechatronics engineers in two different years – 2010 and 2019

Year	Male Mechatronics Engineer (In %)
2010	84.09
2019	83.16

Source: ZIPPIA THE CAREER EXPERT

Table 2. Showing percentage of female mechatronics engineers in two different years – 2010 and 2019

Year	Female Mechatronics Engineer (In %)
2010	15.91
2019	16.84

Source: ZIPPIA THE CAREER EXPERT

Table 3. Showing percentage of Asian mechatronics engineers in two different years – 2010 and 2019

Year	Asian Mechatronics Engineer (In %)
2010	14.77
2019	15.89

Source: ZIPPIA THE CAREER EXPERT

Table 4. Showing average age of Asian mechatronics engineer's gender wise

Gender	Asian Mechatronics Engineer (Average Age)
Male	41
Female	34.5

Source: ZIPPIA THE CAREER EXPERT

Table 5. Showing average annual salary (in 1000 USD) of mechatronics engineers with master's degree

Mechatronics Engineers with Master's Degree (In %)	Average Annual Salary (In 1000 $)
13	93

Source: ZIPPIA THE CAREER EXPERT

Table 6. Showing jobs in two different industries for mechatronics engineers (In %)

Industry	Mechatronics Engineers Jobs (In %)
Technology	19
Start-up	3

Source: ZIPPIA THE CAREER EXPERT

Table 7. Showing unemployment rate of mechatronics engineers (In %) in the year 2010 and year 2019

Year	Unemployment Rate of Mechatronics Engineers (In %)
2010	6.09
2019	1.42

Source: ZIPPIA THE CAREER EXPERT

Table 8. Showing two US states having jobs for mechatronics engineers in the year 2022

US States	Jobs For Mechatronics Engineers in the Year 2022
New York	1,750
California	7,274

Source: ZIPPIA THE CAREER EXPERT

Table 9. Showing market revenue of CAD 3D designs (in Billions USD)

Year	Market Revenue of CAD 3D Designs (In Billions USD)
2018	5.5
2028	10.48

Source: Statista

Table 10. Showing market revenue of CAD 2D designs (in Billions USD)

Year	Market Revenue of CAD 2D Designs (In Billions USD)
2018	1.97
2028	3.35

Source: Statista

Table 11. Showing market revenue share of two popular CAD software vendors in the year 2019 (in %)

CAD Software Vendors	Market Revenue Share In The World In 2019 (In %)
Autodesk	29
Bentley Systems	6

Source: MMR

Figure 2. Showing percentage of male mechatronics engineers in two different years – 2010 and 2019
Source: ZIPPIA THE CAREER EXPERT

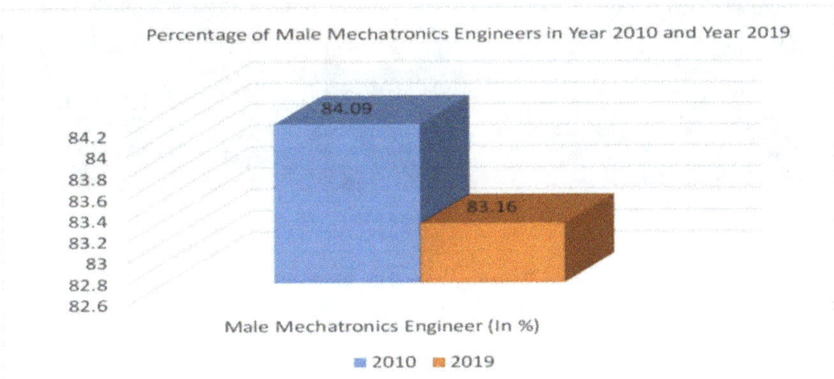

Figure 3. Showing percentage of female mechatronics engineers in two different years – 2010 and 2019
Source: ZIPPIA THE CAREER EXPERT

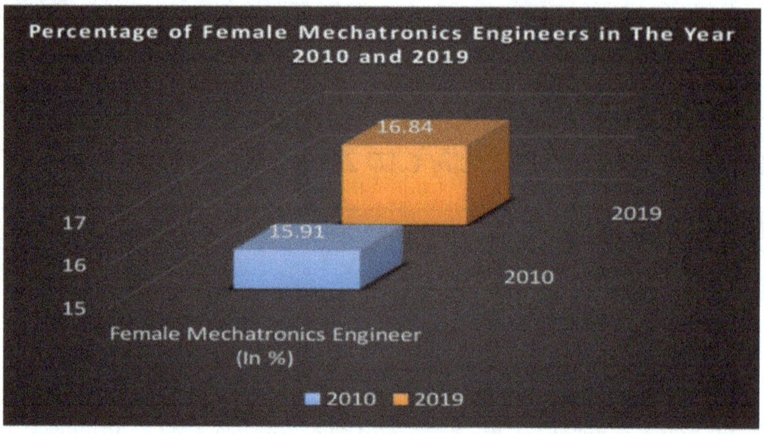

Figure 4. Showing percentage of Asian mechatronics engineers in two different years – 2010 and 2019
Source: ZIPPIA THE CAREER EXPERT

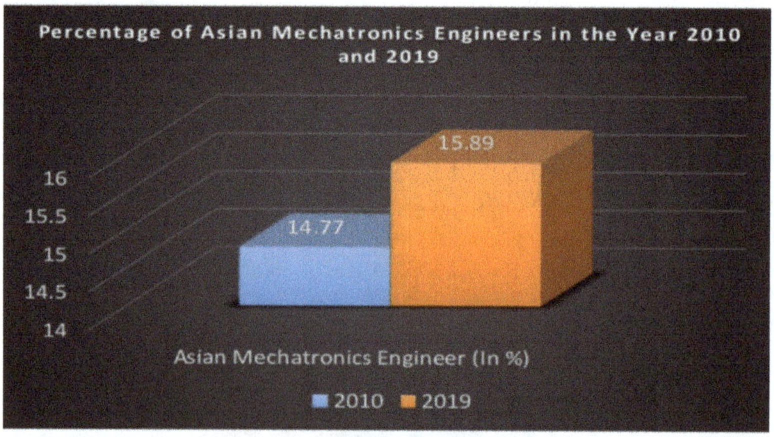

Figure 5. Showing average age of Asian mechatronics engineer's gender wise
Source: ZIPPIA THE CAREER EXPERT

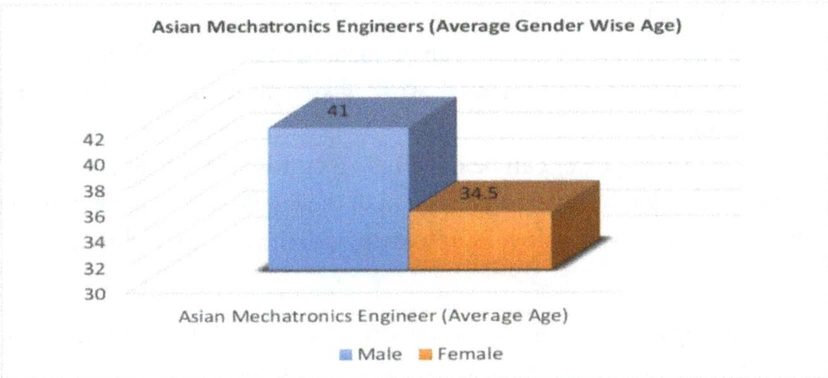

Figure 6. Showing average annual salary (in 1000 USD) of mechatronics engineers with master's degree
Source: ZIPPIA THE CAREER EXPERT

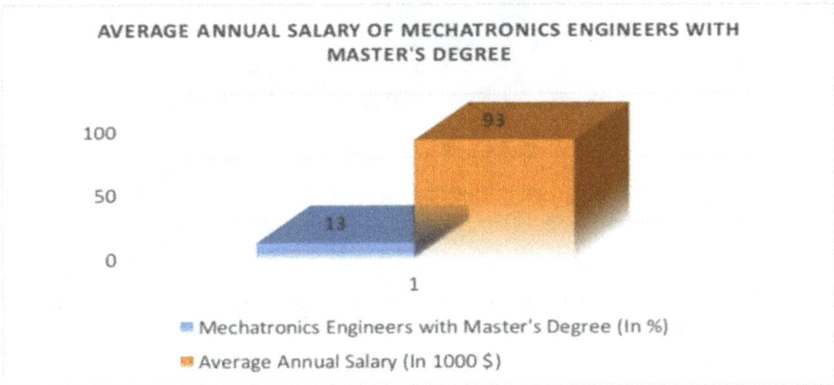

MAJOR ADVANTAGES OF MECHATRONICS ENGINEERING IN THE CONSTRUCTION INDUSTRY

In this section, the major advantages of using applied mechatronics engineering in the construction industry are discussed, which are as follows:

1. The working efficiency of the people increases with the usage of applied mechatronics products or software.
2. Time to complete the construction projects decreases significantly with their usage.

3. Accuracy in design of structures increases considerably.
4. Complex designs are easier to make with their usage.
5. Quality of construction projects increases drastically.
6. Detailed plans of the structures can be made without any difficulty.
7. Significant reduction of manual work takes place during the course of construction projects.
8. Reduction of manpower can be done upto a considerable extent.
9. Technological knowledge improvement of the employees happen while using them and makes the employees upto date with the current technologies.

Figure 7. Showing jobs in two different industries for mechatronics engineers (In %)
Source: ZIPPIA THE CAREER EXPERT

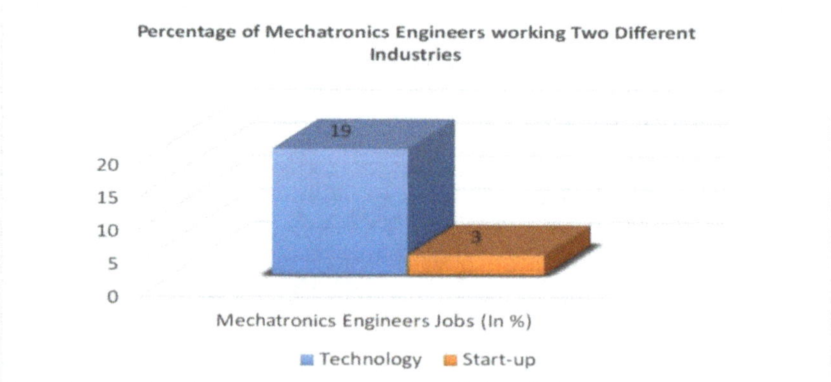

Figure 8. Showing unemployment rate of mechatronics engineers (In %) in the year 2010 and year 2019
Source: ZIPPIA THE CAREER EXPERT

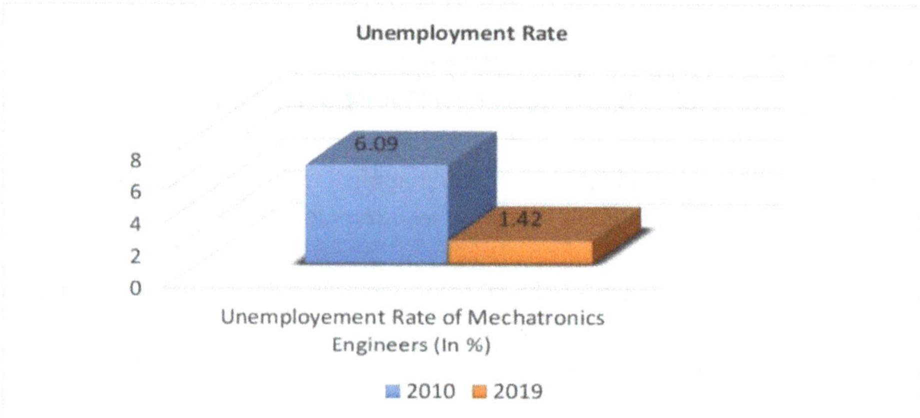

Figure 9. Showing two US states having jobs for mechatronics engineers in the year 2022
Source: ZIPPIA THE CAREER EXPERT

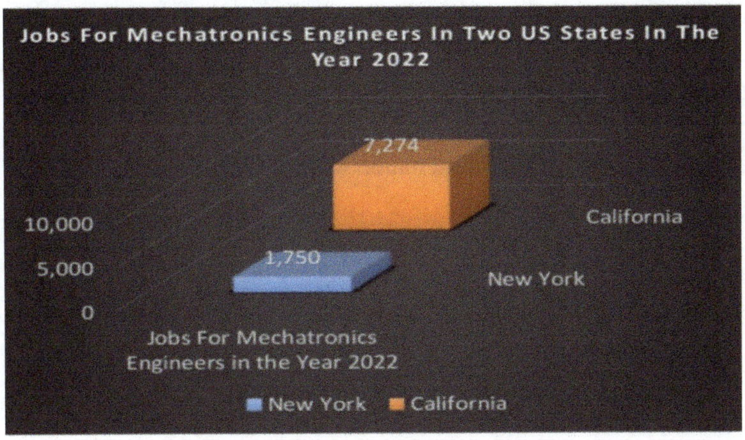

Figure 10. Showing market revenue of CAD 3D designs (in Billions USD)
Source: Statista

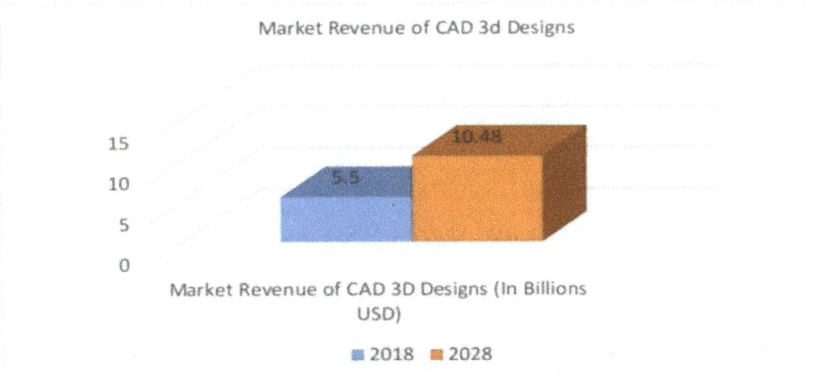

MAJOR DISADVANTAGES OF MECHATRONICS ENGINEERING IN THE CONSTRUCTION INDUSTRY

In this section, the author has discussed some major disadvantages of using applied mechatronics in the field of construction, which are as follows:

1. To utilize applied mechatronics software, systems in a construction project, there is a requirement of highly trained people who can operate it smoothly.

2. Initial cost of such construction projects increases with the modification of designs, requirements of highly sophisticated software, products and so on.
3. It is difficult to detect as well as correct any existing or possible problem which can occur.
4. Engineers who have knowledge of different disciplines and sub-disciplines of engineering are required at least at times when there is a complication in the design as well as execution.
5. The repairing as well as maintenance work are generally costly.

Figure 11. Showing market revenue of CAD 2D designs (in Billions USD)
Source: Statista

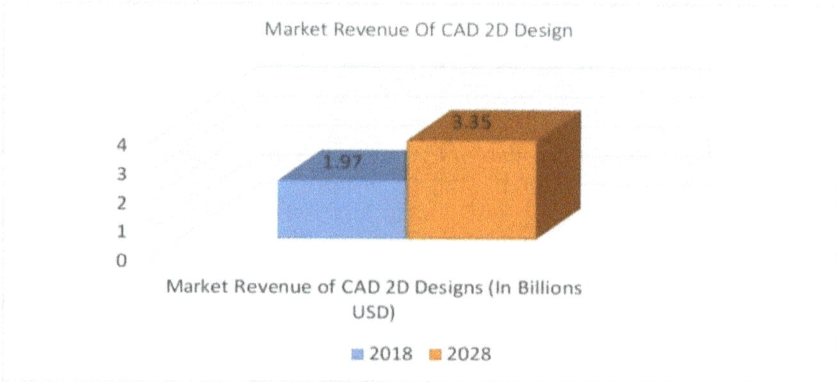

Figure 12. Showing market revenue share of two popular CAD software vendors in the year 2019 (in %)
Source: MMR

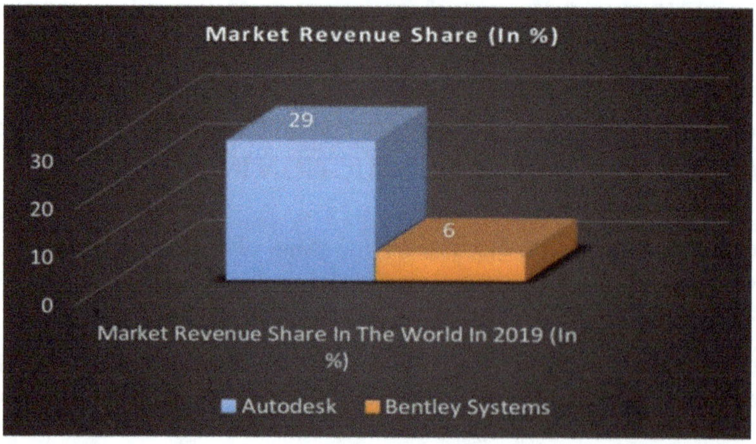

6. Specific applied mechatronics software as well as products are to be chosen carefully in a construction project.

LIMITATIONS

In this section, the limitations of this book chapter has been discussed. The author had focussed in explaining and highlighting the role of mechatronics engineering in the current scenario. A number of scientific and research papers were reviewed by the author, in order to find the gaps in the research. The chapter highlighted history and evolution of mechatronics briefly, not in detail. Further, this chapter focused on applied mechatronics which are and can be used in the construction projects, in short only. The book chapter only mentioned mechatronics engineering utility in the general sense in a limited way. Additionally the usage of applied mechatronics in different stages of construction projects were mentioned briefly.

SOLUTIONS AND RECOMMENDATIONS

Even though mechatronics engineering has a lot of future potential and applied mechatronics can enhance the construction sector greatly, there are still some major disadvantages and challenges of using them. The major problem is the initial cost of using mechatronic and applied mechatronic products as well as the expertise of engineers who need a knowledge of many disciplines of engineering to use them. The author believes that with the technological improvements in the future, the reduction of cost and increase accessibility of mechatronic and applied mechatronic products, systems as well as software will improve tremendously. More focus on mechatronics engineering is required in the universities and colleges to educate the promising engineers of the future. Plus, engineering students from different disciplines in the future must be taught some aspects of other disciplines and sub-disciplines of engineering.

FUTURE RESEARCH DIRECTIONS

With the evolution of mechatronics engineering over the past decades, the applied mechatronics had also emerged in the world. Due to which, the applied mechatronics can be used and are being used upto certain extent in the construction industry. However, to overcome the current limitations and challenges, there is a need to do more research in the area of mechatronics and applied mechatronics in the field of construction. More technological improvement and research on using cost-effective applied mechatronic products are needed to be done. There is a need to do research in the direction to expand the usage of mechatronics as well as applied mechatronics in the construction field, in the future.

CONCLUSION

This book chapter focused on the current status of mechatronics engineering in the present world. It explained mechatronics engineering in brief as well as it highlighted the history in addition to evolution of mechatronics engineering. The author thoroughly reviewed a number of scientific and research papers related to mechatronics engineering and its limited uses in the construction industry, in order to find the gaps in the research, where the author found that not much research work was done in directly using applied mechatronics in the construction projects. The author worked on a methodology in order to show the usage of applied mechatronics in the different stages of a construction project. Further, the chapter talked about the numerous major advantages and also some disadvantages of using applied mechatronics in the field of construction. The book chapter also explained some major applied mechatronics which are used and can be used in the construction industry. The author also used data collected by different organizations in order to create graphs for graphical analysis to support this study.

ACKNOWLEDGMENT

This research received no specific grant from any funding agency in the public, commercial, or not-for-profit sectors.

REFERENCES

Afolalu, I., & Kareem, S. (2021). Enviable roles of manufacturing processes in sustainable fourth industrial revolution – A case study of mechatronics. *Materials Today: Proceedings*, *44*(Part 1), 2895–2901. doi:10.1016/j.matpr.2021.01.099

An, W., Wang, Z., Jiang, T., Chen, P., Liang, X., Shao, J., Nie, J., Xu, M., & Wang, Z. L. (2020). Reliable mechatronic indicator for self-powered liquid sensing toward smart manufacture and safe transportation. *Materials Today*, *41*, 10–20. doi:10.1016/j.mattod.2020.06.003

Baryshev, Y., & Biryokov, S. (2021). Application of the CDIO standards for cyber-physical education in mechatronics and robotics in a research university on the example of development of digital electronic skills. *Procedia Computer Science, 190*, 45–50. doi:10.1016/j.procs.2021.06.006

Bhowmick, P., & Ganguly, A. (2022). Vibration control of a class of mechatronic systems using Negative Imaginary theory. *IFAC-PapersOnLine, 55*(1), 14–19. doi:10.1016/j.ifacol.2022.04.003

Braun, R., & Jazdi, S. (2022). A methodology for the detection of functional relations of mechatronic components and assemblies in brownfield systems. *Procedia CIRP, 107*, 119–124. doi:10.1016/j.procir.2022.04.020

Bula, I., Bula, E., & Hajrizi, E. (2021). Cost Oriented Autonomous Window Cleaning Robot from Mechatronic Scrap – MechCleanBot. *IFAC-PapersOnLine, 54*(13), 233–238. doi:10.1016/j.ifacol.2021.10.451

Carvalho, Hora, & Fernandes (2021). A process for designing innovative mechatronic products. *International Journal of Production Economics,* Volume 231, 107887. *Computer-aided design*, https://en.m.wikipedia.org/wiki/Computer-aided_design

Habib & Komoto. (2021). Comparative analysis of design concepts of mechatronics systems with a CAD tool for system architecting. *Mechatronics, 24*(7), 788–804.

Dybov, A., & Stark, R. (2021). Analysis of common function models for mechatronic products. *Procedia CIRP, 100*, 391–396. doi:10.1016/j.procir.2021.05.091

Geraldes, W., Witvoet, G., & Vermeulen, J. P. M. B. (2022). The mechatronic architecture and design of the High-Dynamic Double-Crystal Monochromator for Sirius light source. *Precision Engineering, 77*, 110–126. doi:10.1016/j.precisioneng.2022.05.009

Goodall, D., Dixon, R., Hamadache, M., & Ward, C. (2020). Railways Discovering Mechatronics and Monitoring - An Overview. *IFAC-PapersOnLine, 53*(2), 8488–8493. doi:10.1016/j.ifacol.2020.12.1427

Gurtner, Z., Heinz, A., & Ophey, T. (2022). Efficient oscillation detection for verification of mechatronic closed-loop systems using search-based testing. *Mechanical Systems and Signal Processing, 163*, 108112. doi:10.1016/j.ymssp.2021.108112

Heimicke, N., Ng, G.-L., Krüger, M., & Albers, A. (2021). A systematic for realizing agile principles in the process of mechatronic systems development through individual selection of suitable process models, methods and practices. *Procedia CIRP*, *100*, 619–624. doi:10.1016/j.procir.2021.05.133

Inci, C., & Bosmans, V. (2022). Offline adaptation of co-simulation time steps by leveraging past co-simulation runs in multi-level mechatronic systems. *Mechanism and Machine Theory*, *171*, 104740. doi:10.1016/j.mechmachtheory.2022.104740

Klein, T. P., & Reinhart, G. (2016). Towards Agile Engineering of Mechatronic Systems in Machinery and Plant Construction. *Procedia CIRP*, *52*, 68–73. doi:10.1016/j.procir.2016.07.077

Komoto, H., & Tomiyama, T. (2012). A framework for computer-aided conceptual design and its application to system architecting of mechatronics products. *Computer Aided Design*, *44*(10), 931–946. doi:10.1016/j.cad.2012.02.004

Mehdi & Boudi. (2021). Towards a sustainable conceptual design of mechatronic systems application to a regenerative braking system. *Materials Today: Proceedings*, *45*(Part 8), 7588–7596.

MMR. (2020). Computer Aided Design (CAD) Market: Global Industry Analysis and Forecast (2021-2027). *Technology Trends*. https://www.maximizemarketresearch.com/market-report/global-computer-aided-design-cad-market/63443/

Numan, J., Jorge M, B., Rodrigo, A., & Esteban, C. (2016). A Computer-Simulated Environment for Modeling and Dynamic- Behavior-Analysis of Special Brushless Motors for Mechatronic Mobile Robotics Systems. *IFAC-PapersOnLine*, *49*(29), 12–17. doi:10.1016/j.ifacol.2016.11.060

Rayyan (2019). *Mechatronics are essential for engineering factories of the future*. Construction Week Online. https://www.constructionweekonline.com/training/257088-mechatronics-are-essential-for-engineering-factories-of-the-future

Sanchez, O., Ordaz, P., & Poznyak, A. (2022). Model based reduced-order observers for a class of mechatronic systems with mitigation of disturbances effects using the Attractive Ellipsoid Method. *Mechatronics*, *84*, 102778. doi:10.1016/j.mechatronics.2022.102778

Shi, Y., & Zhang, K. (2021). Advanced model predictive control framework for autonomous intelligent mechatronic systems: A tutorial overview and perspectives. *Annual Reviews in Control*, *52*, 170–196. doi:10.1016/j.arcontrol.2021.10.008

Smith, B. (2019). *Automation and Robotics for Construction*. AZO Robotics. https://www.google.com/amp/s/www.azorobotics.com/amp/article.aspx%3fArticleID=315

Statista. (2016). *Computer aided design (CAD) market worldwide 2016-2028, by technology*. Statista. https://www.statista.com/statistics/790083/worldwide-computer-aided-design-market-by-technology/

Throat, S. (2022). *Advantages and Disadvantages of Mechatronics System*. Learnmech. https://learnmech.com/advantages-and-disadvantages-of-mechatronics-system/

Tokarev, B., & Klimova, T. (2021). Application of VR instruments in preprofessional education in the area of mechatronics and robotics in a nuclear research university. *Procedia Computer Science*, *190*, 745–749. doi:10.1016/j.procs.2021.06.087

Vogel-Heuser, B., & Bi, F. (2021). Interdisciplinary effects of technical debt in companies with mechatronic products— A qualitative study. *Journal of Systems and Software*, *171*, 110809. https://en.m.wikipedia.org/wiki/Home_automation. doi:10.1016/j.jss.2020.110809

Wang, D., Wang, L., & Wu, J. (2021). Physics-based mechatronics modeling and application of an industrial-grade parallel tool head. *Mechanical Systems and Signal Processing*, *148*, 107158. doi:10.1016/j.ymssp.2020.107158

Wang, R., & Yang, L. & Zhang, C. (2021). Primary failure mode and effects analysis with function-motion-action hierarchy structure for mechatronic systems. *Engineering Failure Analysis, 130*, 105775. https://esub.com/blog/what-is-cad-in-construction/

Xia, Q., & Zhang, Y. (2021). A modular low-cost atomic force microscope for precision mechatronics education. *Mechatronics*, *76*, 102550. doi:10.1016/j.mechatronics.2021.102550

Yakimov, P. I., & Iovev, A. N. (2019). Development of a Modern Learning Environment for Education in Mechatronics and Industrial Automation. *IFAC-PapersOnLine*, *52*(25), 441–444. doi:10.1016/j.ifacol.2019.12.578

Yang, Z., Zhang, T., Hong, J., Zhang, H., Zhao, Q., & Meng, Z. (2021). Research on driving control strategy and Fuzzy logic optimization of a novel mechatronics-electro-hydraulic power coupling electric vehicle. *Energy*, *233*, 121221. doi:10.1016/j.energy.2021.121221

Yuan, H., & Yuan, L. (2022). A design methodology for passive mechatronic vibration absorbers. *Mechanism and Machine Theory, 167*, 104523. doi:10.1016/j.mechmachtheory.2021.104523

Yudin, B., & Moshev, O. (2021). Application of the CDIO standards for cyber-physical education in mechatronics and robotics in a research university on the example of development of 3D-modeling and design skills. *Procedia Computer Science, 190*, 812–816. doi:10.1016/j.procs.2021.06.095

Zeitler, R., & Franke, L. (2016). Computer-Aided Design and Simulation of Spatial Opto-Mechatronic Interconnect Devices. *Procedia CIRP, 50*, 727–732. doi:10.1016/j.procir.2016.04.172

Zippia. (2022). *Best States for a Mechatronics Engineer*. Zippia. https://www.zippia.com/mechatronics-engineer-jobs/best-state s/

Zippia. (2022). *Mechatronics Engineer Demographics and Statistics in the US*. Zippia. https://www.zippia.com/mechatronics-engineer-jobs/demographics/

Chapter 5
COVID-19 Optimization-Based Vibration Controller for a Flexible Manipulator

Mehmet Mert İlman
https://orcid.org/0000-0001-7664-5217
Manisa Celal Bayar University, Turkey

Yılmaz Seryar Arıkuşu
Manisa Celal Bayar University, Turkey

ABSTRACT

Flexible manipulator real-time vibration control methods are effective, but finding the right control gain is difficult. The reason for this is that traditional approaches are not permitted by up-to-date and novel architectures. Current population-based meta-heuristic optimization approaches, on the other hand, can provide solutions for such challenges, as they are inspired by many natural phenomena. Therefore, in the study, the Coronavirus herd immunity optimization (CHIO) method, inspired by the herd immunity mechanism, which is a COVID-19 control method, was used for the optimization of a flexible manipulator control gains. Gray-wolf-optimizer (GWO), another up-to-date population-based algorithm, and traditional particle swarm optimizer (PSO) were used to compare the success of the method. The findings reveal that when it comes to optimizing the vibration controller gains of flexible manipulators, CHIO can outperform its contemporary and traditional competitors.

DOI: 10.4018/978-1-6684-5887-7.ch005

INTRODUCTION

Flexible manipulators created using light and agile links are a field that is currently being studied, as they can work with heavier payloads, consume less energy and are better suited for human-robot interaction compared to traditional rigid manipulators. It has applications in numerous fields such as space, nuclear energy, and medicine as a result of these features. However, its advantageous flexible structure is also very susceptible for vibration. As a result, vibration control of flexible manipulators is required for a successful application. Studies on this topic can be divided into two groups as real-time (closed-loop) and offline (open-loop). In open loop vibration control, Yavuz et al. (2016) succeeded in controlling vibration by establishing a relationship between the natural frequency of the structure and the time parameters of the motor input. Furthermore, this rule association, which was first developed through observation, was turned into a more generalizable state with the use of machine learning and several scenarios in the previous study (Ilman et al., 2022). According to the rule relation, setting the deceleration and acceleration periods of the motor input to be twice the natural vibration period of the structure is sufficient to dampen the vibration by 90%. Despite the fact that this method is convenient, it does not guarantee against disturbing inputs, which is a drawback that should not be overlooked. Fortunately, this is where real-time vibration control methods come into play. However, real-time vibration control also has its own drawbacks and challenges. A downside is the use of costly equipment such as sensors and controllers. Given that the structure's motion control is already been established, integrating vibration control into the motion control while avoiding disruption might be cited as an example of a challenge. While vibration control seeks to manage the structure's flexible movement, motion control aims to control its rigid movement. Since vibration sensors (e.g., accelerometers, strain gauges) do not provide information on the system's rigid movement, studies have underlined the necessity of utilizing more than one sensor in vibration control (Mohamed et al., 2005). However, the control architecture developed in a recent work (Ilman et al., 2022) shown that vibration control with a single sensor can be successfully implemented, ensuring that rigid motion is not disturbed. In this way, vibration control is completely separated from motion control. As a result, vibration control can be simply integrated into motion control. Furthermore, the study found that vibrations that cannot be eliminated by open-loop control can be dampened, which are caused by end effector forces that can occur while holding or releasing loads, such as in pick-and-place equivalent applications.

However, there are issues in real-time vibration control other than disturbing input that need to be addressed. The uncertainty of the payload information that the structure will carry, or the uncertainties that a structure with more than one link will bring, can be stated as prominent challenges. Since the configuration of the system with more than one link will vary throughout the movement, its natural frequency will also change (Karagülle et al., 2017). As mentioned before, since the natural frequency of the flexible manipulator is directly related to the vibration, the controller has to keep up with this situation. As a result, as the configuration changes, the controller gains must update as well. The conventional approaches for obtaining control gains lose their validity as a result of such situations and architectures, while the trial-and-error method becomes extremely time consuming. Fortunately, at this point, population-based meta-heuristic optimization methods (Al-Betar et al., 2021; Dorigo et al., 2006; Karaboga et al., 2014; Mirjalili et al., 2014) come into play. The fundamental advantages of these population-based metaheuristic algorithms are (Mirjalili et al., 2014): (1) Population of candidate solutions share intelligence about the solution field, resulting in unexpected jumps to the important sections. (2) Population of candidate solutions assist each other to avoid local minima. (3) When opposed to single-solution-based algorithms, population-based metaheuristics permit more exploration. Traditional black-box (non-mathematical derivative information) optimization approaches have been applied to flexible manipulators in the literature (Yatim & Mat Darus, 2014; Zain et al., 2006; Zain et al., 2009; Zebin & Alam, 2012). While Zebin and Alam (2012) utilize the classic genetic algorithm (GA) to improve the decision structure of a fuzzy-logic controller for a two-link manipulator system, Zain et al. (2006) and Zain et al. (2009) used it to optimize the gains of a conventional PID-based controller. Yatim and Mat Darus used traditional Particle Swarm Optimization (PSO) to optimize the similar PID-based controller gains. The meta-heuristic optimization methods offer potential for optimizing the controller gains, according to the findings of the researches.

One of the most recent optimization methods is CHIO (Al-Betar et al., 2021). This optimization algorithm was inspired by the COVID-19 outbreak and the concept of herd immunity. Millions of individuals worldwide were infected by the novel severe acute respiratory syndrome coronavirus 2 (SARS-CoV-2) in 2020, resulting in the coronavirus outbreak (COVID-19). In December 2019, the virus was discovered in bats and spread to humans through unidentified gateway species in Wuhan, Hubei Province, China. The virus spreads more swiftly than its two predecessors, SARS-CoV and Middle East respiratory syndrome coronavirus (MERS-CoV), although it is less lethal (Singhal, 2020). Establishing herd immunity within populations constitutes a key concept in dealing with the COVID-19 pandemic (Randolph & Barreiro, 2020). Herd immunity is a condition in which the majority of people in a group are immune to a disease, preventing it from spreading. The herd immunity threshold (HIT) is the level at which those who haven't had the disease and aren't immune are supplied with indirect protection. Using ordinary differential equations, it is feasible to model herd immunity for a variety of diseases. (Randolph & Barreiro, 2020).

Figure 1. Herd immunity concept, a) a model of SIR (susceptible, infectious, recovered) for a completely immunizing infection (R_0 = 4 and HIT = 75%), b) the disease propagation dynamic for naive population, c) the disease propagation dynamic for population with herd immunity (Randolph & Barreiro, 2020)

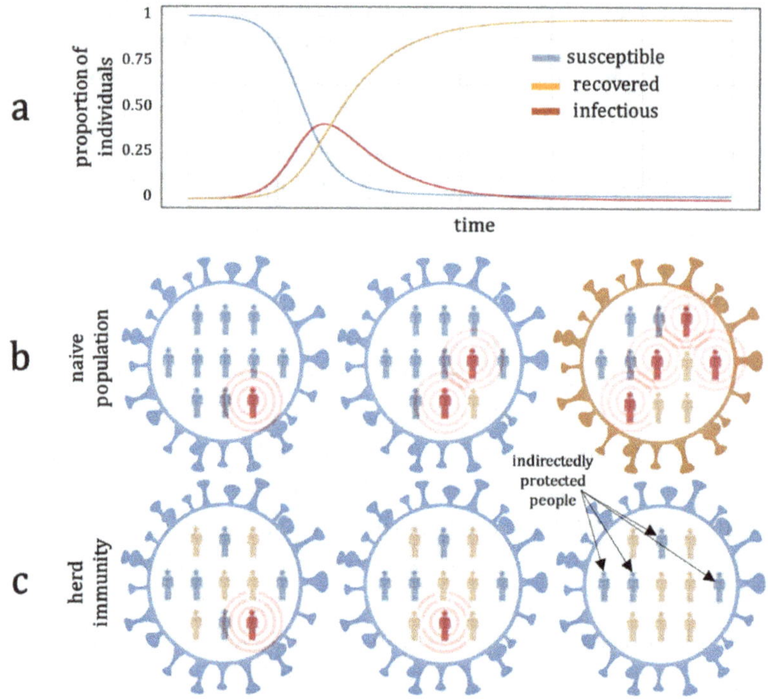

Figure 1 shows the COVID-19 herd immunity model (Figure 1-a) and the spread of the disease for different conditions (Figure 1-b,c). Here, S, I, and R stand for population fraction of susceptible to infection, infected, and recovered from the infection, respectively. It is assumed that people are once recovered, they develop immunity and cannot be infected. Another assumption is that an individual is either susceptible, infected or recovered. However, assumptions are not always valid. Such as a secondary or tertiary infection waves can be occurred as in COVID-19. In this most basic herd immunity model to be created, the HIT value is calculated with the value of $1 - 1/R_0$. Here, the R_0 parameter is called the basic reproduction number and determines the rate of spread of the disease. Using the data, the R_0 value for COVID-19 was determined as 4 (Randolph & Barreiro, 2020) with a corresponding HIT of 0.75. In other words, 75% of them must have a disease in order to achieve herd immunity. Looking at Figure 1-a, when the susceptible proportion of individuals value shown in blue decreases to 0.25 (1-HIT), it is seen that the infectious curve in red peaks, which indicates that herd immunity is provided. In Figure 1-b, on the other hand, it can be observed that in the absence of herd immunity, the outbreak spreads quickly and uncontrollably. In the case of herd immunity, however, the virus does not propagate and vulnerable individuals are indirectly protected, as shown in Figure 1-c.

Despite the fact that this study uses the same real-time controlled flexible manipulator system as the prior one (Ilman et al., 2022), it differs significantly. Unlike the prior approach, which obtains controller gains manually, this study focuses on determining their optimum values by a systematic approach. Also new processes, such as selection and normalization of the error signal, are used to generate the objective function, which allows the optimization approaches to be adapted to the system. As a result the followings are some of the potential novelties that the study could bring to the literature:

- In this study, an up-to-date control architecture (Ilman et al., 2022) was chosen and adapted to optimization methods for which no optimization method has previously been used,
- The ideal controller gains were obtained using CHIO (Al-Betar et al., 2021), one of the most recent population-based meta-heuristic optimization methods,
- The results were reproduced and compared for other population-based meta-heuristic optimization methods such as up-to-date GWO (Mirjalili et al., 2014) and conventional PSO (Kennedy & Eberhart, 1995).

First and foremost, the prior study's system and control architecture are summarized in the following section. The improvements made to the system, as well as the optimization methodologies used, are then detailed. The final section discusses how different optimization strategies can improve the controller gains and thus the system response. Finally, some concluding remarks and future studies are presented in the Conclusion section.

METHODOLOGY

In this section, firstly, the flexible manipulator's closed-loop vibration control diagram is given, followed by an explanation of how the scenario and objective function for optimization are chosen. The population-based meta-heuristic optimization strategies are briefly described after that.

Closed-Loop Flexible Manipulator System

Figure 2 depicts the simulink control diagram of the system utilized in the previous study (Ilman et al., 2022). The plant includes both the flexible manipulator and the servo system model. This is because the manipulator system's motion control (PI control) is already in operation since it is an industrial system. Therefore, the closed loop transfer function (Equation 1) of the servo system and the transfer function of the manipulator system (Equation 2) combine to form the plant transfer function. The strain data measured by the strain-gauge (m/m) is the system's output, while the motor angular velocity (rad/s) is the system's input. When looking at the system's movement, it can be seen that there is a superposition of flexible and rigid body movement (Equation 3). The last term of the equation is not taken into account in industrial systems designed for rigid manipulators, but the second term is taken into account when vibration control is added. Another effect of using two controllers is that the system now has two points of reference. As a result, when the rigid motion reference is the system's motor angular velocity input (Figure 2 – Reference 1), the flexible motion reference is zero since absolute vibration damping is expected (Figure 2 – Reference 2). Two references are presented since vibration control will be applied in the system with motion control. The algorithm will try to reset the control input and disturb the rigid movement if the strain error (Figure 2 – 3rd sum output) is applied straight to the input, therefore the strain error is summed up (Figure 2 – 2nd sum output) with the input error (Figure 2 – 1st sum output). Thanks to the control input loop created as a result of multiplying the input error with the integral controller, it is guaranteed that the position error does not occur. Thus, as promised, it is possible to perform vibration control with a single sensor and independently of motion control. The motion scenario is one of the optimization's prerequisites. In the study, the triangular profile, in which the structure produces the highest vibration, is chosen for the reference of the rigid motion (motor angular velocity input). Angular displacement, maximum angular velocity, acceleration time, deceleration time and movement time of the motion input are 0.5π rad, 0.5π rad/s, 1.935 seconds, 0.065 seconds, and 2 seconds, respectively. The objective function is also required for optimization. The Integral Time Absolute Error (ITAE) approach is utilized for this (Equation 4). Because there are two references in this system, two error signals (motor input error and strain error, respectively) are evaluated concurrently, despite the fact that single error signal is generally selected. Furthermore, because the error signals' physical meanings different (input error is rad/s, strain error is m/m), their order differs as well. As a result, the objective function is normalized so that both signals have the same effect.

$$H_S = \frac{N_S}{D_S} = \frac{\left[689.6s^2 + 19900s + 148800\right]}{\left[1,817.8s^2 + 20300s + 149200\right]} \tag{1}$$

$$H_M = \frac{N_M}{D_M} = \frac{\left[s^4 + 3.2s^3 + 353999.79s^2 + 85644.66s + 4110.12\right]}{\left[85.13s^5 + 272.42s^4 + 29.99e6s^3 + 78e5s^2 + 69.87e9s\right]} \tag{2}$$

$$\delta_{manipulator} = \delta_{rigid} + \delta_{flexible} \tag{3}$$

$$f_{objective}\left(x\right) = \int_0^{\infty} t \left|\frac{e_{rigid}}{e_{rigid}} + \frac{e_{flexible}}{e_{flexible}}\right| \tag{4}$$

Figure 2. Flexible manipulator closed-loop vibration control diagram

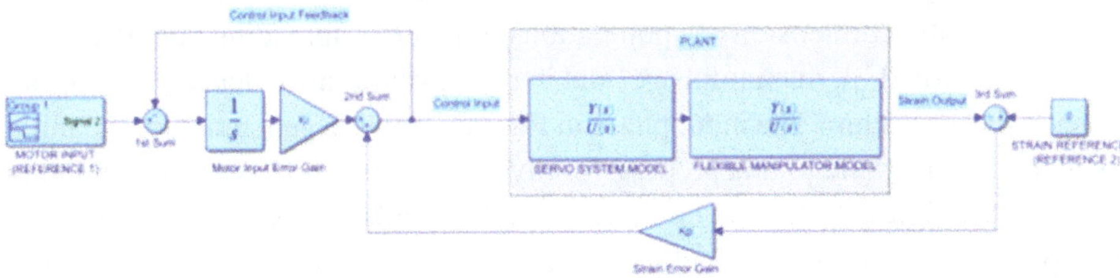

Meta-Heuristic Optimization Algorithms

The proposed CHIO as well as the GWO and PSO algorithms utilized for comparison are briefly presented in this section. These three algorithms have one thing in common: they are all population-based. Figure 3 depicts the main design of these sorts of algorithms, which includes input, initial population, population diversification, and evaluation stages using an objective function. If the evaluation result is negative, the population is updated and the iterations continue; if the evaluation succeeds, the population remains unchanged and finally, convergence occurs.

Figure 3. General diagram of population-based meta-heuristic optimizations

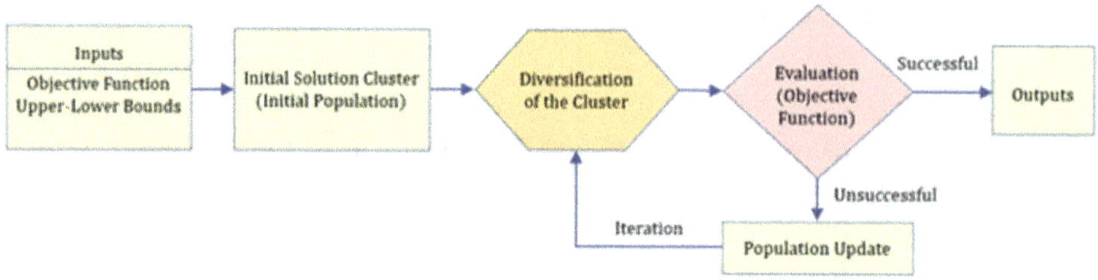

COVID-19 Herd Immunity Optimization (CHIO)

The equivalents of the herd immunity concepts modeled for COVID-19 mentioned in the Introduction section in the optimization algorithm are given in Table 1. The algorithm is based on the idea of protecting the community from disease by turning the majority of the susceptible population that is not contaminated into immune individuals. A customized version of the generalized meta-heuristic diagram displayed in Figure 3 for CHIO was built to visualize the algorithm (Figure 4). As a result, the solution dimension (n), population number (P), spreading rate (R_0), initial infected solution (C_0), maximum age, and iteration values are calculated in the first phase of the algorithm. The input phase in Figure 3 and the generation of the initial population correspond to this process. To diversity the solution set, the population is separated (Figure 4 - before 1st Node) into three independent clusters (infected, susceptible, and immuned) based on the R value (random value between 0 and 1). Diversification functions such as infected (Equation 5), susceptible (Equation 6) and immuned (Equation 7) exploit the relationship between R and R_0 parameters to push the solution closer to a randomly picked element in the cluster it belongs to. In this case, x^C, x^S, and x^I refer to a randomly chosen infected, susceptible, and immuned solution. Furthermore, it can be said that the R-value dictates social distance and diversification will be low in circumstances when it is small. In the following case, if the solution attempting to approach the ideal answer is too far away, a tiny R-value will significantly delay convergence. As a result, the fatality function kicks in and removes from the population the solutions with a minimum R-value (infected) that are not updated above the limit iteration value (Maximum Age). If the diversified solution has a higher fitness value than the prior solution, it will take its place (solution update).

$$\text{infected}\left(x\right) = x + R\left(x - x^{C}\right), \qquad R \in \left[0,\ R_0/3\right] \tag{5}$$

$$\text{susceptible}\left(x\right) = x + R\left(x - x^{S}\right), \qquad R \in \left[R_0/3, 2R_0/3\right] \tag{6}$$

$$\text{immuned}\left(x\right) = x + R\left(x - x^{I}\right), \qquad R \in \left[2R_0/3,\ R_0\right] \tag{7}$$

Table 1. Bridge between COVID-19 and optimization concept (Al-Betar et al., 2021)

No	COVID-19 Context	Optimization Context
1	Population	Population of Solutions (x)
2	Infected, Susceptible, Immuned Case	Solution Set Segmentation Methods for Diversification
3	Immunity Rate	Objective Function: $\min_{x} f\left(x\right), \quad x \in \left[lb, ub\right]$
4	Herd Immunity Treshold (HIT)	Convergence
5	Basic Reproduction Rate ($R_0 = 4$)	Normalized Spreading Rate ($R_0 = [0, 1]$)
6	Social Distancing	Pick random case and rely on the R_0
7	Possibility of Infection	Weak fitness value and inherit COVID-19 features
8	Covid-19 Disease Period	Unchange Iteration No. of Infected Solution
9	Mortality	Exclusion of the solution from the cluster

Figure 4. Flowchart of CHIO algorithm

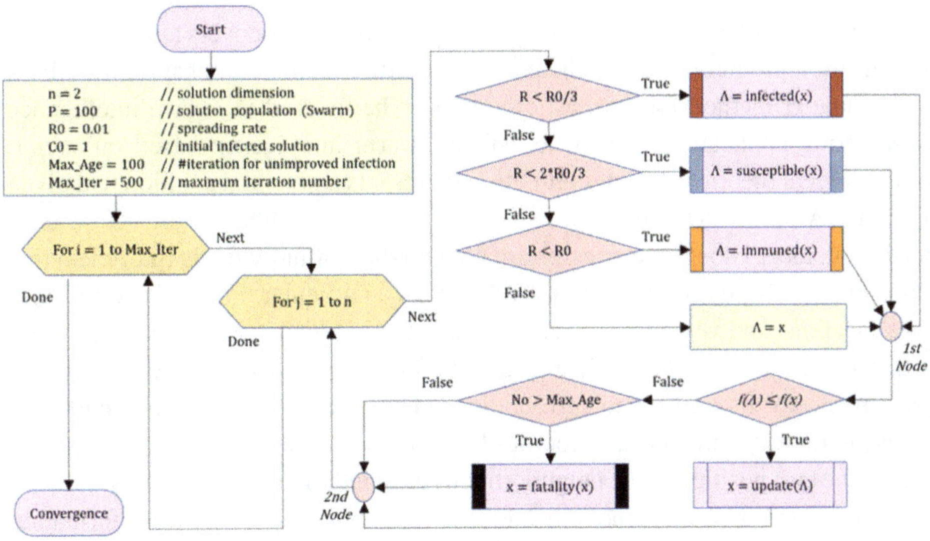

Table 2. Pseudo-code of PSO algorithm(Kennedy & Eberhart, 1995)

Particle Swarm Optimization Algorithm
Set C1, C2, w (max. and min.)
Initialize population of particles that have positions X and velocities V
Set iteration k =1
Compute fitness of particles $F_i^k = f\left(X_i^k\right), \forall i$ and determine the index of the best particle b
Set $pbest_i^k = X_i^k, \forall i$ and $gbest_i^k = X_b^k$
$w = w_{max} - k \times \left(w_{max} - w_{min}\right) / Maxite$ update velocity position of particles as in Equations 8 and 9
evaluate fitness $F_i^{k+1} = f\left(X_i^{k+1}\right), \forall i$ and determine the index of the best particle b1
update Pbest of population ∀i
if $F_i^{k+1} < F_i^k$ then $pbest_i^{k+1} = X_i^{k+1}$ else then $pbest_i^{k+1} = pbest_i^k$
update gbest of population
if $F_{b1}^{k+1} < F_b^k$ then $gbest^{k+1} = pbest_{b1}^{k+1}$ and set b=b1 else $gbest^{k+1} = gbest^k$
if k < Maxite then k = k+1, goto w line else print optimum solution as $gbest^k$

Particle Swarm Optimization (PSO)

In the presence of objectives, a PSO is an algorithm that discovers a solution to an optimization problem in a search space and predicts common behavior. The PSO is a computer method for stochastic processes that is used in swarm intelligence Engineering applications can benefit from swarm intelligence based on social-psychological theories. A fitness function represents a potential solution to a given problem. A social network is described as a set of neighbors with whom each person interacts. Stochastic guesses are a population of individuals (particles) that represent a proposed solution to the problem. At a particular time, t, these particles have locations and velocities. A particle adjusts its velocity to modify its position in quest of a solution. The particle recalls its finest performance as well as the best performance of its immediate surroundings. The system evaluates each particle's fitness iteratively and adjusts velocities based on prior individual best positions as well as the best position in the neighborhood. Each individual keeps its best value as the best "pbest" value in memory. In addition, each individual keeps the information of the best value, gbest information and pbest information (Kennedy & Eberhart, 1995).By moving through the search space, each particle is represented by current velocity and position vectors given by Equations 8 and 9, respectively. Pseudo-code of PSO algorithm is given in Table2.

$$V_i^{k+1} = wV_i^k + C_1 rand_1() \times (pbest_i^k - S_i^k) + C_2 rand_2() \times (gbest_i^k - S_i^k) \tag{8}$$

$$S_i^{k+1} = S_i^k + V_i^{k+1} \tag{9}$$

Grey Wolf Optimization (GWO)

This optimization was inspired by grey wolf predatory hunting. Wolves are social animals who live in packs and have a pyramidal structure The Alpha (α), who is regarded the leader in making crucial decisions, is at the top of the hierarchy. The Beta (β) is the next in the hierarchy, and it assists Alpha in carrying out most of the group's activities while also enforcing discipline. The Delta (δ), which includes hunters, scouts, caretakers, elders, and lookouts, is the third inline. The Omega (ω), the group's last member, can be thought of as a babysitter and is often regarded the group's least significant member. This hierarchy is followed by the hunting optimization. The wolves look for the pray based on their Alpha (best solution candidate), Beta, and Delta positions. In their quest, they move away from each other and then converge to attack. They follow the prey, chasing it down and approaching it. The wolves encircle the pray and update their positions in space near to the prey in any random spot once it is found. The attack usually occurs when the pray is unable to move or flee.The model is represented by Equations 10-16.

$$\vec{D} = |\vec{C}\vec{X}_p(t) - \vec{X}(t)| \tag{10}$$

$$\vec{X}(t+1) = \vec{X}_p(t) - \vec{A}\,\vec{D} \tag{11}$$

$$\vec{A} = 2\vec{a}\vec{r}_1 - \vec{a} \tag{12}$$

$$\vec{C} = 2\vec{r}_2 \tag{13}$$

$$\vec{D}_{\pm} = |\vec{C}_1\vec{X}_{\pm}(t) - \vec{X}|,\ \vec{D}_{,} = |\vec{C}_2\vec{X}_{,}(t) - \vec{X}|,\ \vec{D}_{.} = |\vec{C}_3\vec{X}_{.}(t) - \vec{X}| \tag{14}$$

$$\vec{X}_1 = \vec{X}_{\pm} - \vec{A}_1\vec{D}_{\pm},\ \vec{X}_2 = \vec{X}_{2} - \vec{A}_2\vec{D}_{2},\ \vec{X}_3 = \vec{X}_{.} - \vec{A}_3\vec{D}_{.} \tag{15}$$

$$\vec{X}(t+1) = \frac{\vec{X}_1 + \vec{X}_2 + \vec{X}_3}{3} \tag{16}$$

Where $\vec{A}$ and $\vec{C}$ are the coefficient vectors, t is the present iteration, $\vec{X}$ is the position vector of the grey wolf, $\vec{X}_p$ is the position vector of the prey, $\vec{r}_1$ and $\vec{r}_2$ are random vectors in the range [0,1] and $\vec{a}$ is decreased linearly from 2 to 0 during the iteration. Pseudo-code of GWO algorithm is given in Table 3.

Table 3. Pseudo-code of GWO algorithm (Mirjalili et al., 2014)

Grey Wolf Optimization Algorithm
Initialize the grey wolf population X_i (i = 1,2, …, n)
Initialize A, a and C
Compute the fitness of each search agent
X_α = best search agent
X_β = second best search agent
X_δ = third best search agent
while (t < Max. number of iterations)
for each search agent
update the position of the current search agent by Equation (16)
end for
update A, a, and C
compute the fitness of all search agents
update X_α, X_β, X_δ
compute the fitness of all search agents
update X_α, X_β, X_δ
t= t+1
end while
return X_α

RESULTS AND DISCUSSIONS

The flexible manipulator system is optimized using the methods presented in the previous part in this section. Equation 4 is used to calculate the objective function value for each optimization procedure. In order to make the comparison, the optimization control parameters were set to be as similar as feasible. The maximum population size and iteration value were set to 100 and 500, respectively, for this purpose. Other parameters are chosen based on the research. It is found that when R_0, the most critical control parameter in the CHIO method, is set at 0.01 as suggested by Al-Betar et al. (2021), the best results are obtained. The algorithm's exploration and exploitation rates are also determined by the R_0 parameter. Because meta-heuristic algorithms are part of a stochastic process, random deviations in results may arise when the algorithm is run again. It is expected that by executing each algorithm five times, this effect will be minimized, allowing the findings to be appropriately interpreted. Figure 5 depicts the convergence findings for a total of 15 cases. At first glance, it appears like GWO has the fastest convergence, followed by CHIO, and finally PSO. When we look at Table 4, however, the interpretation changes. CHIO is the fastest to attain the solution space's minimal objective function value (27.9546), while PSO is the second, and then GWO is the third. The reason for this may be that, while having the fastest convergence, GWO becomes stuck in local minima and is unable to achieve the global minima. Looking at the results of the 100th Iteration objective function, the speed order of the algorithms is this time CHIO, GWO and PSO. Therefore, to make a general evaluation among algorithm speeds, GWO gives the best results in the short term (<100 iterations), while CHIO gives the best results in the medium and long term. In addition, when the analysis times are compared, it is seen that CHIO is completed in approximately 10 minutes (12%) longer than the other two algorithms. This establishes a balance between precision and speed. All calculations were performed on the same computer with an I7-7700 K 4.20 GHz CPU and 32 GB RAM.

Table 4. Performance evaluation between the proposed CHIO algorithm and other methods

Test Cases	Kp [0, 1e4]	Ki [0, 1e3]	Average overall computation time (min)	Average no. of iterations for convergence	Average ITAE at 100th iteration
CHIO	3406.72 ± 0.57	22.44 ± 1e-3	94	162 ± 40	27.9548 ± 1e-4
GWO	3406.42 ± 0.16	22.43 ± 6e-4	85	460 ± 76	27.9552 ± 4e-4
PSO	3406.46 ± 0.25	22.44 ± 1e-3	86	396 ± 39	28.6061 ± 63e-2

Figure 5. Convergence plots of CHIO algorithm against the other swarm-based algorithms

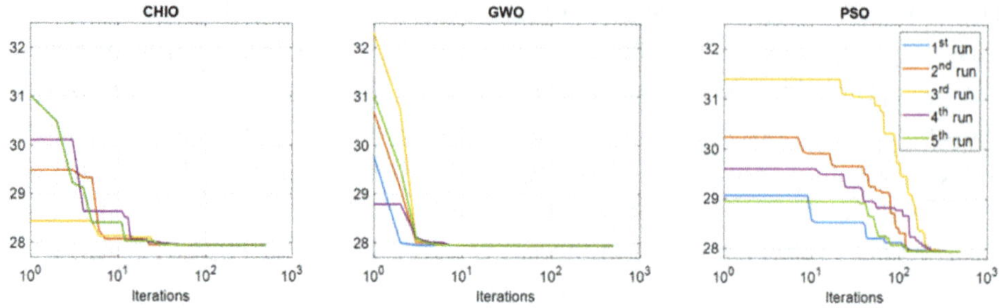

Table 5. Performance criteria results for optimized controller gains (K_p = 3406.72, K_i = 22.44)

System	Overshoot	T_{ss}	Vibration RMS [m/m]	Maximum Deflection
Control-Free	% 1.7298	~ 1 min	9.2624e-05	8.6491 mm
Proposed Controller	% 0.8285	~ 0.5 sec	8.7210e-06	4.1426 mm
Reduction Percentage	% 52.1	~ % 92	% 90.58	% 52.1

Figure 6. Effect of optimized controller gains on control-free flexible manipulator's inputs and outputs

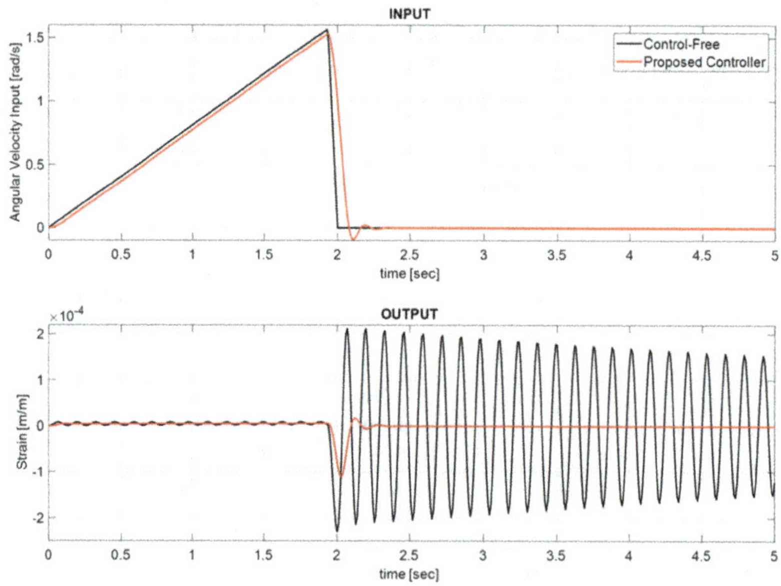

Finally, Figure 6 depicts the effect of optimal control gains on the system. Because all three optimization procedures converged to a similar error value, Kp = 3406.72 and Ki = 22.44 from Table 4 are directly chosen as the optimized control gains. Table 5 shows the results of the performance criteria for the system with these optimized controller gains. Here, it is seen that the maximum deflection and overshoot values are reduced by about half. While calculating the overshoot value, since the end point displacement of the manipulator was given in prior study as 500 mm, the calculated deflection value due to bending is evaluated as the overshoot amount and so the percentage is calculated. Also, the sample frequency in the simulation is 100 Hz. The duration and amplitude of vibrations, on the other hand, show a significant decrease. While the vibration time of the flexible manipulator lasts around one minute, it is reset at the end of 0.5 seconds after the control.

CONCLUSION

The COVID-19 herd immunity optimization (CHIO) approach is used to determine the controller gains for a flexible manipulator with a non-traditional control architecture in this study. Although the control architecture is functional because it does not require additional sensors, the same cannot be said for the computation of control gains. The frequent changing of controller gains makes it more difficult to compute manually, especially in multi-link systems. Fortunately, thanks to the novel objective function created specifically for this architecture, optimization algorithms can be used in this study. In addition to the comparison with the Control-free system, two further optimization approaches were chosen for comparison in order to make the results more interpretable. gray wolf optimization (GWO) and particle swarm optimization are two of these strategies (PSO). These two optimization approaches are both population-based meta-heuristic optimization and of the same kind as the recommended optimization method. Looking at the results, it is seen that each optimization method converges to the similar results. Although the CHIO algorithm took 12% longer to complete than the other two algorithms, it achieved results in less than 50% of iterations. The results show that CHIO holds promise for future studies.

REFERENCES

Al-Betar, M. A., Alyasseri, Z. A. A., Awadallah, M. A., & Abu Doush, I. (2021). Coronavirus herd immunity optimizer (CHIO). *Neural Computing & Applications*, *33*(10), 5011–5042. doi:10.100700521-020-05296-6 PMID:32874019

Dorigo, M., Birattari, M., & Stutzle, T. (2006). Ant colony optimization. *IEEE Computational Intelligence Magazine*, *1*(4), 28–39. doi:10.1109/MCI.2006.329691

İlman, M. M., Yavuz, Ş., Karagülle, H., & Uysal, A. (2022). Hybrid vibration control of an industrial CFRP composite robot-manipulator system based on reduced order model. *Simulation Modelling Practice and Theory*, *115*, 102456. doi:10.1016/j.simpat.2021.102456

İlman, M. M., Yavuz, Ş., & Taser, P. Y. (2022). Generalized Input Preshaping Vibration Control Approach for Multi-Link Flexible Manipulators using Machine Intelligence. *Mechatronics*, *82*, 102735. doi:10.1016/j.mechatronics.2021.102735

Karaboga, D., Gorkemli, B., Ozturk, C., & Karaboga, N. (2014). A comprehensive survey: Artificial bee colony (ABC) algorithm and applications. *Artificial Intelligence Review*, *42*(1), 21–57. doi:10.100710462-012-9328-0

Karagülle, H., Malgaca, L., Dirilmiş, M., Akdağ, M., & Yavuz, Ş. (2017). Vibration control of a two-link flexible manipulator. *Journal of Vibration and Control*, *23*(12), 2023–2034. doi:10.1177/1077546315607694

Kennedy, J., & Eberhart, R. (1995). Particle swarm optimization. *Proceedings of ICNN'95 - International Conference on Neural Networks*, *4*, 1942–1948. 10.1109/ICNN.1995.488968

Kwok, K. O., Lai, F., Wei, W. I., Wong, S. Y. S., & Tang, J. W. T. (2020). Herd immunity – estimating the level required to halt the COVID-19 epidemics in affected countries. *The Journal of Infection*, *80*(6), e32–e33. doi:10.1016/j.jinf.2020.03.027 PMID:32209383

Mirjalili, S., Mirjalili, S. M., & Lewis, A. (2014). Grey Wolf Optimizer. *Advances in Engineering Software*, *69*, 46–61. doi:10.1016/j.advengsoft.2013.12.007

Mohamed, Z., Martins, J. M., Tokhi, M. O., Sá da Costa, J., & Botto, M. A. (2005). Vibration control of a very flexible manipulator system. *Control Engineering Practice*, *13*(3), 267–277. doi:10.1016/j.conengprac.2003.11.014

Randolph, H. E., & Barreiro, L. B. (2020). Herd Immunity: Understanding COVID-19. *Immunity*, *52*(5), 737–741. doi:10.1016/j.immuni.2020.04.012 PMID:32433946

Singhal, T. (2020). A Review of Coronavirus Disease-2019 (COVID-19). *Indian Journal of Pediatrics*, *87*(4), 281–286. doi:10.100712098-020-03263-6 PMID:32166607

Yatim, H., & Mat Darus, I. (2014). Self-tuning active vibration controller using particle swarm optimization for flexible manipulator system. *WSEAS Transactions on Systems and Control*, *9*, 55–66.

Yavuz, Ş., Malgaca, L., & Karagülle, H. (2016). Vibration control of a single-link flexible composite manipulator. *Composite Structures*, *140*, 684–691. doi:10.1016/j.compstruct.2016.01.037

Zain, B. A. M., Tokhi, M. O., & Toha, S. F. (2009). PID-Based Control of a Single-Link Flexible Manipulator in Vertical Motion with Genetic Optimisation. *Third UKSim European Symposium on Computer Modeling and Simulation*, 355–360. doi:10.1109/EMS.2009.86

Zain, M. Z. M., Tokhi, M. O., & Mohamed, Z. (2006). Hybrid learning control schemes with input shaping of a flexible manipulator system. *Mechatronics*, *16*(3), 209–219. doi:10.1016/j.mechatronics.2005.11.004

Zebin, T., & Alam, M. S. (2012). Modeling and Control of a Two-link Flexible Manipulator using Fuzzy Logic and Genetic Optimization Techniques. *Journal of Computers*, *7*, 578–585. doi:10.4304/jcp.7.3.578-585

Chapter 6
Design Optimization Application for a Flexible Robot Manipulator

Mehmet Mert İlman
https://orcid.org/0000-0001-7664-5217
Manisa Celal Bayar University, Turkey

Şahin Yavuz
https://orcid.org/0000-0001-9007-772X
Dokuz Eylül University, Turkey

Doğukan Akgöl
Dokuz Eylül University, Turkey

ABSTRACT

In this chapter, design optimization is performed for a single-link flexible manipulator driven by a standard trapezoidal velocity input. The mass of the original manipulator is reduced by a quarter by design optimization without changing the payload. The finite element model, which is one of the non-rigid models, and topology optimization were utilized for the application. Dynamic stress and natural frequencies of the system are utilized as optimization objectives. The results were compared with the original manipulator in terms of both safety factor and vibration modes. In addition, the system parameters were compared with the tapered beam, and beam with reduced width, which was prepared to have the same mass as the optimized designs. As a result, although the mass of the original manipulator was reduced by one-fourth with the proposed design, it was observed that the dynamic stresses decreased. The study is expected to have significant implications in terms of improving the benefits of flexible robots and providing a contribution to the attenuation of vibrations and stresses.

DOI: 10.4018/978-1-6684-5887-7.ch006

INTRODUCTION

The preference of links with low stiffness in the manufacture of robot manipulators improves various aspects of the robot. The difference in link mass, for example, can be exploited to work with higher payloads as the smaller link mass decreases the load on the actuators and joints. Likewise, as the robot's flexibility improves, so does its ability to work in human-populated environments. Furthermore, lightness improves energy efficiency by lowering energy usage. Moreover, because the robot can travel at higher speeds, shorter cycle durations can be achieved in applications.

Lightness in robot linkages can be achieved in a variety of ways. The utilization of new technological materials is one of these ways. Composite materials such as carbon fiber reinforced polymer (CFRP) and glass fiber reinforced polymer (GFRP) are now often employed. Among the composite materials, it can be said that the carbon fiber reinforced polymer material stands out from the rest. The reason for this is that it is lighter than aluminum, which is one of the lightest metal materials but has higher stability even than steel. Because of its low density, high elasticity modulus, and the optimum ratio these create, it has a high natural frequency. In fact, a robot manipulator with a link composed of carbon fiber reinforced polymer composite was proven in a study to be capable of carrying 16 times its own weight. In addition, cutting-edge additives such as modified (carboxylic acid (COOH) functionalized, fluorinated, or similar) multi-walled carbon nanotubes can be employed to improve composite material characteristics such as natural frequencies and damping ratios. Mechanical parameters such as modulus of elasticity and maximum tensile strength can also be improved using these additions. Despite this, demanding and intricate manufacturing methods, as well as expensive costs, make choosing composite materials over typical metal links difficult.

Aside from all of these benefits, it's important to remember the drawbacks of having flexible links in robot manipulators. As a result of the increased flexibility of the links, the natural frequency of the body decreases, resulting in low frequency and high amplitude oscillations. Furthermore, because the damping ratio is reduced, the vibrations persist for a longer amount of time. A few robot manipulator applications provide an example of the detrimental implications of this circumstance. In a pick-and-place application, for example, it will be essential to wait a lengthy period for the robot to cease vibrating before releasing the workpiece. Because the sensitivity criterion will not be met otherwise. According to studies, residual vibration after 2-3 seconds of movement can easily reach one minute. In this case, the increase in the cycle time by about 30 times makes the flexible robot useless. In addition, the studies focused on the height of vibrations during motion. In this instance, it will be impossible to discuss the effectiveness of robots used for tasks like welding or observation, which require a predetermined path to be followed. The stresses that vibration causes in the system are another adverse consequence. These stresses can cause sudden material damage as well as medium-term issues such as fatigue. It has been emphasized in the studies that flexible manipulators are forced to stress especially in the bending direction. As a result, analyzing these stresses and vibrations is critical for flexible robot efficiency.

Many studies are being conducted today to improve the utility of flexible manipulators by eliminating vibration. These studies can be divided into two groups as open-loop vibration control and closed-loop vibration control. It is fair to state that each group has its own set of advantages and disadvantages. For example, the closed-loop vibration control is effective even at low vibrations and is resistant to the disturbing effects that occur during motion, which is a significant benefit. So much so that disturbing inputs may arise even when robot manipulators are used in planned (deterministic) operations. Examples of these effects are working with payloads of unknown weight in applications such as pick and place, force effects when gripping or releasing the workpiece, contact effects and similar effects. Although it is not possible to talk about such advantages for open-loop control, its effectiveness at high vibrations makes it useful. Furthermore, when considering feedback elements (sensors), controller elements (computer or microprocessor, data acquisition cards), and electronic sensitivity elements for real-time applications (filter and isolation circuits, high sample frequency requirement), all of which are required by closed-loop control architecture, it is important to note that the process's cost increases significantly. So much so that this cost burden can also neutralize the low-cost advantage, which is one of the reasons why the flexible manipulator was chosen in the first place. To minimize these drawbacks, control systems using low-cost sensors such as strain gauges and the robot's own actuators are recommended over expensive piezoelectric sensor-actuator couples, according to the literature. However, since open-loop vibration control requires no equipment, it becomes a very attractive option in conditions where disturbing effects are minimal. Especially in recent studies, it has been observed that vibration can be damped significantly, even with fine adjustments to the robot input times. Thanks to this input shaping method, which is applied by taking into account the natural periods of the robot, high amplitude vibrations can be damped quite effectively.

One of the most important requirements of vibration control methods is the model information of the flexible robot. So much so that a robot model may be needed both in the determination of control gains in closed-loop vibration control applications and in open-loop vibration control methods such as feed-forward control. Furthermore, the model can be utilized to calculate the vibration parameters required in input shaping methods that are described in the literature as model-free. In the literature, it is emphasized that the modeling of flexible robots is more difficult than the modeling of traditional rigid robots. This is because flexible link robot models take into account the robot vibration as well as the kinematic model. So much so that in conventional robot manipulators, vibrations can be neglected thanks to the rigid structure of the system. Three basic parameters can be mentioned in order to achieve a successful non-rigid modeling, these are (i) practicality that can be applied to non-uniform geometries, (ii) reliability and high precision results, (iii) rapid results. Due to its practicality and great sensitivity, the finite element model is a widely preferred method among the several non-rigid modeling techniques in the literature. Reduced-order modeling has also been highlighted as a method for obtaining faster results in various research. In addition to these, the lumped parameter model (LPM) can be preferred especially because of its fast results and practicality. In addition, the assumed mode method (AMM) can also be preferred for its reliability and fast results. Finally, modeling methods based on machine learning can be preferred in terms of their practicality. As a result, it is possible to choose the ideal non-rigid modeling technique in accordance with the intended use.

Design is one of the areas where modeling is applied. So much so that with a successful design, it is possible to consolidate its advantages both by saving more material and by increasing the natural frequency of the system. Today, it is possible to see optimization methods as well as design studies applied to meet different design criteria. Standard gradient-based approaches, as well as new methods such as the genetic algorithm, can be employed to accomplish commonly used topology optimization. Based on provided loads and limitations, the topology optimization method distributes material in the design space by removing unnecessary densities (Liu et al., 2021). However, while optimal design outcomes may appear to be successful at times, they may be meaningless in terms of manufacturability (Lohmeier et al., 2006). On the other hand, the new manufacturing methods such as additive production methods continue to evolve, which provide flexibility in production by removing the constraints that come from conventional production methods (Srinivas and Javed 2020).

Topology optimization of mechanisms is significantly important to reduce their weight and increase the payload/mass ratio, hence improving their operating performance (Guo and Cheng 2010; Deaton and Grandhi 2014; Held et al. 2016). The usage of topology optimization in the robotics field started in early 2000. The stationary part of the ARMAR III humanoid robot is designed by using topology optimization by considering the arrangement of the mechanical and electrical components assembled in this part. All components carried by this part have been subjected to topology optimization together with this stationary part and their placement is also optimized (Albers et al., 2006). Lohmeier et al. (2006) optimized a humanoid robot leg by using the topology optimization technique. The main goal of this research was to obtain a design that can be manufactured with the milling process. The thigh, pelvis, and shank parts of the humanoid robot are complicated parts with geometric limitations. These were created to ensure high strength and reduce the weights of the components by using topology optimization. Xin et al. (2019) conducted size and topology optimizations with static analyses for an industrial tandem manipulator and reduced its weight by 26.5%. The lightweight structure of the manipulator is achieved by using lightweight material and structure.

The weight of the robotic structural components has a significant impact on the industrial manipulator's dynamic performance. A substantial decrease in robotic arm weight, on the other hand, generates stability and precision issues. The stated concerns are first addressed by analyzing 2-3 DOF robotic arms with counterbalancing and modeling of the precision (Hennessey et al. 1987; Oosting and Dickerson 1988). Consideration of manipulator-link dynamic loading is essential for improved design. Because the topology of the manipulator-link is more sensitive to the force's direction. The direction of force is dynamic when the robotic arm revolves around the shaft.

When the studies in the literature on flexible manipulators (Akalla and Fanni, 2021; Sun et al., 2017) examined, it is seen that single-objective optimizations are generally performed. The goal was to maximize the elasticity modulus acquired through static analysis or the natural frequencies obtained through modal analysis. It can be seen that improving these qualities aids the mechanism in its movement. The fundamental goal, however, should be to minimize the greatest transient or residual stresses that the system will experience during its movement. As a result, the absence of these criteria in topology optimization can be considered a gap in the literature.

Figure 1. Multi-objective design optimization framework used in the study

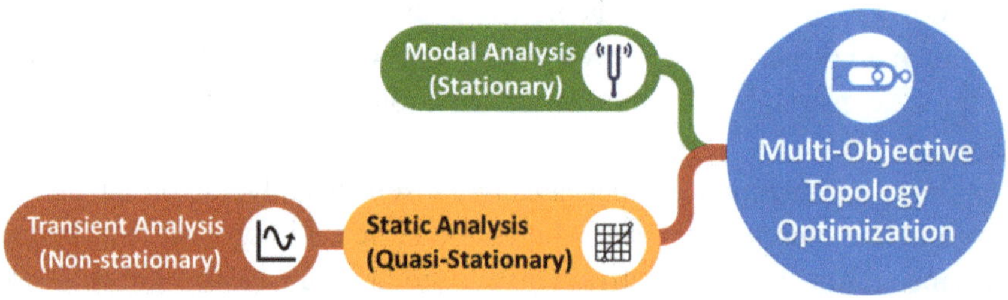

To remedy this problem, the study developed a multi-objective design optimization framework (Figure 1). The D'Alembert principle (Ryu et al., 1992) is used to convert non-stationary transient analyses to quasi-stationary analyses. In this manner, the optimization can be constructed to maximize both the transient and modal analysis criteria.

Figure 2. The original and quarter-mass subtracted designs compared in the study, (middle) original (100% mass) design and boundary conditions for sensor and payload placement, (top left) lightened design by width-reduction, (top right) lightened design by tapering, (bottom) multi-objective topology optimized design

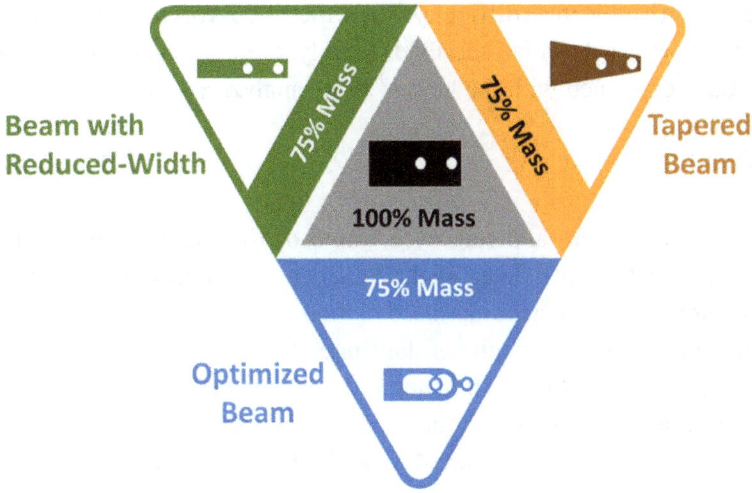

The design optimization of the flexible steel manipulator, the geometry of which was used in the previous study (Yavuz et al., 2020), is discussed in this chapter (Figure 2). Because the study shows the motion scenario in which the system has the most difficulties, the scenario is also incorporated in the design optimization. In this way, it is aimed to reduce the stresses that occur during the movement in the system. The followings are some of the potential novelties that the study could bring to the literature:

(i) Thanks to the new multi-objective topology optimization used, it is possible to include non-stationary criteria among the optimization objectives,
(ii) The dynamic stress results are improved while the mass of the flexible manipulator is reduced,
(iii) The effects of the optimized design and various mass reduction techniques are compared,
(iv) The lightness feature, which is the largest advantage of flexible manipulators, has been increased even more thanks to design optimization.

ANALYSES AND OPTIMIZATION

In this section, firstly, the introduction of the system and finding its natural frequencies are shown. Afterwards, non-stationary analysis is shown for the motion scenario where the system has the most difficulty. Then, the process of converting this analysis to quasi-stationary analysis is described. Finally, the implementation of topology optimization is explained with the help of these analyzes.

Modal Analysis

The experimental system and reference frame of the flexible steel manipulator used in the previous study (Yavuz et al., 2020) are shown in Figure 3. In addition, different parameters and their values used in modeling the system are listed in Table 1. Then, the modal analysis results of the finite element model created using these parameters are shown in Figure 4. When the frequency responses are examined, it is seen that they are consistent with the natural frequency responses in the reference study (Yavuz et al., 2020). In Figure 4, also the frequencies of the first 10 modes of the system are given. These modes will later be used as an objective in topology optimization. In addition, the mode shapes of the first three natural frequencies of the system are given. Because they have small and similar natural frequencies (all three below 50 Hz), the first three modes can be called dominant modes. When the mode shapes are examined, it is seen that the first and third mode shapes are located in the X-Z plane and the second mode shape is located in the Y-X plane. Since the manipulator system moves in the X-Z plane, it can be called a planar manipulator system. Therefore, the first and third modes can be called in-plane modes and the second mode can be called out-of-plane mode. Finally, it can be seen that there is no motion at the starting point due to the boundary condition, and there is no motion at the nodal point in the third mode. This is a special situation arising from the mode shape, and it is important not to place any sensor at this point in order not to lose the relevant frequency information.

Table 1. Properties of the original manipulator system (Yavuz et al., 2020)

System Properties [units]	Values	System Properties [units]	Values
Density – ρ [kg/m^3]	7800	Payload Mass [g]	620
Modulus of Elasticity – E [GPa]	200	Thickness – t [mm]	20
Length – L [mm]	500	Width – w [mm]	3.2
Sensor Length – L_{sensor} [mm]	420	γ, μ (Rayleigh Damping)	3.8e-4
Sensor Mass [g]	54	Joint Stiffness [Nm/rad]	16e3

Figure 3. Experimental setup of the flexible manipulator used in the study (Yavuz et al., 2020)

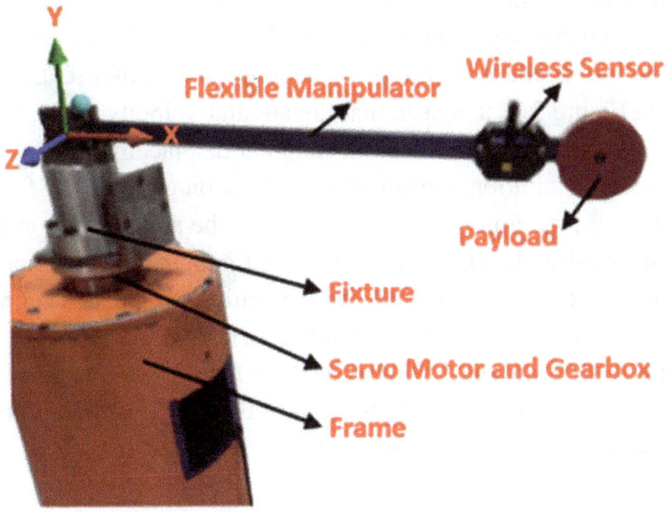

Figure 4. Modal analysis mode shape and mode frequency results, a) first mode shape, b) second mode shape, c) third mode shape

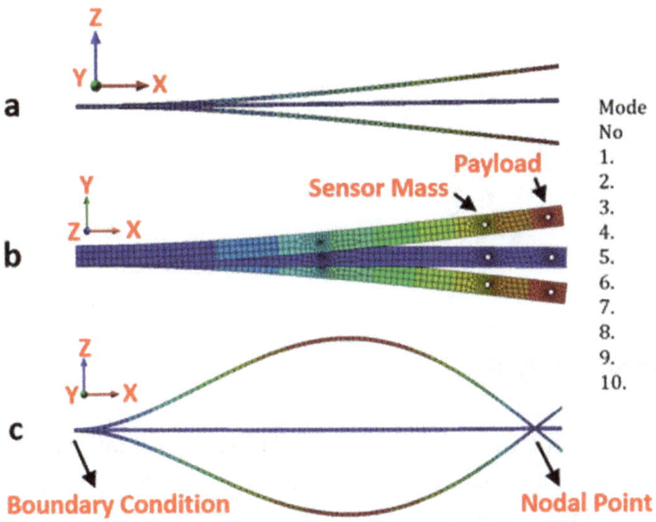

Mode No	Frequency [Hz]
1.	3.1804
2.	19.889
3.	47.680
4.	140.78
5.	287.50
6.	297.17
7.	487.83
8.	511.22
9.	817.32
10.	871.09

Transient Analysis

While performing the transient analysis of the system, the most difficult motion scenario of the flexible steel manipulator used in the previous study (Yavuz et al., 2020) is taken into account. The input functions used in this scenario are shown in Figure 5. In Figure 5.b, it is seen that the angular velocity input is in triangular form and in Figure 5.a, the system has an angular displacement of 90 degrees ($\pi/2$ rad) in 2 seconds. In addition, the angular acceleration graph in the form of a pulse is shown in Figure 5.c. Finally, it can be seen that the angular jerk in Figure 5.d is in the form of impulse, depending on the discontinuous input function used. After the motion input is applied to the system, the equivalent stress (according to von-Mises) response of the system is shown in Figure 6. Accordingly, it is seen that the maximum stress occurs in the 1.99th second (100.66 MPa). Since the tensile yield strength of the material is 250 MPa, the Safety Factor (SF- the ratio of maximum stress to tensile yield strength) of the system is calculated as 2.4837.

Figure 5. Angular motion inputs, a) angular position input (rad), b) angular velocity input (rad/s), c) angular acceleration input (rad/s^2), d) angular jerk input (rad/s^3)

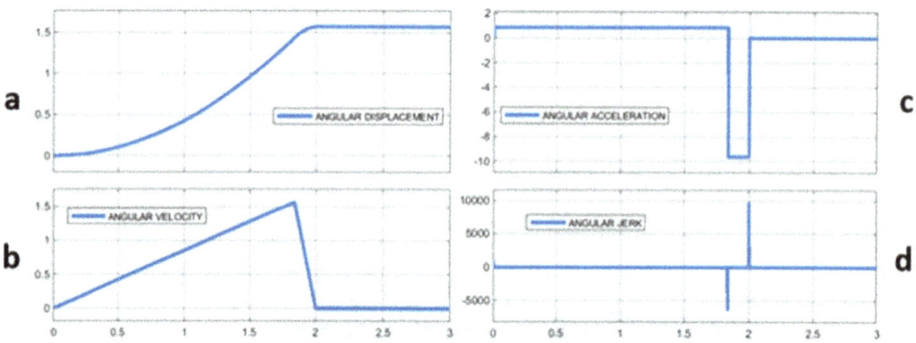

Quasi-Stationary Analysis

As mentioned in the Introduction, it is necessary to convert non-stationary transient analysis to quasi-stationary analysis in order to be used in topology optimization. To make this possible, the D'Alembert principle is applied and the forces and moments of inertia are transferred to the system as virtual inputs, assuming that the system is static (Figure 7). Since it is known that the system is mostly forced in the bending direction, the forces on the Y-axis have been taken into account. In the transient analysis, the input torque, inertial moment, inertial force, sensor inertia and payload inertia values created by considering the torque and accelerations in the 1.99th second when the maximum stress occurs, are 3.447 Nm, 0.2 Nm, 1.03 N, 0.4346 N, 5.9658 N, respectively. The SF value calculated as a result of the quasi-stationary analysis created is 2.4402, which coincides with the result of the transient analysis.

Figure 6. Equivalent stress (according to von-Mises) result (MPa) of the manipulator

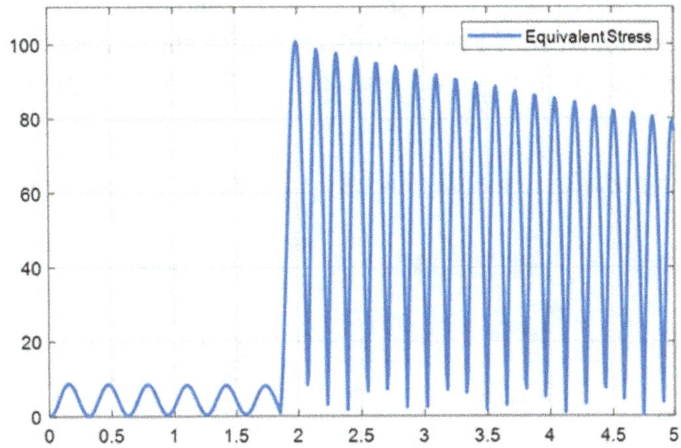

Figure 7. Forces acting on the manipulator according to D'Alembet's principle

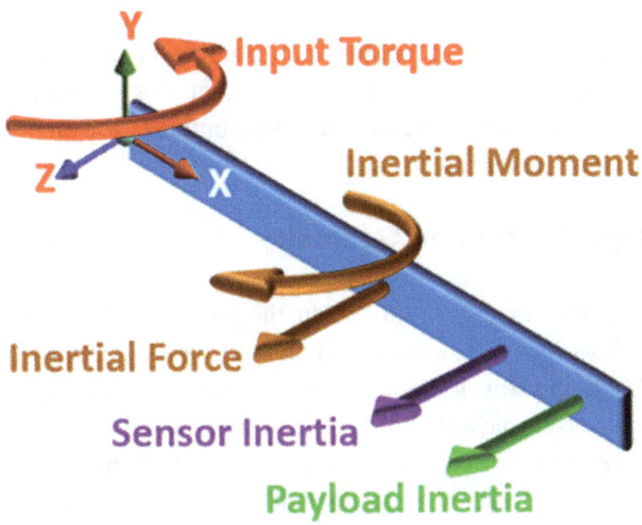

Topology Optimization

The design optimization architecture given in Figure 1 in the introduction is reflected on the program used, as shown in Figure 8. When using the Topology optimization module, both quasi-stationary analysis results and modal analysis results are used. With the selection of the modes to be included from the modal analysis results, multiple topology optimization choices emerge. In the optimizations, while reducing the manipulator mass to 75%, at the same time, the condition that the design should be symmetrical around the X-axis is added. In addition, the joint, sensor and payload connection points, which are called boundary conditions, are kept unchanged.

Figure 8. Design optimization framework on the finite element analysis program

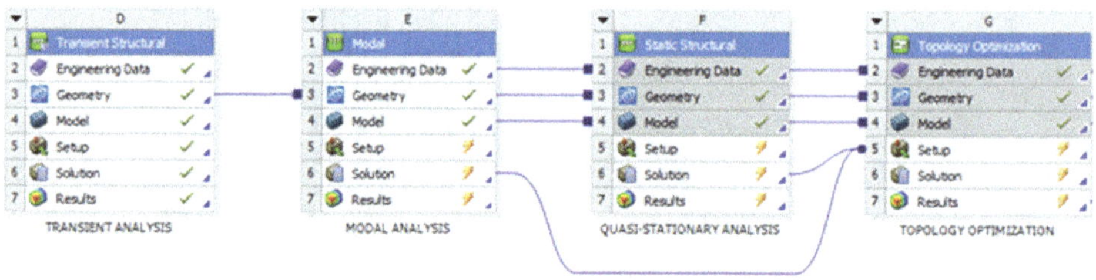

RESULTS AND DISCUSSIONS

In this section, first the effects of topology optimization on geometry are examined. The different designs created according to different mesh quality and objective selection are shown. Then, the effects of topology optimization on the safety factor and vibration modes are examined.

Effects of Optimization on Geometry

While performing the topology optimization, the parts where the stress is minimum are reduced with various iterations according to the analysis results (Bendsoe and Sigmund, 2003). In this way, it is aimed to reduce the mass of the system by getting rid of the parts that do not have much effect on the mechanical strength. The precision of this process depends on the mesh quality in the finite element model. In Figure 9, it is shown that the effects of the change of mesh quality on the created design and the optimization may fail when the mesh quality is low. While determining the mesh quality, mesh relevance value (limited between -100 and 100) and refinement order (maximum 3rd level) parameters are used. Fine mesh structure is utilized in the optimizations.

Figure 9. Optimization results for different mesh levels, (1st from top) zero relevance, (2nd from top) maximum relevance, (3rd from top) maximum relevance and 1st order refinement, (2nd from bottom) maximum relevance and 2nd order refinement, (1st from bottom) maximum relevance and maximum refinement

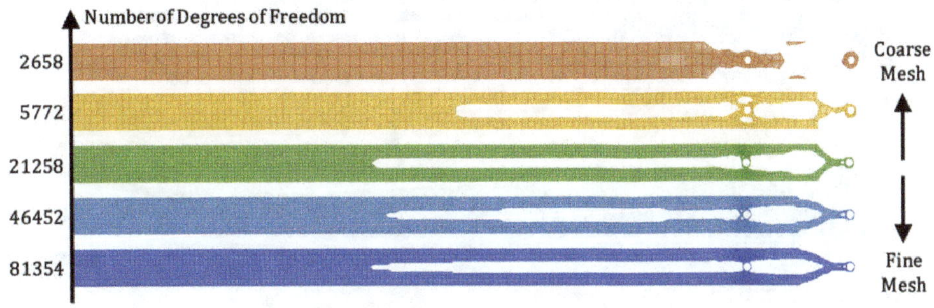

In the previous section, it has been explained that quasi-stationary and modal analysis results are taken into account while determining the objectives during topology optimization. From the modal analysis results, the mode shape of each considered mode is included in the optimization. The number of modes considered and the corresponding design results are shown in Figure 10. Accordingly, in the first three figures, it is seen that the results are not completely converged and there are parts that are not connected to each other in the final version of the designs. Although it is seen that the optimization results converge better when it goes to the bottom of the figure, it is seen that the geometry becomes too complex. This is because mode shapes corresponding to high vibration modes are also included in the optimization. However, since high vibration modes are not dominant, optimizing them doesn't have much effect on the systems dynamic behavior. Therefore, the optimization that includes the first three dominant vibration modes, shown in Figure 4, can be selected as the most ideal option. This design is also the first result to be converged in Figure 10 and has the simplest geometry among the other designs.

Figure 10. The designs created according to the number of modes taken into account as obcevtives in the topology optimization, e.g., the first result is carried out only by considering the quasi-stationary analysis and vibration modes are not included. In the final design, both the quasi-stationary analysis results and the mode shape results of the first 10 vibration modes are taken into account.

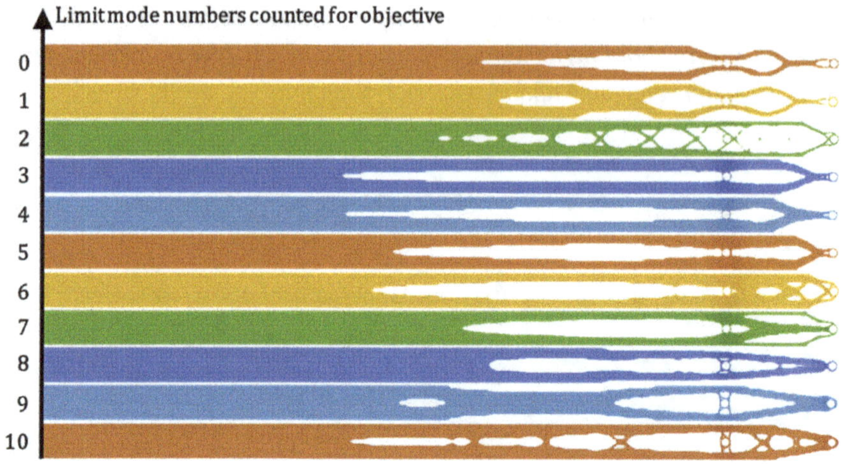

Effects of Optimization on Safety Factor and Vibration Modes

Among the designs created in the previous section, the most ideal design (4th from top in Figure 10) is chosen as the proposed design and the design that includes all vibration modes (at the bottom of Figure 10) is chosen as the most complex design. For comparison with these designs, the original beam, tapered beam and beam with reduced-width shown in Figure 2 are selected. The original beam has a mass of 249 grams, while the other four designs have a mass of about 187 grams.

These five designs, whose analysis results are compared in the section, are shown in Figure 11, respectively. Comparison of vibration mode frequencies of these designs is given in Figure 12. Accordingly, for the original beam, proposed design, complex design, tapered beam and beam with reduced-width: (i) the first natural frequencies are, 3.1865 Hz, 3.1097 Hz, 3.125 Hz, 3.0279 Hz and 2.7897 Hz, (ii) the second natural frequencies are, 19.876 Hz, 16.987 Hz, 19.783 Hz, 16.09 Hz, 13.062 Hz, and (iii) the third natural frequencies are 47.769 Hz, 47.854 Hz, 45.992 Hz, 47.596 Hz, 46.608 Hz, respectively. Therefore, it can be seen that the designs whose natural frequencies are closest to the original design are the designs produced by topology optimization.

Figure 11. Manipulator designs for analysis results comparison, a) proposed design, b) complex design, c) orjinal beam, d) tapered beam, e) beam with reduced-width

Figure 12. Comparison of the first three vibration mode frequencies for different designs (Figure 11)

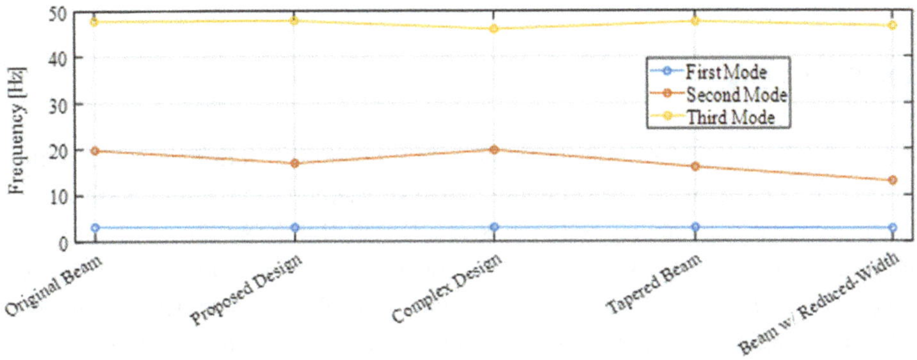

The comparison of the Safety Factor values calculated as a result of the transient analyses created with the velocity profile in Figure 5 is given in Figure 13. Accordingly, the comments about the results can be listed as follows:

- It is seen that the design, whose thickness is reduced by the direct mass reduction ratio, without any optimization, gives the lowest SF value (below 2),
- On the contrary, the two topology-optimized designs reached higher SF values (above 2.5) than both the tapered beam and the original beam,
- Finally, the proposed design has the simplest geometry among the optimized designs, as well as the highest SF value.

Figure 13. Comparison of transient analysis Safety Factor (SF) values for different designs (Figure 11)

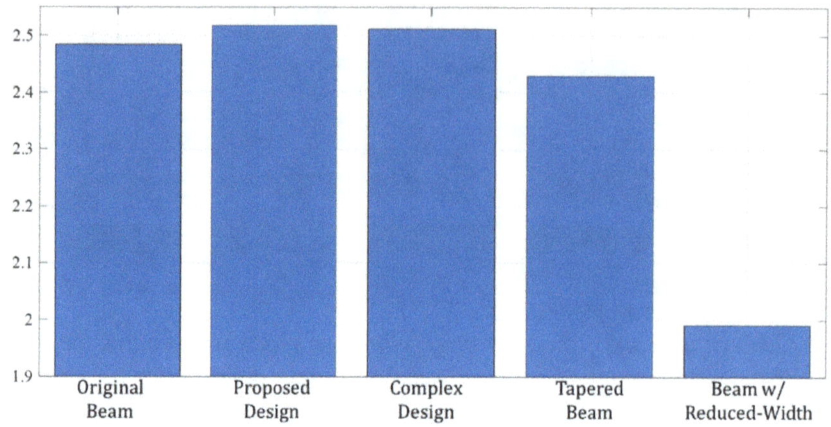

CONCLUSION

In the study, design optimization of a flexible manipulator with industrial servo motion control was carried out by finite element analysis method. As the motion scenario, the scenario that was stated to be the most challenging for the system in the previous study was chosen. During the optimization, both modal analysis and transient analysis converted to quasi-stationary analysis were utilized. While comparing the results, both natural frequencies and transient analysis safety factor values were employed. In addition, the results were compared with tapered beam and beam with reduced width, in which the mass reduction (one in forth) was realized. Considering the first three natural frequencies and quasi-stationary analysis, the optimized design has been the proposed design in the study, as it has both the simplest geometry among the optimized designs and the highest safety factor among the designs including original manipulator design.

It is possible to use the design obtained in the study for links made of different materials and in multi-link systems. It is thought that combining design optimization with materials with high strength and lightness, such as carbon-fiber-reinforced-polymer composite, will further increase the efficiency of flexible multibody systems.

REFERENCES

Albers, A., Brudniok, S., Ottnad, J., Sauter, C., & Sedchaicharn, K. (2006, December). Upper body of a new humanoid robot-the design of ARMAR III. In *6th IEEE-RAS International Conference on Humanoid Robots,* (pp. 308-313). IEEE.

Alkalla, M. G., & Fanni, M. A. (2021). Integrated structure/control design of high-speed flexible robot arms using topology optimization. *Mechanics Based Design of Structures and Machines*, *49*(3), 381–402. doi:10.1080/15397734.2019.1688170

Bendsoe, M. P., & Sigmund, O. (2003). *Topology Optimization.* Springer Science & Business Media. https://link.springer.com/book/10.1007/978-3-662-05086-6

Deaton, J. D., & Grandhi, R. V. (2014). A survey of structural and multidisciplinary continuum topology optimization: Post 2000. *Structural and Multidisciplinary Optimization*, *49*(1), 1–38. doi:10.100700158-013-0956-z

Guo, X., & Cheng, G.-D. (2010). Recent development in structural design and optimization. *Lixue Xuebao*, *26*(6), 807–823. doi:10.100710409-010-0395-7

Held, A., Knufer, S., & Seifried, R. (2015). *Topology Optimization of Members of Dynamically Loaded Flexible Multibody Systems using Integral Type Objective Functions and Exact Gradients*. 6.

Held, A., & Seifried, R. (2012). Topology Optimization of Members of Elastic Multibody Systems. *Proceedings in Applied Mathematics and Mechanics*, *12*(1), 67–68. doi:10.1002/pamm.201210025

Hennessey, M., Priebe, J., Huang, P., & Grommes, R. (1987, March). Design of a lightweight robotic arm and controller. In *Proceedings of the IEEE International Conference on Robotics and Automation,* (Vol. 4, pp. 779-785). IEEE. 10.1109/ROBOT.1987.1087976

Liu, Y., Yang, C., Wei, P., Zhou, P., & Du, J. (2021). An ODE-driven level-set density method for topology optimization. *Computer Methods in Applied Mechanics and Engineering*, *387*, 114159. doi:10.1016/j.cma.2021.114159

Lohmeier, S., Buschmann, T., Schwienbacher, M., Ulbrich, H., & Pfeiffer, F. (2006, December). Leg design for a humanoid walking robot. In *6th IEEE-RAS International Conference on Humanoid Robots,* (pp. 536-541). IEEE.

Oosting, K., & Dickerson, S. L. (1988, April). Simulation of a high-speed lightweight arm. In *Proceedings. 1988 IEEE International Conference on Robotics and Automation,* (pp. 494–496). IEEE. doi:10.1109/ROBOT.1988.12100

Ryu, J., Sung-Soo, K., & Sang-Sup, K. (1992). An efficient computational method for dynamic stress analysis of flexible multibody systems. *Computers & Structures, 42*(6), 969–977. doi:10.1016/0045-7949(92)90107-B

Srinivas, G. L., & Javed, A. (2020). Optimization approaches of industrial serial manipulators to improve energy efficiency: A review. *IOP Conference Series. Materials Science and Engineering, 912*(3), 032058. doi:10.1088/1757-899X/912/3/032058

Sun, J., Tian, Q., & Hu, H. (2017). Topology optimization based on level set for a flexible multibody system modeled via ANCF. *Structural and Multidisciplinary Optimization, 55*(4), 1159–1177. doi:10.100700158-016-1558-3

Yavuz, Ş., İlman, M. M., & Binici, B. (2020). Modeling approach and analysis time comparison of single-link flexible steel- and epoxy-glass/carbon-fiber composite manipulators. *Structures, 26*, 396–405.

Chapter 7
System Approaches to the Problem of Dynamic Damping of Vibrations of Technical Objects of Transport and Technological Purpose:
Reduced Stiffness, State Sets, and Form Interactions

Andrey Vladimirovich Eliseev
https://orcid.org/0000-0003-0222-2507
Irkutsk State Transport University, Russia

Nikolay Konstantinovich Kuznetsov
Irkutsk National Research Technical University, Russia

Artem Sergeevich Mironov
Irkutsk State Transport University, Russia

ABSTRACT

The methodological basis for solving the problems of assessment, control, and formation of dynamic states of technical objects of transport and technological purposes in a state of vibration loading is being developed. Within the framework of structural mathematical modeling, mechanical oscillatory systems used as design schemes of technical objects are compared with schemes of dynamically equivalent automatic control systems. New results are presented in the field of technology for

DOI: 10.4018/978-1-6684-5887-7.ch007

evaluating dynamic states and forms of interactions of elements of mechanical oscillatory systems with the reduced elastic characteristics. For mechanical oscillatory systems, an interpretation of dynamic states and forms of interactions in the form of an oriented graph has been developed, creating prerequisites for the classification of a set of dynamic states in which a system can be located under the condition of various types of force excitation

INTRODUCTION

Many technical facilities, including transport and technological purposes, are operated under conditions of increased dynamic loads. Such conditions imply the need for calculations at the stages of preliminary research and the adoption of constructive and technical solutions that ensure the necessary level of reliability and safety of operation of technical means (see DeSilva, 2000; Harris & Cred, 2002; Iwnicki, 2006). Technical objects can be quite complex systems that include not only mass-inertial elements, but a significant number of other elements that implement both elastic interactions and the dissipation of oscillation energy, as well as functional ties. Functional ties are capable of creating various dynamic states and specific forms of dynamic interactions of machine and equipment elements. Moreover, functional connections have the potential of new dynamic effects of practical significance.

Considerable attention is paid to the development of system approaches to the tasks of machine dynamics, ensuring operational safety, including protection from vibrations of technical objects of transport and technological purposes, vibration isolation of equipment, devices and equipment under the influence of vibration loading (see Banakh & Kempner, 2010; Karnovsky & Lebed, 2016; Panovko, 2006). The problems of machine dynamics are very diverse. However, it would be advisable to single out research areas related to the tasks of vibration protection and vibration isolation of machines, devices and equipment, as well as control modes of dynamic vibration damping, special forms of coherence of movement of system elements, as well as other specific manifestations of forms of interaction of elements (see Bykhovsky, 1969; Galiev, Nekhaev, & Nikolaev, 2010; Goncharevich & Frolov, 1981; Rocard, 1949). A wide range of research in the field of dynamics of mechanical oscillatory systems is aimed at creating a scientific and methodological basis (see Eliseev, Lukyanov, Reznik, & Khomenko, 2006; Eliseev, Reznik, & Khomenko, 2011; Eliseev, Reznik, Khomenko, & Zasyadko, 2008). Within the framework of such a theoretical foundation, the problems of the dynamics of machines, equipment and mechanisms operating under periodic (usually harmonic) external influences of a forceful or kinematic nature are solved, taking into account a certain degree of connectivity between external disturbances. Structural mathematical modeling occupies a significant place among the common technologies for assessing, monitoring and forming dynamic states of technical objects under vibrational loading of a power or kinematic nature (see Belokobylsky, Eliseev, & Kashuba, 2013; Eliseev, 2018; Eliseev & Artyunin, 2016).

Within the framework of structural approaches, mechanical oscillatory systems used as design schemes of technical objects are compared with schemes that are equivalent in dynamic terms to automatic control systems. Frequency characteristics of systems are essential in mathematical modeling based on structural methods. Such characteristics give an idea of the features of the characteristic dynamic states of the system in the form of resonant phenomena and the effects of dynamic damping of vibrations along a fixed coordinate.

Dynamic states and forms of dynamic interactions of system elements are of great importance for the formation of dynamic quality and ensuring the safety conditions for the operation of technical facilities, etc. The need to determine the features of the dynamic states of systems and forms of dynamic interactions presupposes the preliminary construction of certain mathematical models that display the steady state motion of elements. Mathematical models make it possible, along with the natural frequencies of the system, to determine the frequencies of dynamic damping of fixed coordinate oscillations on the basis of analytical or approximate approaches, which is usually used in engineering practice. At the same time, in many cases there is a need for detailed representations of dynamic states and forms of dynamic interactions not of one coordinate, but of a set of coordinates of mechanical oscillatory systems. The analytical approach to solving such problems encounters certain difficulties.

In the proposed work, within the framework of structural mathematical modeling methods, the methodological basis of a systematic approach to assessing the totality of dynamic states and forms of dynamic interactions of elements of mechanical oscillatory systems under conditions of kinematic or force perturbation is developed.

GENERALITIES. PROBLEM STATEMENT

The considered mechanical oscillatory system (Figure 1) is formed by three mass-inertial elements m_1, m_2, m_3 connected by means of elastic elements with stiffness k_1–k_5, kinematic disturbances $z_1(t)$, $z_2(t)$ act from the limiters; in addition, external harmonic synchronous disturbances $Q_1(t)$, $Q_2(t)$, $Q_3(t)$ act on the mass-inertial elements. The system makes small fluctuations relative to the static equilibrium position. The displacements y_1, y_2, y_3 of the mass inertial elements relative to the position of the static equilibrium are considered as generalized coordinates.

Figure 1. Design scheme of a technical object represented by a mechanical oscillatory system with three degrees of freedom

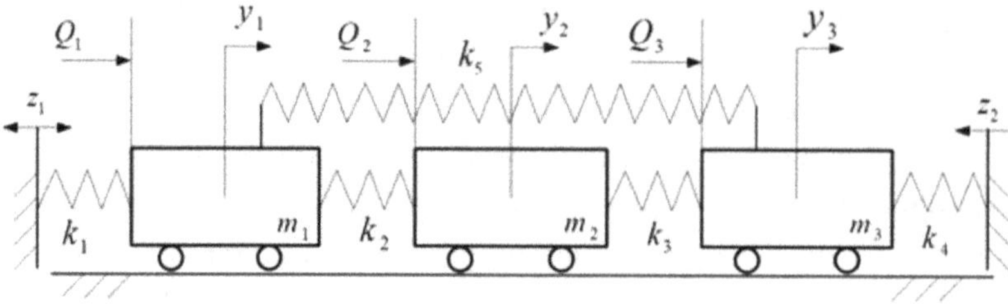

To construct a mathematical model of the system, Lagrange equations of the 2-nd kind are used, compiled on the basis of expressions for kinetic and potential energy (see Eliseev, Kuznetsov, & Moskovskikh, 2019; Eliseev, & Eliseev, 2020; Eliseev, Eliseev, Bolshakov, & Khomenko, 2021):

$$T = \frac{1}{2}m_1\dot{y}_1^2 + \frac{1}{2}m_2\dot{y}_2^2 + \frac{1}{2}m_2\dot{y}_2^2 \tag{1}$$

$$\prod = \frac{1}{2}k_1(y_1 - z_1)^2 + \frac{1}{2}k_2(y_2 - y_1)^2 + \frac{1}{2}k_3(y_3 - y_2)^2 + \frac{1}{2}k_4(y_3 - z_2)^2 + \frac{1}{2}k_5(y_3 - y_1)^2, \tag{2}$$

After composing a system of differential equations taking into account zero initial conditions and integral Laplace transformations, we write down an algebraic system:

$$\bar{y}_1\left(m_1p^2 + k_1 + k_2 + k_5\right) - \bar{y}_2k_2 - \bar{y}_3k_5 = \bar{Q}_1 + k_1\bar{z}_1, \tag{3}$$

$$-\bar{y}_1k_2 + \bar{y}_2\left(m_2p^2 + k_2 + k_3\right) - \bar{y}_3k_3 = \bar{Q}_2, \tag{4}$$

$$-\bar{y}_1k_5 - \bar{y}_2k_3 + \bar{y}_3\left(m_3p^2 + k_3 + k_4\right) = \bar{Q}_3 + k_4\bar{z}_2. \tag{5}$$

It is assumed that external force perturbations have connectivity properties of the form:

$$\bar{Q}_2 = \alpha\bar{Q}_1, \tag{6}$$

$$\bar{Q}_3 = \beta \bar{Q}_1 , \quad (7)$$

where α and β are constant values taking positive, negative and zero values.

The task is to develop graph-analytical methods for assessing the dynamic features of mechanical oscillatory systems with three degrees of freedom, which are under the influence of inphase external disturbances of a force and kinematic nature.

FEATURES OF TAKING INTO ACCOUNT THE COEFFICIENTS OF CONNECTIVITY OF EXTERNAL FORCE EXCITATIONS

The system of coefficients of the equations (3)-(5) of motion in coordinates $\bar{y}_1, \bar{y}_2, \bar{y}_3$ (3)-(5) under the condition (z_1=0, z_2=0) are presented in table 1.

Table 1. The coefficients of the algebraic system

a_{11}	a_{12}	a_{13}
$m_1p^2+k_1+k_2+k_5$	$-k_2$	$-k_5$
a_{21}	a_{22}	a_{23}
$-k_2$	$m_2p^2+k_2+k_3$	$-k_3$
a_{31}	a_{32}	a_{33}
$-k_5$	$-k_3$	$m_3p^2+k_3+k_4+k_5$
	Generalized forces	
$\bar{Q}_1$	$\bar{Q}_2$	$\bar{Q}_3$

The algebraic system of equations in Laplace images (table 1) are used to construct structural schemes that are dynamically equivalent to automatic control systems. Figure 2 shows a structural model of the system, which allows to obtain a detailed description of the connections between the elements of the system and to build the transfer functions of the system ($p=j\omega$ is a complex variable, the «¯ » - icon above the variable means a Laplace image).

Based on table 1, expressions can be constructed for the relationships between the variables y_1, y_2, y_3 and the impact of $\bar{Q}_1$, taking into account the connectivity coefficients:

$$W_1(p) = \frac{\bar{y}_1}{\bar{Q}_1} = \frac{\left(a_{22}a_{33} - a_{23}a_{32}\right) + \alpha\left(a_{13}a_{32} - a_{33}a_{12}\right) + \beta\left(a_{12}a_{23} - a_{13}a_{22}\right)}{A_0}, \tag{8}$$

$$W_2(p) = \frac{\bar{y}_2}{\bar{Q}_1} = \frac{\left(a_{23}a_{31} - a_{31}a_{33}\right) + \alpha\left(a_{11}a_{33} - a_{13}a_{31}\right) + \beta\left(a_{13}a_{21} - a_{11}a_{23}\right)}{A_0}, \tag{9}$$

$$W_3(p) = \frac{\bar{y}_3}{\bar{Q}_1} = \frac{\left(a_{21}a_{32} - a_{22}a_{31}\right) + \alpha\left(a_{12}a_{31} - a_{11}a_{32}\right) + \beta\left(a_{11}a_{22} - a_{12}a_{21}\right)}{A_0}, \tag{10}$$

where

$$A_0 = a_{11}a_{22}a_{33} - a_{11}a_{23}a_{32} - a_{33}a_{12}a_{21} - a_{22}a_{13}a_{31} + a_{13}a_{21}a_{32} + a_{12}a_{23}a_{31}, \tag{11}$$

where the characteristic equation A_0=0 allows us to determine the natural frequencies.

Figure 2. Structural (a) and detailed (b) schemes of a mechanical oscillatory system with three degrees of freedom

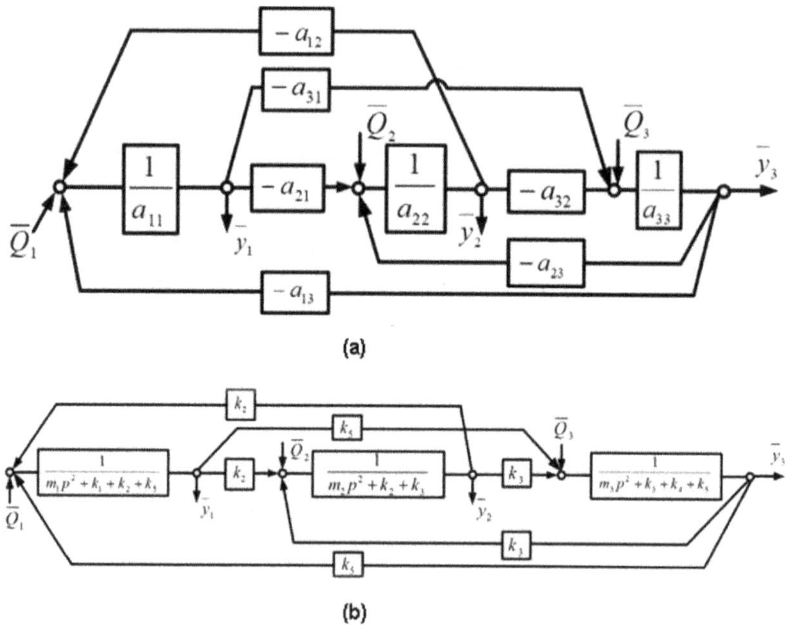

Further, it is assumed that the force effects disturbances $\bar{Q}_1, \bar{Q}_2, \bar{Q}_3$ are applied to the mass-inertial elements m_1, m_2, m_3 respectively, without regard to connectivity (α=0, β=0).

METHOD CONSTRUCTION OF THE REDUCED CHARACTERISTICS OF THE SYSTEM BASED ON THE FREQUENCY CHARACTERISTIC EQUATION

To study systems in a generalized form reduced to one coordinate $\bar{y}_1$ by excluding coordinates $\bar{y}_2, \bar{y}_3$, taking into account that the force disturbance $\bar{Q}_1$ is applied to the element m_1, transformations of the characteristic frequency equation (11) are considered in the form:

$$a_{11} - \frac{a_{31}\left(a_{31}a_{22} - a_{21}a_{23}\right)}{\left[m_2 m_3\right]} - \frac{a_{12}\left(a_{12}a_{33} - a_{23}a_{31}\right)}{\left[m_2 m_3\right]} = 0\,, \tag{12}$$

where

$$\left[m_2 m_3\right] = a_{22}a_{33} - a_{23}a_{32} = \left[m_2 p^2 + k_2 + k_3\right] \cdot \left[m_3 p^2 + k_3 + k_4 + k_5\right] - k_3^{\,2}\,. \tag{13}$$

The block diagram of the system (Figure 2, a, b) after exception of the coordinates $\bar{y}_2, \bar{y}_3$, takes the following form (Figure 3, a, b).

Figure 3. Block diagram of the system reduced to one coordinate of motion of the mass-inertia element m₁, taking into account the disturbing force $\bar{Q}_1$: a) - generalized view; b) - detailed view ($\bar{Q}_1 \neq 0, \bar{Q}_2 = 0, \bar{Q}_3 = 0$)

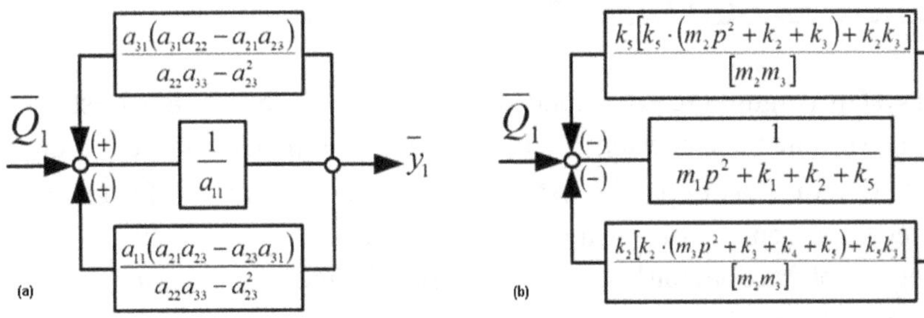

Transformation of the original system (Figure 2) to a generalized system (Figure 3) suggests the possibility of interpreting individual parts of the system using representations of quasi-springs, the stiffness of which depends on the frequencies of external excitation:

$$\overline{k'}_{\text{rd.1}}(p) = \frac{k_5\left[\left(m_2p^2 + k_2 + k_3\right)k_5 + k_2k_3\right]}{\left[m_2m_3\right]}; \tag{14}$$

$$\overline{k''}_{\text{rd.2}}(p) = \frac{k_2\left[\left(m_3p^2 + k_3 + k_4 + k_5\right)k_2 + k_3k_5\right]}{\left[m_2m_3\right]}. \tag{15}$$

As a result of transformations related with the exception of the coordinate $\bar{y}_2, \bar{y}_3$, characteristic frequency equation (12) can be represented as:

$$m_1p^2 + k_1 + k_2 + k_5 - k_{\text{rd.1}} - k_{\text{rd.2}} = 0. \tag{16}$$

After excluding the coordinates $\bar{y}_2, \bar{y}_3$, the equation of the system (Figure 2) with respect to $\bar{y}_1$ is equal to zero at the point of application of force Q_1, the sum of all forces applied to the mass-reaction element m_1, including inertia and bond reaction forces:

$$m_1p^2\bar{y}_1 + k_1\bar{y}_1 + k_2\bar{y}_1 + k_5\bar{y}_1 - k_{\text{rd.1}}\bar{y}_1 - k_{\text{rd.2}}\bar{y}_1 - \bar{Q}_1 = 0, \tag{17}$$

which is a representation of the Dalembert principle.

A METHOD FOR EVALUATING THE FORMS OF TIED MOVEMENTS BASED ON PARTIAL SYSTEMS

The model system (Figure 1) can be reduced to a chain system by zeroing the stiffness parameter k_5=0. When considering $\bar{y}_1$ coordinates as a kinematic perturbation, by analogy with the concept of a partial system of the usual form (all coordinates except one are fixed), the concept of a generalized partial system representing a system with two degrees of freedom can be considered. The schematic diagram and structural model of the generalized partial system is shown in Figure 4.

Figure 4. Generalized partial system: a) - schematic diagram ($\bar{y}_1$ is considered as a kinematic excitation), b) - block diagram ($\bar{y}_1$ - input excitation, $\bar{y}_2$ and $\bar{y}_3$ - output movements)

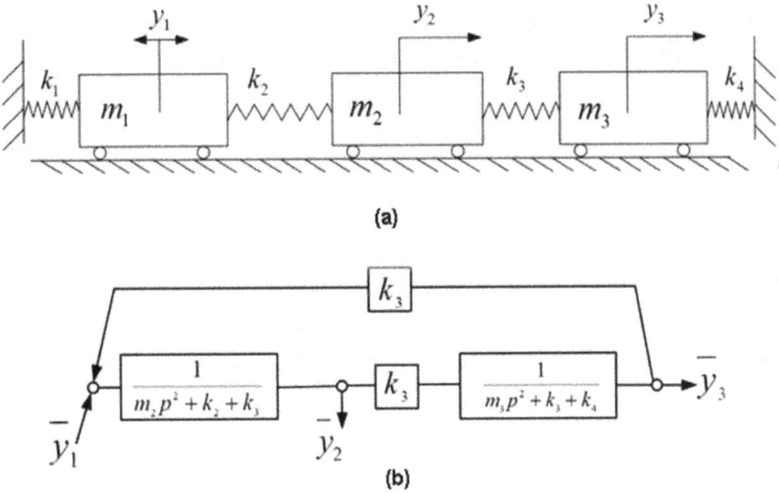

The dynamic properties of a generalized partial system in the form of a chain mechanical oscillatory system moving under the influence of a kinematic disturbance in the form of a $\bar{y}_1$ coordinate can be characterized by the so-called transfer functions of interpartial connections:

$$W_1''(p) = \frac{\bar{y}_2}{\bar{y}_1} = \frac{\left(m_3p^2 + k_3 + k_4\right)\cdot k_2}{\left[m_2m_3\right]}, \tag{18}$$

$$W_2'''(p) = \frac{\bar{y}_3}{\bar{y}_1} = \frac{k_2k_3}{\left[m_2m_3\right]}. \tag{19}$$

Generalized partial system (Figure 4, a), considered as an isolated mechanical vibrating system with two degrees of freedom, has two frequency «natural» oscillations and determined from the equation $[m_2m_3]=0$ (13).

The characteristic equation of the system (16) is provided $k_5=0$ taking into account expressions (14) and (15) for the dynamic stiffness $k_{rd.1}(p)$ and $k_{rd.2}(p)$ takes the form:

$$m_1p^2 + k_1 + k_2 - k_{\text{rd.1}} - k_{\text{rd.2}} = m_1p^2 + k_1 + k_2 - 0 - \frac{k_2^{\,2}\left(m_3p^2 + k_3 + k_4\right)}{\left[m_2m_3\right]}. \tag{20}$$

After replacing p=ωj, j=Ö-1, the expression (20) can be represented as an equation with respect to the natural frequency of the original system ω:

$$-m_1\omega^2 + k_1 + k_2 = \frac{k_2^{\,2}\left(-m_3\omega^2 + k_3 + k_4\right)}{\left(-m_2\omega^2 + k_2 + k_3\right)\cdot\left(-m_3\omega^2 + k_3 + k_4\right) - k_3^2}. \tag{21}$$

The graphical solution of equation (16), which determines the natural oscillation frequencies of the initial system (Figure 1), can be represented by the intersection points of the graphs of the functions of the left $D(\omega)$ and right $D_1(\omega)$ parts:

$$D\left(\omega\right) = k_1 + k_2 - m_1\omega^2; \tag{22}$$

$$D_1\left(\omega\right) = k_{rd.2}\left(\omega\right). \tag{23}$$

The frequencies $n_2^{'}, n_3^{'}$, found from equation $[m_2m_3]=0$ as the «natural» frequencies of the partial system viewed in isolation (Figure 4), deliver infinitely large values to the reduced stiffness $k_{\text{rd.2}}^{'}\left(\omega\right)$. The equality of the dynamic stiffness $D_1(\omega)$ of the quasi-spring to infinitely large values determines the modes of dynamic quenching of vibrations for an element with mass m_1.

FREQUENCY CHARACTERISTICS IN THE EVALUATION OF FORMS OF DYNAMIC INTERACTIONS

The amplitude-frequency characteristics of the transfer functions of the system along various coordinates depend significantly on the points of application of external force actions. The key features of the graphs of amplitude-frequency characteristics are the signs of the frequencies of dynamic damping of oscillations, natural frequencies and partial frequencies. The family of frequency response graphs (Figure 5) is constructed for a mechanical oscillatory system with a force disturbance $\bar{Q}_1 \neq 0$ ($\bar{Q}_2 = 0, \bar{Q}_3 = 0$).

Figure 5. Amplitude-frequency characteristics of the transfer functions of the system (according to Figure 1): $1-\frac{\bar{y}_1}{\bar{Q}_1}; 2-\frac{\bar{y}_2}{\bar{Q}_1}; 3-\frac{\bar{y}_3}{\bar{Q}_1}$*; pp.1-4 - determine the frequencies of dynamic quenching at* $\bar{Q}_1 \neq 0, \bar{Q}_2 = 0, \bar{Q}_3 = 0$*, pp.5-7 - partial frequencies, pp.8-10 - natural frequencies, p.11 - determines the zero frequency*

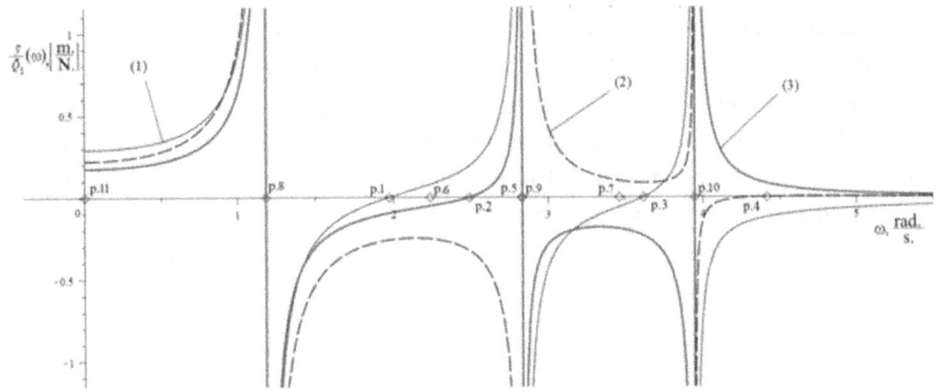

The determination of the frequencies of natural oscillations of $\omega_{1nat}, \omega_{2nat}, \omega_{3nat}$ by expressions (22) and (23) can be interpreted as a method of finding the frequencies of special modes depending on the object to which an external force disturbance is applied, in frequency, from the element m1. The peculiarity of this method in this case is the construction of an expression for the frequencies of natural oscillations, ensuring that the dynamic stiffness of the system is equal to zero in the form

$$D(\omega) = D_1(\omega) \tag{24}$$

The form of expression for dynamic stiffness (24) allows us to consider its zero, positive, and at the same time, its negative values.

Within the framework of the developed concepts, resonant modes in mechanical oscillatory systems reflect the absence of resistance from the system elements to external force influences of a certain frequency applied to mass-inertial elements.

FEATURES OF DISPLAYING THE DIRECTIONS OF MOVEMENT OF ELEMENTS BASED ON AMPLITUDE-FREQUENCY CHARACTERISTICS

By means of assessing the forms of dynamic interactions, the amplitude-frequency characteristics of the system, the amplitude-frequency characteristics of interpartial connections and the given dynamic stiffness can be selected.

The essential features of the dynamic interactions of the system elements, displayed on the frequency response graphs (Figure 5), can be presented in tabular form (table 2). As a demonstration of the developed method, part of the dynamic states and features of the forms of movement of the elements can be built on the basis of the amplitude-frequency characteristics considered in the frequency range $\omega \in (0, \omega_{2nat})$ are selected to evaluate the dynamic states and features of the forms of movement of the elements (table 2, column 1, rows II-IV).

Table 2. States and forms of dynamic interactions ($\bar{Q}_1 \neq 0, \bar{Q}_2 = 0, \bar{Q}_3 = 0$)

	1	2	3	4	5	6	7	8	9	10	11	12	13	14
I		0		ω_{1nat}		ω_{dyn1}		ω_{par1}		ω_{dyn3}		ω_{par1}		ω_{2nat}
II	$\frac{\bar{y}_1}{\bar{Q}_1}$		+	∞	–	0	+	+	+	+	+	+	+	∞
III	$\frac{\bar{y}_2}{\bar{Q}_1}$		+	∞	–	–	–	–	–	–	–	–	–	∞
IV	$\frac{\bar{y}_3}{\bar{Q}_1}$		+	∞	–	–	–	–	–	0	+	+	+	∞
V	Point	p.11		p.8		p.1		p.6		p.2		p.5		p.9
VI	State	000		∞∞∞		0II		I*I		II0		*II		∞∞∞
VII	Form		+++		---		+--		+--		+-+		+-+	

The set of characteristic frequencies of external force action is represented by natural frequencies, frequencies of dynamic vibration damping and partial frequencies; at the same time, characteristic frequencies determine the frequency intervals at which the amplitude-frequency characteristics retain a certain sign (table 2, row I). The modes of dynamic vibration damping at the corresponding coordinate («0II» – vibration damping mode at the $\bar{y}_1$ coordinate, «II0» – vibration damping at the

$\bar{y}_3$ coordinate), vibration modes at partial frequencies («I*I», «*II» – oscillation at the second and first partial frequencies) and resonance modes («∞∞∞») are indicated as characteristic modes, using symbols.

In the intervals between characteristic frequencies, the signs of amplitude-frequency characteristics (table 2, columns 3,5,7,9,11,12,13, rows II-IV) denoting the direction of movement of individual coordinates of the system are determined. As forms of dynamic interactions, the placements of triple sets formed by pluses and minuses are indicated in accordance with the signs of amplitude-frequency characteristics in frequency intervals.

The totality of states and forms of dynamic interactions of a mechanical oscillating system under the forced force action $\bar{Q}_1$, presented in table 2, can be represented by a graph in which the vectors of values of amplitude-frequency characteristics in characteristic frequencies form vertices, and the frequency intervals of the sign of the amplitude-frequency characteristics, displayed by sets of directions, form arcs.

The vertex «000» (Figure 6) is a point with zero frequency (Figure 5, p.11). The arc 0 corresponds to the form of movement of the system elements with the same directions, which is depicted by a set of «+++» next to the arc; the end of the arc (Figure 6, p.8) «points» to the vertex «∞∞∞» corresponding to the resonance mode (Figure 5, p.8). Arc 1, displaying the form of dynamic interaction of the same orientation, but opposite to the directions displayed by arc 0. Arc 1 indicates the vertex «0II» (Figure 6, p.1), corresponding to the dynamic quenching mode at the y_1 coordinate (Figure 5, p.1). Arcs 2-9 successively «pass» through the vertices, in particular, oscillation modes at the third partial frequency, the mode of dynamic damping of oscillations along the y_1 coordinate, resonance, the mode of dynamic damping of oscillations along the y_1 coordinate and the dynamic locking effect, identified with the «zero» mode within this consideration. In general, this mechanical oscillatory system «passes» through eight different forms of dynamic interactions.

Along with amplitude-frequency characteristics, amplitude-frequency characteristics of interpartial connections and dynamic stiffness depending on the frequency of external disturbance can be used to assess the states and forms of dynamic interactions.

CONCLUSION

For a special case of joint application of external force disturbances, an approach has been developed to assess the states and forms of dynamic interaction of elements of mechanical oscillatory systems with three degrees of freedom based on the use of amplitude-frequency characteristics of the system, dynamic stiffness of the system and its individual fragments based on transformations of the frequency equation.

Figure 6. Graph of states and forms of dynamic interactions of a mechanical oscillatory system in figure 1 for conditions $\bar{z}_1 = 0, \bar{z}_2 = 0, \bar{Q}_1 \neq 0, \bar{Q}_2 = 0, \bar{Q}_3 = 0$

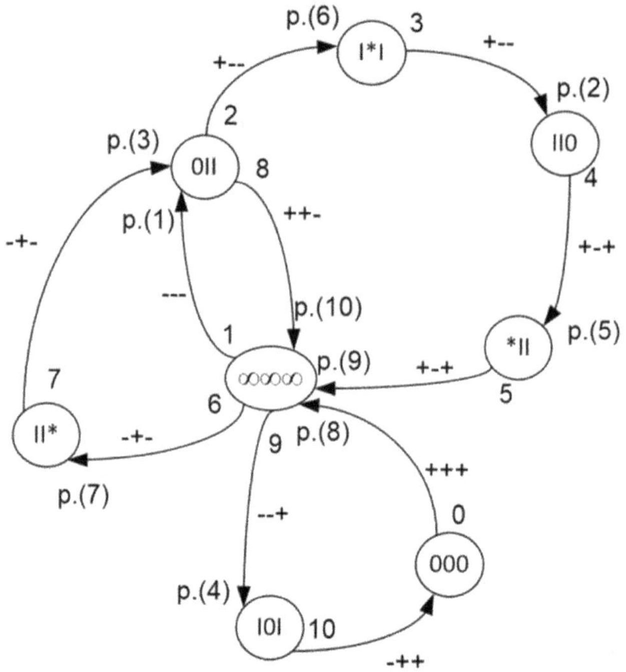

It is shown that the resonant mode of the system «as a whole» displays the state of zeroing of the dynamic stiffness of the system. Zeroing the dynamic stiffness of the quasi-spring implements a certain form of joint movement of the system elements. At the same time, the modes of dynamic damping of vibrations are realized at infinitely large values of dynamic stiffness at the selected point.

To evaluate the variants of dynamic modes and forms of interaction of elements of mechanical oscillatory systems with several degrees of freedom, a graphical-analytical approach has been developed that creates prerequisites for determining the characteristics of the influence of the application of external influences of a forceful and kinematic nature, taking into account the choice of an object whose dynamic state is evaluated.

REFERENCES

Banakh, L., & Kempner, M. (2010). *Vibrations of mechanical systems with regular structure*. Springer. doi:10.1007/978-3-642-03126-7

Belokobylsky, S. V., Eliseev, S. V., & Kashuba, V. B. (2013). *Prikladnye zadachi strukturnoj teorii vibrozashchitnyh sistem* [Applied problems of the structural theory of vibration protection systems]. Polytechnic.

Bykhovsky, I. I. (1969). Osnovy teorii vibracionnoj tekhniki [Fundamentals of the theory of vibration technique]. *Mechanical Engineering (New York, N.Y.).*

DeSilva, C. W. (2000). *Vibration. Fundamentals and practice.* CRC Press.

Eliseev, A. V., Kuznetsov, N. K., & Moskovskikh, A. O. (2019). *Sistemnye predstavleniya, strukturnye skhemy i svyazi elementov* [Dynamics of machinery. System representations, block diagrams and connections of elements]. Innovative Mechanical Engineering.

Eliseev, S.V. (2018). *Prikladnoj sistemnyj analiz i strukturnoe matematicheskoe modelirovanie (dinamika transportnyh i tekhnologicheskih mashin: svyaznost' dvizhenij, vibracionnye vzaimodejstviya, rychazhnye svyazi)* [Applied system analysis and structural mathematical modeling (dynamics of transport and technological machines: connectivity of movements, vibration interactions, lever connections)]. Publishing house of Irkutsk State Transport University.

Eliseev, S. V., & Artyunin, A. I. (2016). Prikladnaya teoriya kolebanij v zadachah dinamiki linejnyh mekhanicheskih sistem [Applied theory of oscillations in problems of dynamics of linear mechanical systems]. *Science.*

Eliseev, S. V., & Eliseev, A. V. (2020). *Theory of oscillations. Structural mathematical modeling in problems of dynamics of technical objects.* Springer International Publishing. doi:10.1007/978-3-030-31295-4

Eliseev, S. V., Eliseev, A. V., Bolshakov, R. S., & Khomenko, A. P. (2021). Metodologiya sistemnogo analiza v zadachah ocenki, formirovaniya i upravleniya dinamicheskim sostoyaniem tekhnologicheskih i transportnyh mashin [Methodology of system analysis in the tasks of assessment, formation and management of the dynamic state of technological and transport machines]. *Science.*

Eliseev, S. V., Lukyanov, A. V., Reznik, Y. N., & Khomenko, A. P. (2006). *Dynamics of mechanical systems with additional ties.* Publishing House of Irkutsk State University.

Eliseev, S. V., Reznik, Y. N., & Khomenko, A. P. (2011). Mekhatronnye podhody v dinamike mekhanicheskih kolebatel'nyh sistem [Mechatronic approaches in the dynamics of mechanical oscillatory systems]. *Science.*

Eliseev, S. V., Reznik, Y. N., Khomenko, A. P., & Zasyadko, A. A. (2008). *Dinamicheskij sintez v obobshchennyh zadachah vibrozashchity i vibroizolyacii tekhnicheskih obektov* [Dynamic synthesis in generalized problems of vibration protection and vibration isolation of technical objects]. Publishing House of Irkutsk State University.

Galiev, I.I., Nekhaev, V.A., & Nikolaev, V.A. (2010). *Metody i sredstva vibrozashchity zheleznodorozhnyh ekipazhej* [Methods and means of vibration protection of railway railway train]. Publishing house of the Educational and Methodological Center for Education on railway transport.

Goncharevich, I. F., & Frolov, K. V. (1981). Teoriya vibracionnoj tekhniki [Theory of vibration technique]. *Science*.

Harris, S. M., & Cred, C. E. (2002). *Shock and vibration handbook*. McGraw – Hill Book Co.

Iwnicki, S. D. (2006). *Handbook of railway vehicle dynamics,* (1st ed.). CRC Press. doi:10.1201/9781420004892

Karnovsky, I. A., & Lebed, E. (2016). *Theory of vibration protection*. Springer International Publishing. doi:10.1007/978-3-319-28020-2

Panovko, G. Y. (2006). Dinamika vibracionnyh tekhnologicheskih processov [Dynamics of vibration technological processes]. *Regular and Chaotic Dynamics*.

Rocard, Y. (1949). *Dynamique generale des vibrations* [General dynamics of oscillations]. Masson.

Chapter 8
Application of Robots for Welding Inspection in the Shipbuilding Industry

Duong Vu
Duy Tan University, Vietnam

ABSTRACT

In certain industries, to conduct the inspection of construction such as storage tankers (big volume), bridge bars, ship hulls, or other heavy infrastructure, inspectors often need to climb up at high altitude, or slip into limited space. Due to the hazardous conditions (elevated temperature and presence of pollution), this type of operation is hard to achieve, or simply not accessible. Moreover, the inspectors are unable to conduct the operation with a high level of accuracy and productivity. Recent achievements in the study and design of robotic systems have encouraged the application of magnetic climbing wheeled robots to conduct many tasks instead of physical labor from humans. This chapter introduces the design process of the wheeled climbing robot for the welding inspection in shipbuilding industry.

INTRODUCTION

In certain industries, to conduct the inspection of construction such as storage tankers (big volume), bridge bars, ship hulls or other heavy infrastructure, inspectors often need to climb up at high altitude, or slip into limited space. Due to the hazardous conditions (elevated temperature and presence of pollution), this type of operation is hard to achieve or simply not accessible. Moreover, the inspectors are unable to conduct the operation with a high level of accuracy and productivity. Recent

DOI: 10.4018/978-1-6684-5887-7.ch008

achievements in the study and design of robotic systems have encouraged the application of magnetic climbing wheeled robots to conduct many tasks instead of physical labor of humans. This chapter introduces the design process of the wheeled climbing robot for the welding inspection in shipbuilding industry.

ADHESIVE FORCE AND LOCOMOTION

Since this robot can be adhesive to the ferromagnetic surface in all space positions, such as vertical wall, ceiling, roof (see Figure 1), it is seen as an attractive alternative in industry deal with the inspection task, because it is expected to enhance efficiency in several times and decrease the risk injury for technical staff (Jaise Jose et al, 2018). The materials used in ship's hull are mostly steel plates and the welding seams are almost flat. In this case, the wheel locomotion is the most effective among the three most common locomotion methods of using magnetic climbing robot: wheeled locomotion, tracked (stair) locomotion and legged locomotion.

Figure 1. Space position of robot in testing

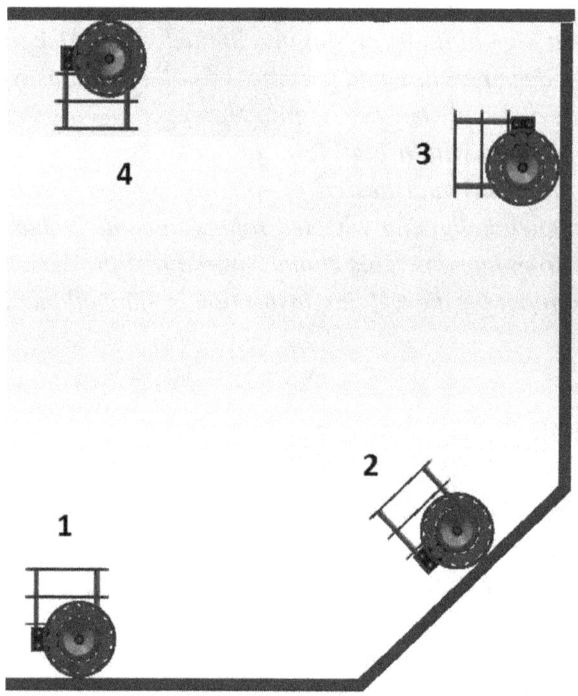

With dynamic stability, the robot can be firmly adhesive to the ferromagnetic surface of construction and freely move in all space positions. From one side, the magnetic and friction forces must be reasonable to provide balance between them, otherwise the robot will overcome the exceeding adhesive (magnetic) force leading to overload against the actual payload. On the other hand, it is needed to reach strong enough magnetic force when the robot is positioned in special places such as vertical or ceiling directions. Magnetic adhesive force, friction force and surface conditions were the subjects in recent publications (Abdulkader et al, 2019; Ravindra et al, 2018; Jiao et al, 2022; Olivier, 2018; Zhong et al, 2021). The magnetic force is created by using a permanent magnetic or electro – magnetic field. Regarding the permanent magnetic field, it does not require the external electric field to maintain adhesion. The intensity of this magnetic force depends on magnetic flux density. This attractive force changes accordingly to the gap between the magnet and the ferromagnetic surface (Welch and Mondal, 2019) (see Figure .2).

$$F_{\text{air gap}} = \frac{F_{\text{magnetic}}}{1+S^3}$$

Where:

- $F_{\text{air gap}}$ is magnetic force depending on the air gap (N).
- S is the distance between the magnet and the steel surface (mm).

Figure 2. Relation between magnetic force and air gap

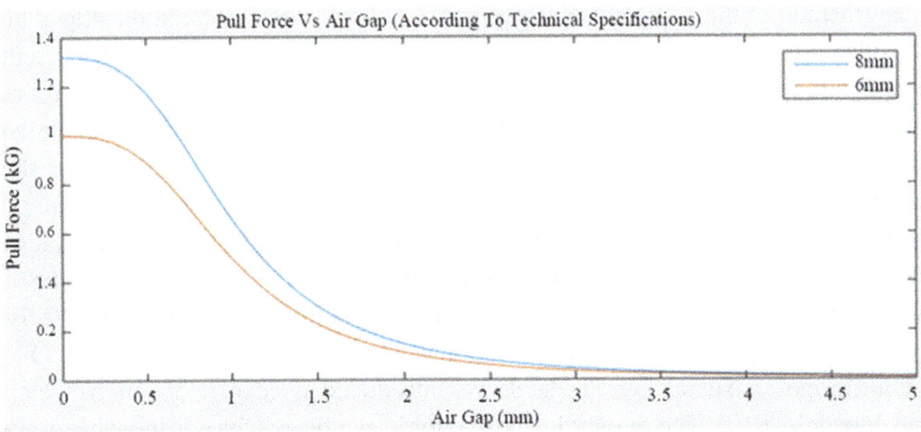

There is one drawback to using permanent magnet: in some instances, it is difficult to release the robot from the ferromagnetic surface when the task is near completion. To avoid this situation, a dedicated device for removing the robot from the surface is provided.

The advantage of using electromagnet is that the adhesive force can be varied according to the current through the coil. But unlike the permanent magnet, the electromagnet requires a constant supply of power. By variation of power, the adhesion can be adjusted in a wide range, although the configuration of this design seems more complex. In comparison with the permanent magnetic, the electromagnet consumes more energy. Given the drawback the designers need to consider using reasonable magnetic force in certain settings.

There are two main locomotion methods: legged and wheeled locomotion. With legged locomotion, sometime the stability is affected due to the dislocation of its center of gravity. The problem has arisen in the design for re-orientation of the center of gravity with proper control tools. In this case, for better adhesion to the dislocated center of gravity while in motion each type of robot needs a specific way of placing magnets to obtain the mutual variation of magnetic strength for detachment and stable motion on the surface. The robot with legged locomotion can spontaneously overcome any obstacle on the way, but its speed is slower.

For wheeled locomotion, the robot can move at a relatively high and stable speed. In some Non- Destructive Examination (NDE) processes, it is important to keep the movement of the robot at a stable speed. When compared with any other locomotion modes, wheeled locomotion shows a competitive advantage due to its ability to maintain a smooth average motion. The disadvantage of wheeled locomotion is the weak obstacle avoidance. By the looks of the noted irregularities, the gap between the magnet and surface changes and this could lead to the weakened or strengthening of the magnetic adhesive force. This observed correlation should be considered in deriving at a solution. If this adhesive magnetic force is insufficient to attach to the robot, in the worst situation the sliding or rolling over of the robot could occur. In a situation where a robot was climbing onto the ship's hull whilst carrying an expensive NDE device if the robot failed to adhere to the surface the expenses would not be compensable. The different methods of locomotion can be used to achieve the target of motion with continuous changing of its direction of motion alongside the testing object's surface. There are different configurations of wheels and magnet placement, such as two independently active controlled wheels, four-wheel design and an Omni - directional wheel design (see Figure 3.4). One unique proposal of using hybrid wheels was introduced in recent publication (Welch and Mondal, 2019). The required adhesive force can be achieved if the wheels are embedded with magnets, magnetic rollers or separate magnets placed in the frame close to the surface. Nevertheless, the author of this publication prefers the wheels integrated with permanent magnets design.

Omni wheels have rollers arranged tangentially around the circumference of the wheels. Meconium wheels are similar to omni-wheels except the rollers are arranged at an angle of 45° to the main wheel axis. This configuration supports combining forces to maintain holonomic motion.

Figure 3. Omni wheel

Figure 4. Meconium Wheel

STRUCTURE COMPONENT OF SYSTEM

In the shipbuilding industry, there is a huge volume of metallic construction undergoing different manufacturing processes to build the main unit – the hull. During this manufacturing chain, the welding process is used to connect the steel plates and contractors need to control the quality of welding joints. A typical NDE inspection scheme on a ship requires a chain of checkpoints to examine the quality of welds as shown in Figure 5.

Figure 5. Objective of shipbuilding NDE main component

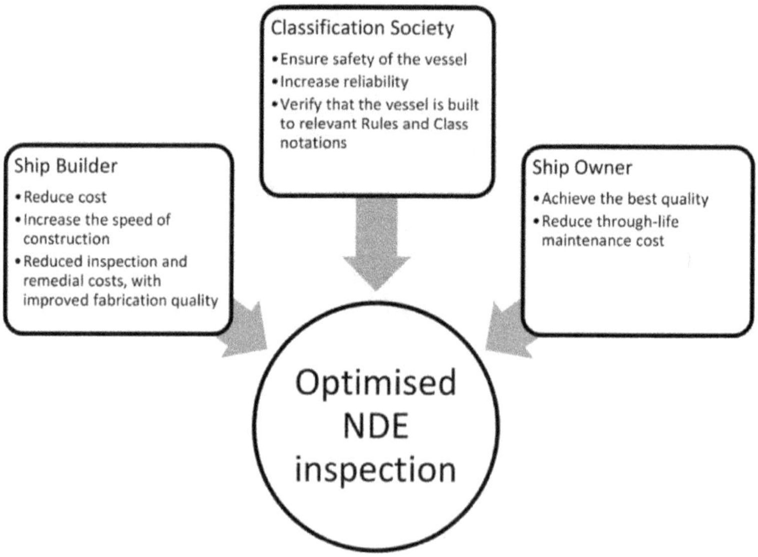

There are many different techniques to conduct an examination of the welding joint, but the ultrasonic method is considered the most effective and reliable method. The NDE technique is the recommended technique used to detect and localize welding defects. In this process, the robotic carrier or scanning system can help speeding up data collection and improving detection accuracy as compared to human labor. The scanning arm should be able to carry the sensor. The increasing interest in developing new modern robotic devices for NDE evaluation of welding joints presents considerable benefits of using robots over manual inspection. These benefits include the removal of the repetitiveness aspect of the inspection, human fatigue, robotic resilience in hash conditions which by turn will increase the inspection productivity. In common practice for conducting the NDE control of welding seams in ship's hull, huge scaffolding needs to be erected alongside the whole ship. The invention of a robot device carrying the ultrasonic detector will reduce expenses associated with erecting scaffolding. The use of robotic device in inspection will save time as well as material costs. Even when the costs for robotic NDE inspection are high, this type of inspection will provide assurance that welding discontinuities would be detected prior to a catastrophic incident. In some circumstances, this inspection is a favorable solution to extending the equipment' lifetime which could offer potential economic benefits in savings by not replacing the equipment as often. The following are some challenges faced in a robotic inspection:

- The ability to achieve the required positional accuracy in comparison with the experience of human inspector.
- The consistency of evaluation due to the challenging environments and the accurate, reliable motivation for detecting defects can be affected.
- Data collection during an inspection must be transmitted in real time for future analysis and stored in a database for the lifetime of the ship in case of future repairs.
- The mobility of the robot to smoothly move accordingly to the complex trajectory of welding seam where the metallic surface is not flat.
- In the case of transmitting the acquired data via a Wi-fi network, there are many different conditions which impact negatively on the error accumulated.

The general configuration of a robot used in a NDE inspection in Figure. 6.

Figure 6. Principal design of robot for inspection

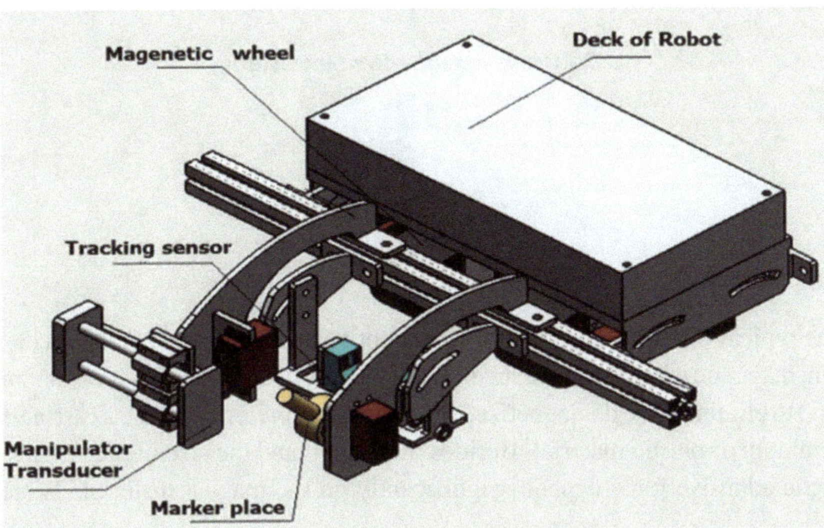

DESIGN OF MAGNETIC CLIMBING ROBOT

Mechanical Design

Based on the goal and scope of an inspection, below is the thought process to create the principal design (see Figure 7).

Figure 7. Design philosophy

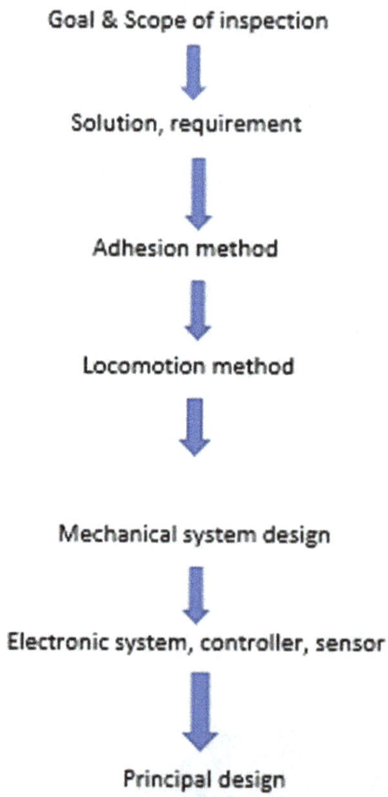

The typical mechanical structure of a robot consists of a body frame, a driving system, magnetic adhesion mechanism and an inspection system. To keep the weight low and to eliminate the demagnetized influence, it is preferred to use a carrier made of aluminum or plastic material. Besides the air gap and the size of the magnet, the magnetic adhesive force depends significantly on the magnet material. There are four main types of permanent magnets: Neodymium iron boron magnet (NDFeB); Samarium cobalt magnet (SmCo); Ferrite magnet and Alnico magnet (AlNiCo).

The magnetic property of NDFeB type has the strongest coercive force as its material is composed of rare earth elements. The high magnetic energy facilitates the design of the magnet to be compact in size.

Electronic Design

The schematic structure of controlling a robot is shown in Figure 8 and Figure 9.

Figure 8. Block scheme of central control unit

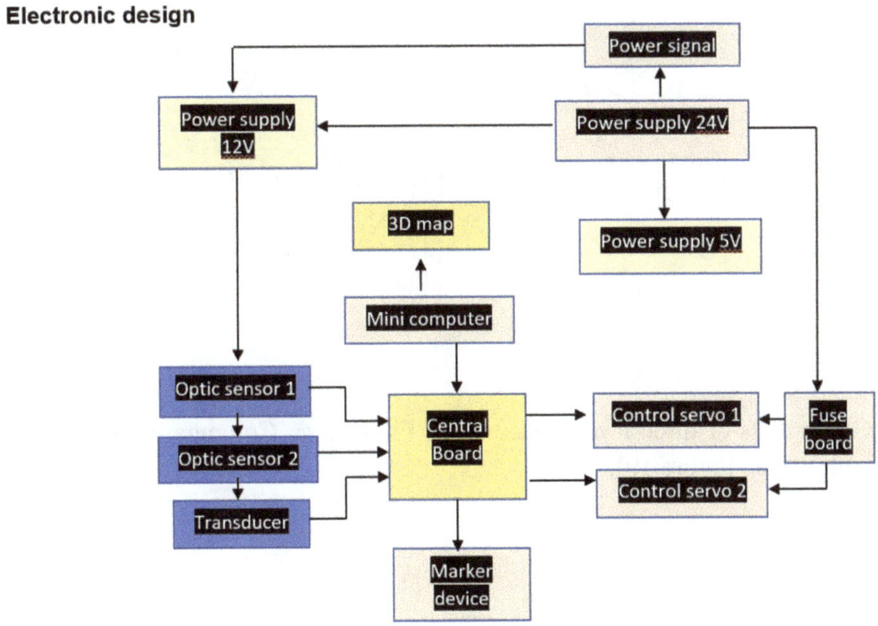

Robot Operating System (ROS)

Any robot needs an operating system to fulfil the applications. An operating system consists of software programs and tools. ROS is designed as a loosely coupled system consisting of nodes. Each node is responsible for one task. Nodes communicate with each other by passing information via topics. The development principal in ROS is that any developer can establish a complex system by connecting existing solutions for small tasks.

Figure 9. Micro - controller Atmega 2560

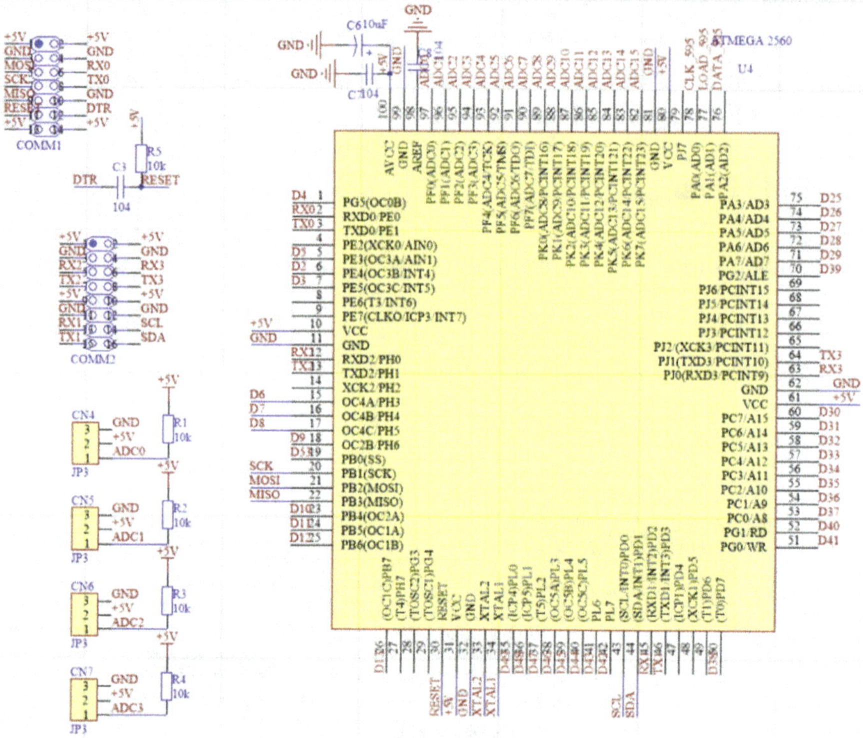

ROS consists of three layers of concept: Filesystem, Computation Graph and Community (see Figure 10).

Figure 10. Basic Model of ROS

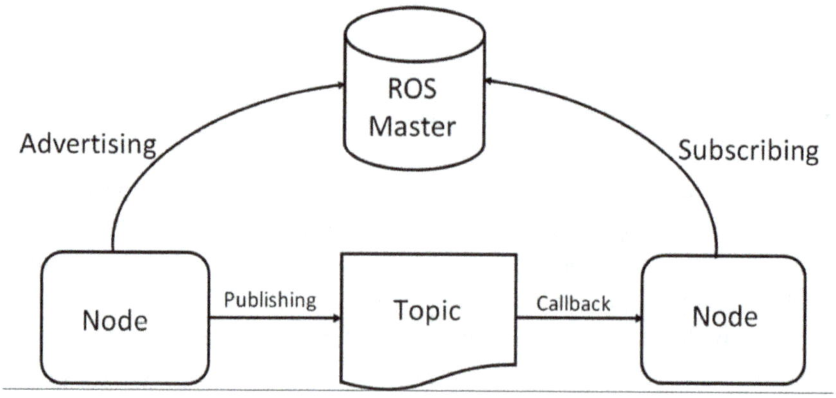

Filesystem in ROS is used to centralize the build process of a project whilst providing relative flexibility and tools to decentralize its dependencies. The structure of filesystem is shown in Figure 11.

Figure 11. Structure of ROS Filesystem

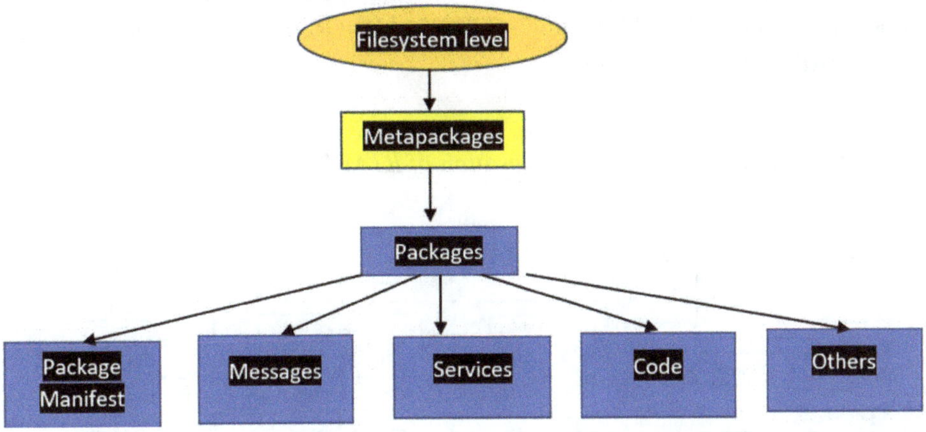

The computation graph model is a specific structure that consists of nodes, connected by edges called topics. Nodes can pass messages to one another through topics. A node makes service calls to another nodes. Messages and service calls do not pass through the master. Each node represents one process running the ROS graph. Every node is assigned a name and this name registers with the ROS master before it can accept a new command. Topics are names of buses over which they send and receive messages. A service is an action that a node can take which will have a single result. ROS community concepts are the ROS resources that enable the exchange of software and knowledge between separate communities. The following resources are provided through ROS:

- Visualization
- Object recognition
- Navigation
- Manipulation/ Grasping
- Plugging in

3D Tracking Development for Robot

The Simultaneous Localization and Mapping – SLAM is applied to define the position in real time and helps to create a 3D mapping. This technique in robotics has been widely studied in recent time and found a successful application in UAV area. For shipbuilding, because of the limited space without GPS support, this method is recommended for establishing a 3D map on a small scale based on an available Real -Time Appearance - Based Mapping - RTABMAP in ROS. The principal operation of RTABAP is shown in Figure 12.

Figure 12. General structure and operation of RTABMAP

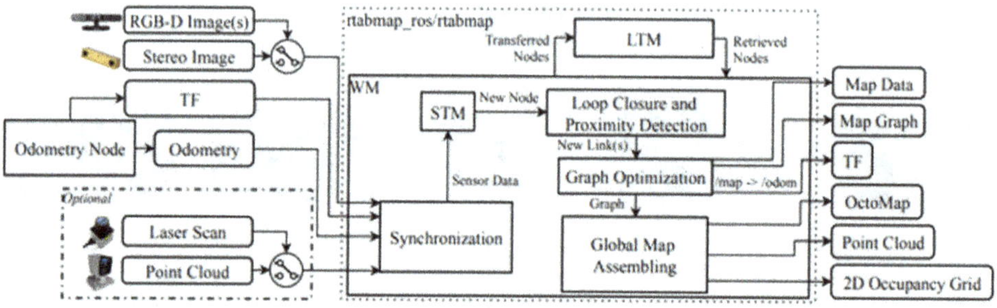

Figure 13. General scheme of connection with scan laser

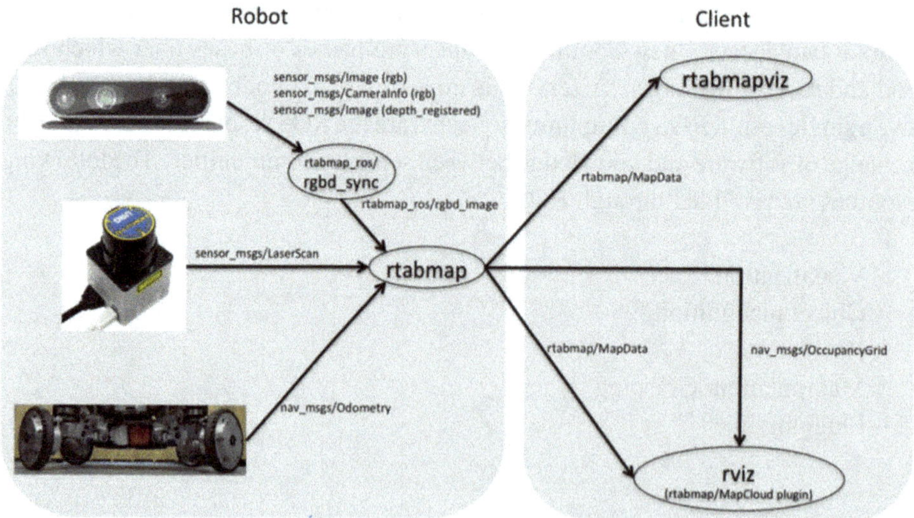

TF is a term frequency for definition the position of sensor. STM is short term memory. LTM is long term memory. WM is working memory. The total scheme of connection using a scan laser is described in Figure 13.

A 3D laser seam tracking system is an effective method which has a laser beam shooting out of the unit reaching a surface, reflecting off the surface, transmitting back into the sensor and the sensor picks up where the beams scan.

A robot may perform desired motions without human intervention using programming. This programming can be online or offline. In offline programming, the robot needs a motion program generated through motion trajectories and motion simulation in a virtual environment. The drawback of offline programming is the availability of the model in advance, which could limit the influence of model correction on program development. In the operating conditions, the generated program faces the question of adaptability. In welding production, the programming often deals with 3D space curved welding. In this case, the manual teaching programming is considered insufficient due to the complex surface and the trajectory of welding that need to be taught (L. Alonso Ferreira et al, 2017). The offline programming cannot be adapted if there is no 3D model available. To overcome this challenge, the robot for NDE and for welding also, must be a "smart perceptive" in obtaining information that is needed for robot motion and activation. This process involves the development of a robot with intelligence and flexibility. A model using a linear structured-light sensor is a reasonable solution to creating a robot that has the 3D modeling perception of welding seam in 3D space with high precision. The image processing method is the basic background for designing this robot. This method builds a visual system for varied scenarios. The advanced 3D programming method is used to improve precision and deliverability. The reliability and effectiveness of this method has been proven through experiments in comparison with the simulation (Bo Zhou et al, 2021).

Stability of Wheeled Climbing Robot

Stability is a necessary property of any mobile robot, including the wheeled climbing robots. Stability in this case, can be static or dynamic. Static stability helps the robot stand without rolling over and can be achieved through mechanical design. Dynamic stability is the ability of a robot not rolling over while in motion. The dynamic stability enables a robot to stand while moving and this requires active control. The topic of balance and dynamic stability is a major topic in control and robotics.

Static Stability

The wheeled robot with four magnetic wheels is chosen for the analysis of stability (see Figure 14), where:

- θ is the inclination angle of the surface on which the wheeled robot is moving.
- W is the total weight of the robot (including NDE device attached).
- N is the total reaction force from the surface, directed to the robot.
- O is the center of gravity of the robot.
- F_m is the magnetic force directed to the surface.
- D is the diameter of the wheel.
- L is the distance between the two (front and rear) wheels.
- D is the distance from the frame's body's center of gravity to the surface.
- μ is the friction coefficient.

Figure 14. Conditions of stability for climbing robot

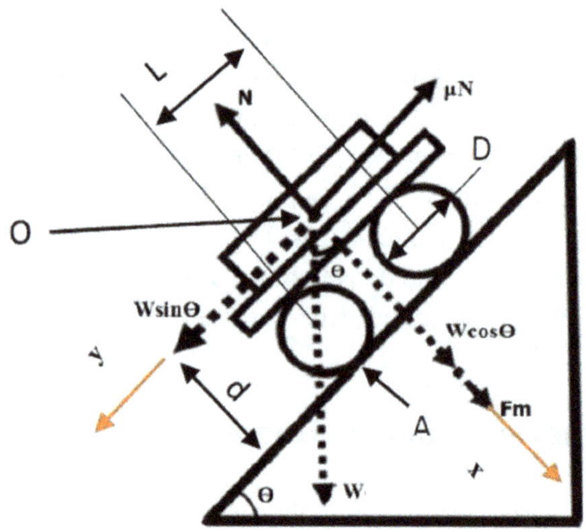

The static stability of a climbing robot depends on turnover avoidance (Figure 15, a), sliding avoidance (Figure 15, b) and rollover avoidance (Figure 15, c) as shown in Figure 15.

In designing a climbing robot, the static analysis helps to define the geometry and force parameters. In real conditions, this climbing robot should move on surfaces with different and varied slopes.

Figure 15. Static stability of wheeled climbing robot

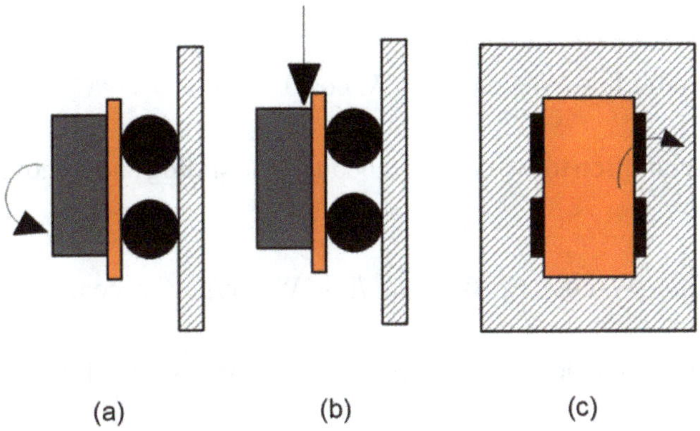

From the statics, is it known that:

$$W.\cos\theta + F_m = \frac{W\sin\theta}{\mu} \quad (1)$$

$$\text{Therefor: } F_m = \frac{W\sin\theta}{\mu} - W.\cos\theta \quad (2)$$

From the equation (2), the condition for sliding avoidance is:

$$F_m \geq \frac{W\sin\theta}{\mu} - W.\cos\theta \quad (3)$$

Considering the total moment for this system (see Fig 14), according to point A as the center of rotation:

$$W.\ d.\ \sin\theta + R/2.\ L - W.\ L/2.\ \cos\theta - F_m.\ L/2 = 0 \quad (4)$$

From this equation, the value of force F_m is equal to: (2. W. d. sin θ)/L + N – W. cos θ

By substitution N = W. sin θ / μ, the equation (4) will be as follow:

$$F_m = (2.\ W.\ d.\ \sin\theta) / L + W.\ \sin\theta / \mu - W.\ \cos\theta \quad (5)$$

The equation (5) is the condition of the turnover avoidance for the climbing robot, namely

$$F_m^{3} \, (2.\ W.\ d.\ \sin\theta) / L + W.\ \sin\theta / \mu - W.\ \cos\theta \qquad (6)$$

The general condition to avoid simultaneously the sliding and turnover, is (Duong Vu et al, 2021):

$$F_m^{3} \, \max\ [(W.\ \sin/\mu\text{-}W.\ \cos);\ (2.\ W.\ d.\ \sin /L + W.\ \sin/\mu - W.\ \cos)] \qquad (7)$$

The transverse force acts perpendicular on the robot's wheel (see Figure 16)

Figure 16. Transverse force acting on robot

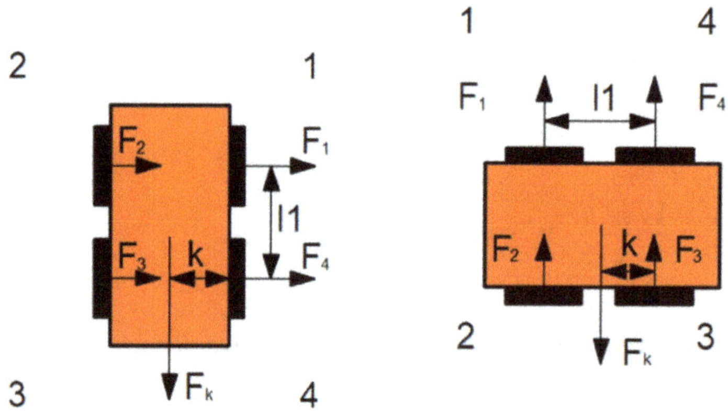

Where:

- k is the distance between the center of gravity and the center of the robot's wheel.
- F_i (I = 1,2,3,4) is the transverse force acting on each wheel.
- 1_1 is the moment arm (distance between the two wheel's centers).
- M_r is the moment that caused the rollover of the robot.
- F_k is the total drive force acting parallel to the robot's body

$$M_r = (F_k \text{ x } k) - 2\,(F_i \text{ x } l_1) = 0, \text{ therefore: } F_i = \frac{F_{k \times K}}{2\, l_1} \tag{8}$$

From the equation (8), the condition of rollover's avoidance is:

$$F_i^3 \frac{F_{k \times K}}{2\, l_1} \tag{9}$$

Dynamic Stability (see Figure 17)

Figure 17. Torque calculation

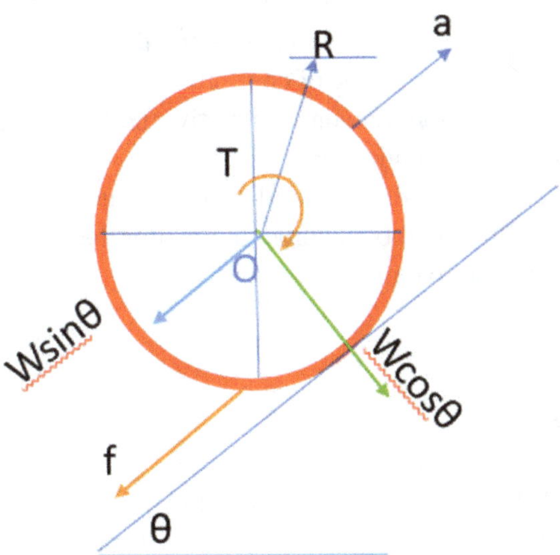

The moment of the robot's weight will deter the rotation of each wheel. This moment is nominated as 'M_w' the radius of the wheel is noted by 'R':

$$M_w = (W.\ \sin\theta / 4).\ D/2 \tag{10}$$

Moment M_f from the friction force and magnetic adhesive force is calculated as:

$$M_f = ((W.\ \cos\theta / 4) + F_m/4).\ \mu.\ D/2 \tag{11}$$

From (10) and (11), it can give the formula for calculation motor drive torque:

$$M_d = M_f - M_w = [-W. \sin\theta + \mu. (W. \cos\theta + F_m)]. D/8 \quad (12)$$

Substituting the value of the magnetic adhesive force from equation (7) with equation (12), the final torque value can be calculated. Note that, while inspecting the quality of welding (ISO 17640:2017), the robot must be moving in all space positions, while the inclination angle, ‘θ’, changes from 0° to 180°. Using equations (7) and (12) will evaluate the magnetic adhesive force and torque in all critical positions.

CONCLUSION

- The magnetic wheeled climbing robot is a competitive device to perform the NDE tasks in the shipbuilding industry.
- To improve the accuracy and reliability of this method, it is required to study more extensively on the adhesive analysis and the tracking of locomotion for 3D trajectory in real time mode of controlling.
- Climbing robot operations remain an attractive field of research and realization in a widened scope of applications.
- In furure research, it will be useful to carry out the static and dynamic stability of the climbing robot in order to monitor the real payload with better manoeuvrability and traction.

REFERENCES

Duong, V., Mikhail, M., & Thu, Q. L. (2021). Study on the Stability of a Wheeled Climbing Robot, *Proceedings of the 2nd Annual International Conference on Materials, Machines and Methods for Sustainable Development (MMMS 2020)*, 638-645. doi:10. 1007/978-3-030-69610-885.

Enjikalayin, A. R., Prabakaran, V., Htet, L. N., Selva, K., Raj, V. S., & Elara, M. R. (2020). A Magnetic Climbing Robot for Autonomous Thickness Measurement in Ship Hull Maintanance. *Journal of Marine Science and Engineering, 469.* doi:10.3390/jmse8060469,8

Ferreira, L., Figuera, Y., Iglesiaz, I., & Souto, M. (2017). Offline CAD-based robot programming and welding parametrization of a flexible and adoptive robotic cell using enriched CAD/CAM system for shipbuilding. *Procedia Manufacturing, 11*, 215–223. doi:10.1016/j.promfg.2017.07.228

Gao, X., Cui, D., Gou, W., Mu, Y., & Li, B. (2017). Dynamic and stability analysis on stairs climbing of wheel-track mobile robot. *International Journal of Advanced Robotic Systems, 14*(4), 1–13. doi:10.1177/1729881417720783

ISO. (2017). Non-destructive testing of weld - Ultrasonic testing - Techniques, testing levels, and assessment. *ISO Standards.*

Jiao, S., Zhang, X., Zhang, X., Jia, J., & Zhang, M. (2022). Magnetic, Circuit Analysis of Halbach, Array and Improvement of Permanent, Magnetic Adsorption, Device for Wall- Climbing Robot. *Symmetry, 14*(2), 429. doi:10.3390ym14020429

Jose, J., Dinakaran, D., Ramya, M. M., Samuel, D. G. H. (2018). A survey on magnetic wall-climbing robots for inspection, *International Journal of Mechanical and Production Engineering Research and Development, 8*(6), 59-68.

Mansi S. Chabukswar, R., Nanwatkar, K., & Aparna, M. B. (2019). Design and Experimental Analysis of Magnetic Climbing Robot, *International Journal of Research in Engineering, Science and Management, 2*(12), 418-425.

Manzoor, M., Wu, Q., & Masood, R.. (2014). Qinghe Vu & Rana Javed Masood. Stability, Trajectory Following and Formation Control of Wheel Mobile Robot. *The Open Automation and Control Systems Journal, 6*(1), 1798–1809. doi:10.2174/1874444301406011798

Olivier, K. A. (2018). A magnetic climbing robot to perform autonomous welding in the shipbuilding industry. *Robotics and Computer-integrated Manufacturing, 53*, 178–186. doi:10.1016/j.rcim.2018.04.008

Ravindra, S. B., Pushparaj, M. P., & Saroj, K. P. (2018). Experimental investigation on permanent magnet based wheel mechanism for safe navigation of climbing robot. *Procedia Computer Science*. doi:10.1016/j.procs.2018.07.046,133:377-384

Welch, H., & Mondal, S. (2019). Analysis of Magnetic Wheel Adhesion Force for Climbing Robot. *Journal of Mechatronics and Robot, 3*(1), 534–541. doi:10.3844/jmrsp.2019.534.541

Zhong, Z., Xu, M., Xiao, J., & Lu, H. (2021). Xu Minh, Xiao Junhao & Lu Huimin. Design and Control of an Omnidirectional Mobile Wall-Climbing Robot. *Applied Sciences (Basel, Switzerland), 11*(22), 11065. doi:10.3390/app112211065

Zhou,B., Liu, Y., Xiao, Y., Zhou, R., Gan, Y., & Fang, F. (2021). . Intelligent Guidance Programming of Welding Robot for 3D Curved Welding Seam, *Digital Object Identifier.*

Chapter 9
Application of TOPSIS Optimization Technique in the Micro-Machining Process

Boopathi Sampath
https://orcid.org/0000-0002-2065-6539
Muthayammal Engineering College, India

Sasikumar C.
Bannari Amman Institute of Technology, India

Sureshkumar Myilsamy
Bannari Amman Institute of Technology, India

ABSTRACT

Multi-criteria optimization techniques have been applied to predict the best alternative solutions to meet the best results for all attributes. In the micro-machining processes, the various process parameters in the micro-machines, special experimental setup, tool profiles, workpiece materials, and working mediums are considered. Sometimes, the response attributes conflict with each other, making it difficult to estimate the best-influenced process parameters. The multi-criteria optimization methods are applied to solve the abovementioned conditions. In the chapter, the multi-criteria optimization technique: TOPSIS (technique for order of preference by similarity to ideal solution) implementation procedures, flowchart, and result interpretation have been demonstrated to optimize or predict the best micromachining process parameters. A case study on the application of the TOPSIS technique in electrical discharge machining (EDM)) has been illustrated.

DOI: 10.4018/978-1-6684-5887-7.ch009

INTRODUCTION

The micro-machining processes are generally playing an important contribution in the various industries to machine complicated, contour, hard, and micro-sized components. a one-of familiar type of non-conventional machining process. In this aspect, the micro-machining parameters optimization of response parameters is essential to select the required process parameters for the various machining conditions: materials, environment, and minimizing the environmental impacts. Material removal rate, surface roughness, dimensional accuracy, ovality, roundness errors, tool wear rate, tool selection, and work material are all factors to consider. Input parameters such as power, current, voltage, chemical concentration, fluid requirements, fluid pressure, flow rate, material hardness, and parameters related to non-renewable energy sources are also considered. The material removal process is done by applying periodic intervals of an electric pulse during the machining process. The selection of suitable process parameters for the various industrial requirements is essential to minimize the workload of machine operators, improve quality and enhance productivity. Thus, the optimization techniques play a very important role in selecting the optimum parameters in the micromachining processes. The utilization of optimum resources in the manufacturing industry leads to reducing the waste and cost of manufacturing. The different type of optimization techniques was applied to micro-machining processes to improve the machining characteristics. The multi-criteria TOPSIS optimization method plays a very important role in the optimization of the various micro-machining processes. In this chapter, the electrical discharge machining process is selected as a case study to illustrate the implementation, analysis, and prediction procedures of the TOPSIS technique.

The Technique for order of preference by similarity to ideal solution (TOPSIS) is one of the multi-response decision-making techniques that is utilized to forecast the optimum solution to the multiple responses. In this study, the TOPSIS technique has been applied to estimate the best combination of operating parameters (Boopathi, 2013; Boopathi & Sivakumar, 2013, 2016a). In the 1980s, it was established by Hwang and Yoon for optimization problems. It is based on the Euclidean distance method, which finds the relative closeness of each factor to the ideal solutions. Initially, it converts the dimensional factors to unidimensional or dimensionless factors. Two alternative solutions have been made for the maximization and minimization of responses with respect to ideal solutions. Positive Ideal Solutions (PIS) were obtained for the expectations of maximising the responses. In this case, Negative Ideal Solutions (NIS) were obtained with the expectation of minimizing the responses. If the obtained solutions have a short distance from PIS and a long distance from NIS, it is selected as the best solution. In TOPSIS, the beneficial output is related to maximizing the responses, and non-beneficial output is related to minimizing the responses(Nilesh T. Mohite, Pratik P.Shinde & Amar B. Shirage, Shubhash T. Vhagade, 2020; Sahoo et al., 2019).

GENERAL OPTIMIZATION PROCEDURE IN MICRO-MACHINING PROCESS

The general procedure of optimization in micromachining processes is illustrated in Figure 1. The general steps are explained as follows.(Boopathi, 2021b; Boopathi et al., 2012; Boopathi, Thillaivanan, et al., 2021; Haribalaji, Boopathi, & Asif, 2021a)

Step 1: Understand the machining process, and input and output parameters.
Step 2: Define the controlled and uncontrolled parameters. The controlled parameters are called measurable parameters, and the uncontrolled parameters are also called noise or non-quantified parameters.
Step 3: Select the suitable design of the experiment method (Factorial/ Taguchi / Response surface methods) for the micro-machining processes.
Step 4: Define the number of trails or experiments to be performed.
Step 5: The experiments are performed.
Step 6: The experimentally observed data has been examined.
Step 7: To predict the best and optimum process parameters, use single and multi-criteria optimization techniques.
Step 8: Predicted results (quality characteristics) have been validated by confirmation experiments.
Step 9: Apply/ Utilize the best process parameters in the component manufacturing process.

The parameters are individually controlled to obtain the best response parameters and outputs. Based on the problem description, different types of techniques have been developed and applied in machining processes to control the quality and cost of the product.

The full factorial, partial factorial, Taguchi, and Response Surface design of experiments are applied to conduct the experiments and single objective optimization processes. In the single objective methods, there are some design and analysis limitations to predicting the combined effects of all parameters on all the responses simultaneously. It can only be applied to find the optimum value for individual response and only be employed within the defined levels. However, multi-objective optimization methods (fuzzy, neural network, TOPSIS, Grey rational Analysis, etc.) have additionally been used to find the best alternative solutions for all selected attributes(Bhuyan et al., 2015; Kumar & Rai, 2020; Mohanty et al., 2018).

Figure 1. Micro-machining Process General Optimization Procedure

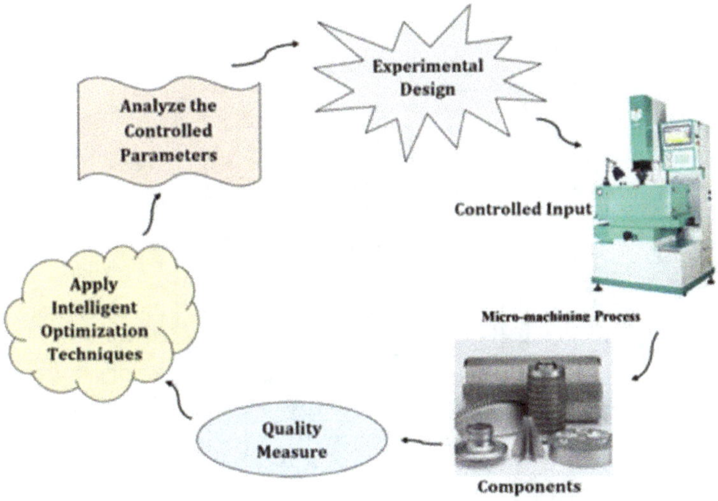

In this chapter, the data collected from the conventional design of experiment methods is utilized as a source for multi-criteria analysis. In the micro-machining and material processing research activities, multi-objective optimization techniques were incorporated to optimize the two or more conflicting responses(Kumar & Rai, 2020; Mao et al., 2018; P. Wang et al., 2015). For rough and finish cutting operations, the design of experimental methods is not only suitable to obtain full results. Some of the techniques used to develop the polynomial regression analysis were also applied. These equations are used as input sources for multi-objective genetic algorithms, non-sorting genetic algorithms-II, particle swarm algorithms, multi-objective evolutionary algorithms, Simulated Annealing, artificial neural networks, and artificial bee colony algorithms. These are based on Pareto-front and Pareto-set concepts to find the best concatenated solutions. The Fuzzy logic technique, grey rational analysis, and TOPSIS are working based on source/experimental data obtained from the design of experiments that have been directly utilized(Mandal & Sarkar, 2012; C. C. Wang et al., 2007).

Applications of Various Optimization Methods in Micro-Machining Processes

The various optimization techniques applied in micro-machining processes are illustrated in Figure 2. The various process/input parameters that are usually applied in micro-machining processes are shown in Figure 3.

Figure 2. Application of various optimization methods in micromachining processes

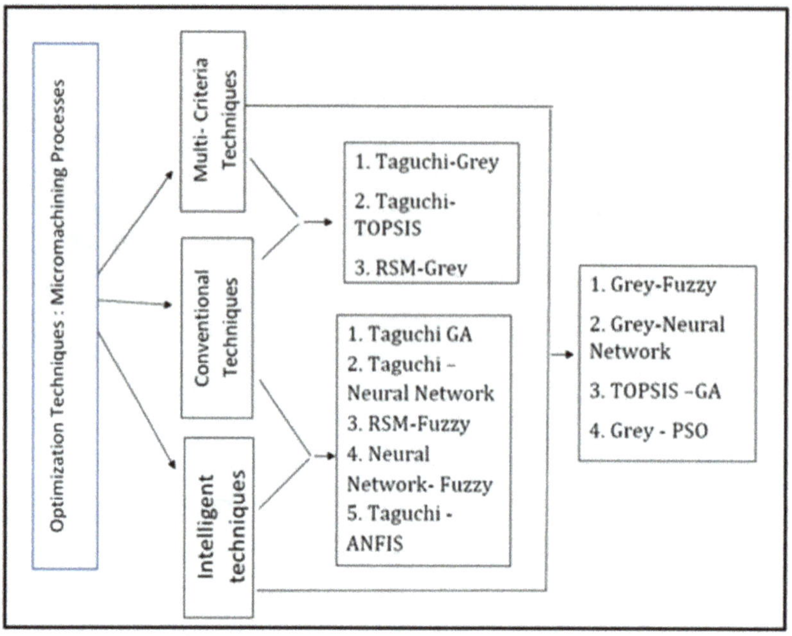

Figure 3. Various process parameters in micro-machining processes

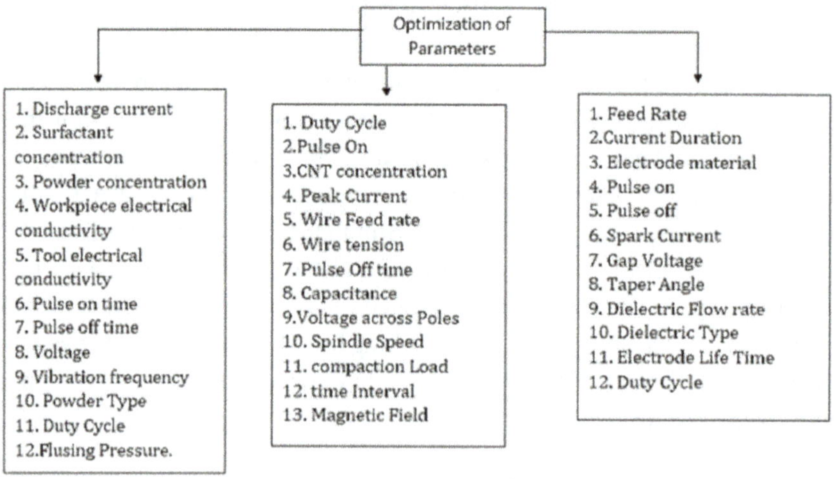

Conventional Techniques: All design of experiments methods were used, including factorial, Taguchi, response surface, and mixture methods. The Taguchi and response surface methods are applied in micro-machining processes.

Intelligent Techniques: It is also called non-traditional optimization techniques. These methods work based on natural inspirational behaviour and population-based algorithms. Examples: multi-objective genetic algorithm, non-sorting genetic algorithm-II, particle swarm algorithm, multi-objective evolutionary algorithm, simulated annealing, artificial neural network, and artificial bee colony algorithm.

Multi-Criteria Optimization Techniques: The weightages have been separately assigned to the selected attributes and the experimental raw data is directly utilized to perform the multi-criteria optimization processes. TOPSIS, GRA (Grey rational method)(C. C. Wang et al., 2007), MOORA (multi-objective optimization based on ratio analysis)(Tansel & Yldrm, 2013) and WSM (Weighted sum method), and WPM (weighted product method)(Rana & Patel, 2020).

Integrated optimization techniques: Two or more optimization techniques are integrated or joined together to obtain the best solution to the output criteria and are called integrated techniques. These types of methods are emerging to predict the solutions to complex multi-objective problems (Figure 2).

TOPSIS IMPLEMENTATION PROCEDURE

The implementation mechanism of TOPSIS in the micromachining processes is illustrated in Figure 4 and explained as follows.

Step 1: Experimentation:

- **Define the Process parameters and response variables**: It can be varied based on the selected micromachining technology. The process parameters are divided into two major categories: **Control factors**: The output can be changed based on certain variations of factors. The control factors had been allowed to modify their levels depending upon the nature of the experimentation setup. The responses were affected by control factors (Boopathi & Sivakumar, 2012, 2016a, 2016b).
  - **Noise factors**: Some factors would not have majorly contributed to the changes in the response factors of the system. These parameters were not included in the systematic tests (Boopathi et al., 2022; Haribalaji et al., 2022; Yupapin et al., 2022).
  - **Response variable:** It is called output, criteria, attributes, or machining characteristics. It is modified based on the changes in the process parameters. For example, the material removal rate can be varied by changing the voltage and pulse duration of the electrical discharge machining process.

- **Selection of Design of experiment (DOE) method:** There are various designs of experiments to decide the experimental design and its values for an experiment. These experiment numbers have been decided by the number of process parameters and the DOE method. Factorial design, the Taguchi method, and response surface methods are widely applied techniques in DOE. In the factorial design, the number of trial experiments has been increased by increasing variables. Thus, it cannot be applied to a large number of variable experiments. The Taguchi method has a similar working mechanism as factorial design, conversely, partial factorial design, i.e., only a few experiments can be conducted to obtain the same results. Response surface methods are employed to design the experiments based on the 5 different levels of each variable in the experimentation (Boopathi, 2021c; Boopathi, Thillaivanan, et al., 2021; Haribalaji, Boopathi, & Asif, 2021b; Haribalaji, Boopathi, Asif, et al., 2021; Sampath et al., 2021).

Figure 4. Flow chart for TOPSIS Technique

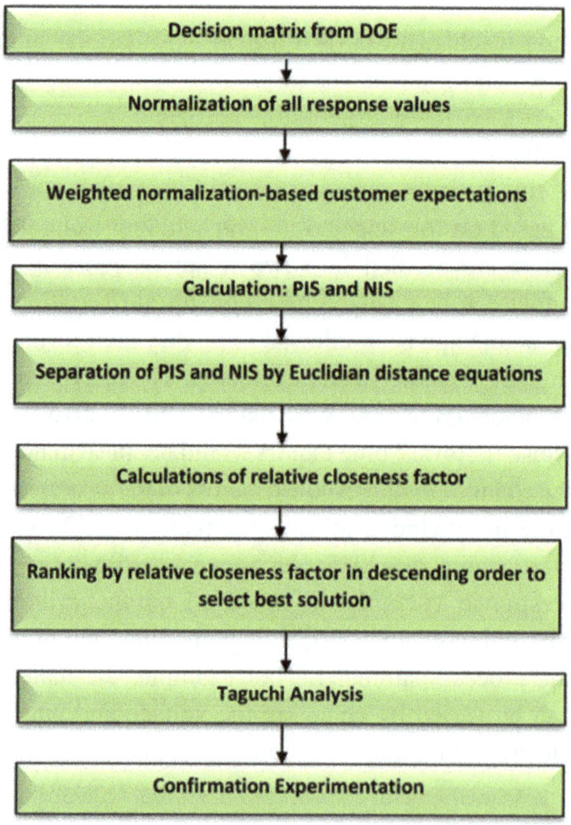

- **Conducting the Experiments**: The experiments are performed based on the DOE representations, and response observations have also been noted for the statistical analysis. Many experiments have randomly been conducted to overcome the bias errors. Sometimes, an experiment is two or three times repeated or replicated to confirm the experimental readings (Boopathi & Sivakumar, 2016a).
- **Statistical Analysis/ ANOVA:** Statistical analysis has been performed to analyse the experimental results using the formulae: sum of the square, variance, F-value, and total sum of the square. It has been varied based on the selected DOE method. In this chapter, the ANOVA of the Taguchi method has been illustrated in the case study session(Haribalaji, Boopathi, & Asif, 2021b).
- **Single Objective Optimization**: The best process parameter settings are obtained for the single response variable based on the ANOVA methods only. It is a very simple method to obtain single-objective results.
- **Multi-Objective Optimization:** In the case of multi-response optimization processes, the response variables might conflict with each other. Few response variables are expected to increase their values, while others are expected to decrease their values. In this situation, conventional statistical analysis is not enough to find the best results for both conflicting attributes. TOPSIS, GRA, MOORA, and other non-tractional optimization techniques have been used to solve multi-objective experimental problems.

Step 2: The collected data from experiments is transformed into normalization to calculate the uniform statistical calculation. Hence, all response data is standardised between 0 and 1 for easy implementation. Equation (1) represents the matrix format of experimental data. Equation (2) is applied for the normalisation of the matrix from equation (1)(Kannan et al., 2022).

$$N = \begin{bmatrix} n_{11} & n_{12} & \cdots & n_{1x} \\ n_{21} & n_{21} & \cdots & n_{21} \\ \vdots & \vdots & \vdots & \vdots \\ n_{y1} & n_{y1} & \cdots & n_{xy} \end{bmatrix} \tag{1}$$

$$Nm_{ij} = \frac{N_{ij}}{\sqrt{\sum_{i=1}^{y} N_{ij}^2}} \tag{2}$$

Where Nm_{ij} is the normalized response value for i=1, 2, ..., x and j=1, 2, …, y.

Step 3: The total weight for all the attributes is equal to one. The weights are equally divided based on the number of attributes. Sometimes, the total weight is unequally separated based on the customer's expectation. However, it was not recommended due to some usage difficulties. The weighted normalised equation is obtained by multiplying the weights of each response. The weight assigned to the responses is based on customer expectations or output requirements (Equation 3). However, the sum of all weights (WS_{ij}) must be equal to 1(Sahoo et al., 2019).

$$WS_{ij} = Wt_{j} \times Nm_{ij} \tag{3}$$

Where $\sum_{j=1}^{p} Wt_{j} = 1$

Step 4: The equations (4) and (5) have been applied to calculate the Positive ideal (μ^{+}) and negative ideal (μ^{-}) solutions using weighted normalized values, respectively.

$$\mu^{+} = \left(\mu_{1}^{+}, \mu_{2}^{+}, \ldots\ldots.\mu_{n}^{+}\right) = \begin{cases} \left(\max \delta\mu_{ij} \middle| j \in C_{1}\right) \\ \left(\min \mu_{ij} \middle| j \in C_{2}, i = 1,2,\ldots\right) \end{cases} \tag{4}$$

$$\mu^{-} = \left(\mu_{1}^{-}, \mu_{2}^{-}, \ldots\ldots.\mu_{n}^{-}\right) = \begin{cases} \left(\min \mu_{ij} \middle| j \in B_{1}\right) \\ \left(\max \mu_{ij} \middle| j \in B_{2},\right) \end{cases} \tag{5}$$

Where C_1 and C_2 are called beneficial and non-beneficial attributes respectively.

Step 5: Positive and negative Euclidian distance ($AS_{i}^{+/-}$) of positive and negative alternative solutions are found by equations (6) and (7) respectively.

$$AS_{i}^{+} = \sqrt{\sum_{j=1}^{1} \left(\mu_{ij} - \mu^{+}\right)^{2}} \tag{6}$$

$$AS_{i}^{-} = \sqrt{\sum_{j=1}^{1} \left(\mu_{ij} - \mu^{-}\right)^{2}} \tag{7}$$

Step 6: Relative Closeness Factor (RCF) is determined by equation (8). It can be used to find the best alternative solution and its respective best attributes.

$$RCF = \frac{AS_i^-}{AS_i^+ - AS_i^-} \tag{8}$$

Step 7: RCF values are sorted by descending order, from highest to lowest. The best solution is connected to rank one with the best attributes and solutions. The last ranked solutions and corresponding attributes are called "worst data," which means they could not be applied for further processes.

Step 8: In some of the research activities, the best solutions and criteria are obtained based on the experimental methods. However, most of the research articles show the same result as compared with the ranking method. Thus, this step may be omitted by many researchers.

Step 9: All the best results derived from TOPSIS have been verified by conducting confirmation experiments through micro-machining experiments. Two or more replications of confirmation trials are recommended to examine the results.

CASE STUDY: IMPLEMENTATION OF TOPSIS IN MICRO-MACHINING PROCESS

In the session, the Taguchi method was used to conduct the methodical design of experiments. The observations of data from the near-dry wire electrical discharge machining experiments have statistically been examined to select important factors for each response. The Technique for Order of Preference by Similarity to Ideal Solution (TOPSIS) has been used to forecast the combinations of optimal process parameters on three conflicting response parameters for the expectations of machine operators. In this chapter, the basic theories and important formulae of the Taguchi and TOPSIS approaches have been discussed(Kannan et al., 2022).

(1) Research Objectives

- To investigate the cryogenically treated molybdenum wire electrode used in near-dry machining processes.
- To conduct experiments using the Taguchi L27 orthogonal array.
- To identify the influencing controlled input process parameters and their contribution towards wire wear ratio and material removal rate using a single objective optimization technique: Taguchi analysis.

- To predict the best combinations of controlled input parameters levels to optimize the wire wear ratio, and material removal rate using a cryogenically cooled wire near-dry machining process by the TOPSIS technique.

(2) Experimental Part

In a near-dry machining process, liquid nitrogen is used to cool the molybdenum wire electrode during machining. Experimental arrangements are developed to conduct the experiments on cryogenically cooled wire used in the near-dry WEDM process. The molybdenum wire electrode cooled on the top and bottom sides of the wire, and the air-mist dielectric was sprayed in the machining zone in the WEDM process. The wire wear ratio and material removal rate are optimized.

(3) Design of Experimentation

Systematic experimentation is used to conduct the tests with the minimum number of tests to obtain optimal parameter settings for qualitative and quantitative analysis. The Taguchi method is one of the best experimental design methods applied in the design and manufacturing fields for successful statistical analysis. It is based on partial factorial design concepts with the lowest number of trails. Taguchi suggested an orthogonal array design for experimentation. The control and noise factors are considered to select the orthogonal array table.

The steps of Taguchi's design and analysis procedures are shown in Figure 5 and explained as follows.

- **Selection of Orthogonal Array:** An orthogonal array is selected based on the number of control variables, range of each variable, levels, and interaction effect requirements(Boopathi, 2019, 2021c; Boopathi et al., 2012; Sampath & Myilsamy, 2021).
- Orthogonal array for Two-Level factors: L_4, L_8, L_{12}, L_{16}, L_{32}
- Orthogonal array for Three-Level factors: L_9, L_{27}
- Orthogonal array for mixed-level factors: L_{18}

In this study, all-controlling parameters have three levels. Thus, the L27 orthogonal array has been selected and is shown in Table 1.

- **Assigning parameters:** Assigning parameters to the column of orthogonal array-based requirements of interaction effects to be discussed. The linear graphs and triangular tables are used to allocate the factors to certain columns. In this study, there were no interaction effects studied. All five factors are assigned sequentially from the first to the fifth column. The remaining column of OA would not be taken for the collection of data and analysis.

Figure 5. Flow chart for Taguchi design and analysis

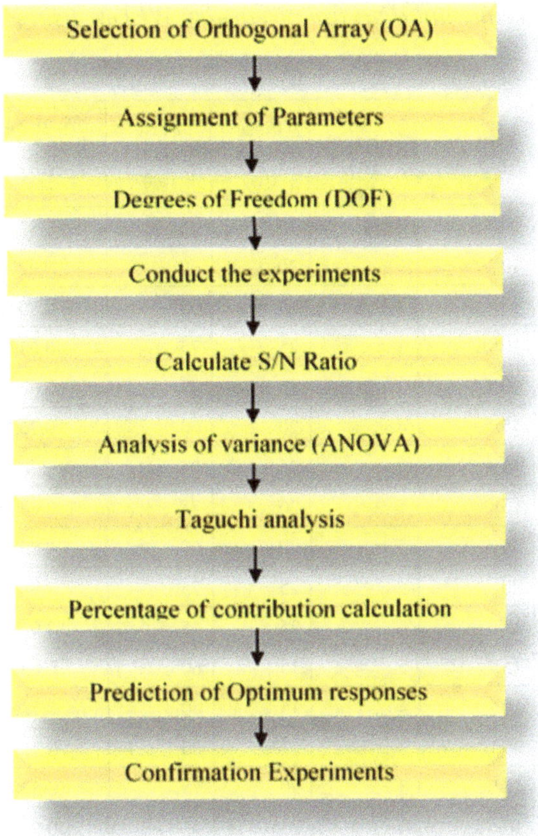

- **Degrees of Freedom (DOF):** Total DOF should be greater than the sum of DOF of parameters and DOF of interaction factors. It is also essential to select the OA(Boopathi, Thillaivanan, et al., 2021).
  Total DOF = number of trails -1 = 27-01 =26
  DOF of each parameter = number of levels -1 = 3-2 =2
  Total DOF for 5 controlling parameters = 5×2 =10
  Thus, the total DOF is greater than the total DOF of all parameters. Hence, L27 OA is a well-stable method for this research work.

- **Conducting Experiments:** As per Taguchi principles, two or three repetitions/ replications of trails should be conducted to avoid noise error. If the deviation of results from reputations is more than 5%, the number of trials should be increased to more than two times.

Table 1. Mean Experimental data from Near-dry WEDM using the cryogenically cooled wire tool

Exp. No.	Pressure (bar)	Flow rate (ml/min)	Current (A)	Pulse width (μs)	Pulse interval (μs)	Mean value of Observations	
						WWR x10^{-3}	MRR (mm^3/ min)
1	5	20	5	15	45	0.8250	10.6196
2	3	10	3	20	60	0.6292	3.7586
3	4	15	4	15	45	0.7047	6.9166
4	3	10	3	25	75	0.6430	4.9262
5	3	10	3	15	45	0.6272	3.7030
6	4	15	4	20	60	0.7629	7.3614
7	4	15	4	25	75	0.7879	7.4838
8	3	15	5	20	75	0.6669	7.3281
9	5	20	5	20	60	0.8577	10.8865
10	5	20	5	25	75	0.8882	11.1534
11	3	15	5	25	45	0.9280	10.4528
12	3	15	5	15	60	0.6364	7.1946
13	5	10	4	15	60	0.7260	7.5838
14	4	20	3	15	60	0.4558	5.8158
15	4	20	3	20	75	0.5085	5.9381
16	5	10	4	25	45	0.9676	11.0978
17	4	20	3	25	45	0.8078	9.9190
18	5	10	4	20	75	0.7565	7.0946
19	3	20	4	25	60	0.6553	10.4862
20	3	20	4	15	75	0.3854	5.8825
21	5	15	3	20	45	0.6808	10.7086
22	4	10	5	25	60	0.9187	11.2646
23	3	20	4	20	45	0.7137	10.2749
24	5	15	3	15	75	0.4818	6.1382
25	4	10	5	15	75	0.6624	7.5394
26	5	15	3	25	60	0.6635	11.2090
27	4	10	5	20	45	0.9861	10.9643

- **Calculation of Signal to Noise Ratio:** After conducting research experiments, the mean value of the observed data is calculated. And the signal-to-noise ratio is calculated based on customer requirements. There are three signal to noise ratio equations (9), (10), and (11) illustrated as follows (Boopathi, 2021b; Boopathi, Balasubramani, et al., 2021; Myilsamy & Sampath, 2021; Sampath & Haribalaji, 2021; Sampath & Myilsamy, 2021; Serrano Cardona & Muñoz Mata, 2013).

S/N Ratio: Higher the better condition

$$S/N_{HB} = -10\log_{10}\left(\frac{1}{n}\sum_{i=1}^{n}\frac{1}{Y_i^2}\right) dB \tag{9}$$

S/N Ratio: Nominal the better

$$S/N_{NB} = -10\log_{10}\left(\frac{1}{n}\sum_{i=1}^{n}(Y_i^2 - Y_0)\right) dB \tag{10}$$

S/N Ratio: Smaller the better

$$S/N_{LB} = -10\log_{10}\left(\frac{1}{n}\sum_{i=1}^{n}Y_i^2\right) dB \tag{11}$$

Where, Y_i-Response value of ith experiment,
Y_{0-}Target response value,
N- Number of replications.

For 'Nominal the better' expectations, the moderate values of responses are elected and parameters are also predicted. For 'higher the better' expectations, maximizing the response factors is based on the best level of controlling parameters. For the 'Smaller the Better' expectations, minimizing the response is based on the best level of control parameters.

In this study, the MRR is assigned to the 'Higher the better' condition, and it should be maximized based on the best levels of each control parameter. The wire wear ratio and surface roughness are connected to "the smaller the better' and these should be minimized based on different levels of parameters.

- **Analysis of Variance (ANOVA) test:** The analysis of variance tests has been performed by collecting data from experiments and S/N ratio values. The formulae of the sum of the square, variance, F-value, and total sum of the square are shown in equations (12), (13), (14), and (15) respectively(Abdul Haq et al., 2019; Uthayakumar et al., 2019)
  The sum of square of the factor

$$SS_{factor} = \frac{factor_1^2}{N_{A1}} + \frac{factor_2^2}{N_{A2}} + \frac{factor_3^2}{N_{A3}} - \frac{\sum_{i=1}^{27} y_i}{N} \tag{12}$$

$$\text{Variance } V_{factor} = \frac{SS_{factor}}{Dof_{factor}} \tag{13}$$

$$\text{F-Value } F_{calculated} = \frac{V_{factor}}{V_{error}} \tag{14}$$

$$\text{The total sum of square } S_{Total} = \sum_{i=1}^{27} y_i - \frac{\sum_{i=1}^{27} y_i}{N} \tag{15}$$

Where, N - Total number of experiments

y_i- number experiments with respect to factor

$factor_1^2$ – Number of experiments in the First level of factor,

$factor_3^2$ – Number of experiments in the Third level of factor

- **Taguchi Analysis:** At this stage, the significant contributing factors were identified based on the F-test. Both mean values and S/N values are considered for conducting the test. Non-contributed factors were pooled, i.e., they would be taken into account for predicting the result. The calculated F value has been compared with F-Table data based on degrees of freedom and several factors. The significant parameters are identified based on the 'F' test and the percentage of contribution calculations. If variance is less than the F value, then the factor is insignificant. If the percentage of contribution is less than 5%, then the factor is also called an insignificant factor.
- **Percentage of Contribution:** The percentage contribution of each parameter to the responses was calculated using the following formula. The ratio of the individual factor sum of the square to the total sum of the square is called the percentage of contribution of the factor (Equation (16)). Based on this value, the significant parameters are sequentially ranked.

$$\text{Percentage of contribution } P_{\%} = \frac{SS_{factor}}{SS_{Total}} \tag{16}$$

- **Prediction of Optimal Parameters:** The best values of responses are predicted using a combination of parameter levels. It could be obtained through a combination of highly contributed individual and interactive factors. In this research, the best levels of five process factors were selected from the Taguchi analysis. The combinations of parameter levels for each response are calculated using equation (17). The mean values of each parameter and total mean values have been taken to predict the optimum responses (Boopathi et al., 2012; Sivaiah & Chakradhar, 2017).

$$Optimum\,response = \overline{Total} + \left(\overline{factor_1} - \overline{Total}\right) + \left(\overline{factor_2} - \overline{Total}\right) \ldots \left(\overline{factor_n} - \overline{Total}\right) \tag{17}$$

where, $\overline{Total}$ is the overall mean all the trails, $\overline{factor_1}, \overline{factor_2}, and\, \overline{factor_n}$ are the mean response value of the first factor, second factor, and nth factors respectively.

- **Confirmation Experiments:** The estimated results obtained from Taguchi analysis are verified by conducting validation tests. The best levels of each parameter are accounted for, directing the experiments with interaction terms. In these tests, the noise factors were not considered for performing the tests. The tests have been replicated two times to avoid experimental errors. The mean value of combination tests has been compared to validate whether the predicted values have coincided with the test result or not. If the results did not match the tests, all of the experiments should be repeated.

Based on the responses from pilot experimentation, the pressure, flow rate, current, pulse width, and pulse interval are considered as alternative solutions. The Wire Wear Ratio (WWR), and Material Removal Rate (MRR), are called criteria or attributes. These are obtained from experiments using changing alternative solutions. All the experimentation has been carried out using Taguchi's L27 orthogonal array. It guides the combination of alternative solutions and experimental directions. All the experimental observations are shown in Table 1. These have been used to analyse single and multi-criteria optimization problems.

(4) Multi-criteria Optimization Using TOPSIS

Step 1: The mean values of experimental observations of near-dry WEDM are taken as a decision matrix of the TOPSIS Approach. The 27 sets of decision matrices are shown in Table 1. In the decision matrix, the attributes represent output responses, such as MRR, and WWR, and alternate solutions represent test trails.

Step 2: The units of MRR and WWR are dissimilar and have different units of measurement. As a result, these values are normalised between 0 and 1 to facilitate inter-calculation. The normalised response values of 27 trails are shown in Table 2.

Step 3: MRR and WWR are considered for this study. These are equally treated and the weight assigned to each response is 1/3. The weighted normalized values of each response are shown in Table 3. However, the sum of weights is equal to one.

Step 4: Positive ideal solutions (PIS) and negative ideal solutions (NIS) are calculated from weighted normalized values. PIS and NIS values of 27 trails are shown in Table 3.

Step 5: The Euclidian distance of positive and negative alternative solutions to each response from ideal solutions is determined. In this study, the $\mu+$ and $\mu-$ values are varied based on maximisation and minimization of responses. Euclidian distances AS+ and AS- for response variables were obtained from those equations.

Step 6: The Relative Closeness Factor (RCF) of each response is determined by the Euclidian distance between AS+ and AS-. It is calculated for 27 trials as shown in Table 4.

Experiment number five is the first ranked (1st) alternative, which is called the best ideal solution for this problem. The 26th experiment trail is the last (27) ranked alternative, which is called the worst ideal solution. The best and worst alternative solutions are shown in Table 4. These are also illustrated in Figure 6.

(5) Integrated TAGUCHI -TOPSIS Method

This method is used to predict or confirm the obtained results from the TOPSIS approach. Sometimes, it has been performed to improve the prediction values using DOE analysis. In this session, the RCF values obtained from the TOPSIS approach are considered to perform the Taguchi analysis to predict the combined significant parameter percentages of contributions, and effects of each parameter on the RCF. It was done by statistical software such as Minitab, STATCASE- Design Expert, and STATISTICA and MATLAB. The variations of each operating parameter on

RCF are calculated or estimated. The S/N ratio, residual graphs, normality, random variations, and variance test results were performed to find the statistical verification. The significant parameters are graded based on the percentage of the contribution of the ANOVA test. In the ANOVA test, 5% experimental error is acceptable with a 95% confidence level.

Table 2. Normalized Responses from Experiments

Exp. No.	Response square		Normalized response	
	WWR^2	MRR^2	WWRn	MRRn
1	0.6806	112.7759	0.1722	1.1178
2	0.3959	14.1271	1.2577	1.2577
3	0.4966	47.8394	0.8826	0.9611
4	0.4134	24.2674	1.1447	1.1508
5	0.3934	13.7122	1.2623	1.2623
6	0.5820	54.1902	0.8126	0.9295
7	0.6208	56.0073	0.7928	0.9224
8	0.4448	53.7010	0.8170	0.9307
9	0.7356	118.5159	0.1212	1.1748
10	0.7889	124.3983	0.0887	1.2373
11	0.8612	109.2610	0.2178	1.0873
12	0.4050	51.7623	0.8383	0.9396
13	0.5271	57.5140	0.7751	0.9161
14	0.2078	33.8235	1.0377	1.0616
15	0.2586	35.2610	1.0217	1.0496
16	0.9362	123.1612	0.1116	1.2254
17	0.6525	98.3866	0.3246	0.9983
18	0.5723	50.3333	0.8554	0.9478
19	0.4294	109.9604	0.1923	1.0902
20	0.1485	34.6038	1.0289	1.0550
21	0.4635	114.6741	0.1424	1.1346
22	0.8440	126.8912	0.0915	1.2652
23	0.5094	105.5736	0.2424	1.0525
24	0.2321	37.6775	0.9947	1.0300
25	0.4388	56.8426	0.7819	0.9180
26	0.4402	125.6417	0.0408	1.2486
27	0.9724	120.2159	0.1315	1.1947
Sum	**14.4511**	**2011.1184**	**Sum of Square**	

Table 3. Weighted responses, Positive and Negative ideal solutions

Exp. No.	Weighted Responses		Ideal solutions	
	WWRn/2	MRRn/2	Positive (PIS)	Negative (NIS)
1	0.0895	1.2574	0.0601	0.0758
2	0.0521	0.1575	0.0598	0.0733
3	0.0653	0.5334	0.0527	0.0695
4	0.0544	0.2706	0.0484	0.0576
5	0.0517	0.1529	0.0523	0.0527
6	0.0766	0.6042	0.0565	0.0482
7	0.0817	0.6244	0.0662	0.0556
8	0.0585	0.5987	0.0732	0.0553
9	0.0968	1.3214	0.0749	0.0567
10	0.1038	1.3870	0.0516	0.0522
11	0.1133	1.2182	0.0526	0.0510
12	0.0533	0.5771	0.0840	0.0505
13	0.0693	0.6412	0.0426	0.0756
14	0.0273	0.3771	0.0442	0.0692
15	0.0340	0.3931	0.0624	0.0526
16	0.1231	1.3732	0.0456	0.0605
17	0.0858	1.0970	0.0512	0.0549
18	0.0753	0.5612	0.0847	0.0550
19	0.0565	1.2260	0.0438	0.0753
20	0.0195	0.3858	0.0674	0.0550
21	0.0610	1.2785	0.0630	0.0592
22	0.1110	1.4148	0.0478	0.0558
23	0.0670	1.1771	0.0775	0.0554
24	0.0305	0.4201	0.0788	0.0568
25	0.0577	0.6338	0.0402	0.0764
26	0.0579	1.4008	0.0462	0.0663
27	0.1279	1.3403	0.0493	0.0678
μ+	**0.0195**	**1.4148**	**IDEAL SOLUTIONS**	
μ-	**0.1279**	**0.1529**		

Table 4. Calculations for Relative closeness coefficient and ranking

Exp. No.	Pressure (bar)	Flow rate (ml/min)	Current (A)	Pulse width (μs)	Pulse interval (μs)	RCF	Rank	Remarks
1	5	20	5	15	45	0.1335	20	
2	3	10	3	20	60	0.5000	2	
3	4	15	4	15	45	0.4787	8	
4	3	10	3	25	75	0.4987	3	
5	3	10	3	15	45	0.5000	1	Best Solution
6	4	15	4	20	60	0.4664	12	
7	4	15	4	25	75	0.4622	13	
8	3	15	5	20	75	0.4675	11	
9	5	20	5	20	60	0.0935	23	
10	5	20	5	25	75	0.0669	26	
11	3	15	5	25	45	0.1669	18	
12	3	15	5	15	60	0.4715	10	
13	5	10	4	15	60	0.4583	15	
14	4	20	3	15	60	0.4943	4	
15	4	20	3	20	75	0.4933	6	
16	5	10	4	25	45	0.0835	24	
17	4	20	3	25	45	0.2454	16	
18	5	10	4	20	75	0.4744	9	
19	3	20	4	25	60	0.1500	19	
20	3	20	4	15	75	0.4938	5	
21	5	15	3	20	45	0.1115	21	
22	4	10	5	25	60	0.0674	25	
23	3	20	4	20	45	0.1872	17	
24	5	15	3	15	75	0.4913	7	
25	4	10	5	15	75	0.4600	14	
26	5	15	3	25	60	0.0317	27	Worst Solution
27	4	10	5	20	45	0.0991	22	

Figure 6. TOPSIS: Identification of Best and Worst ideal solution by Relative closeness coefficient

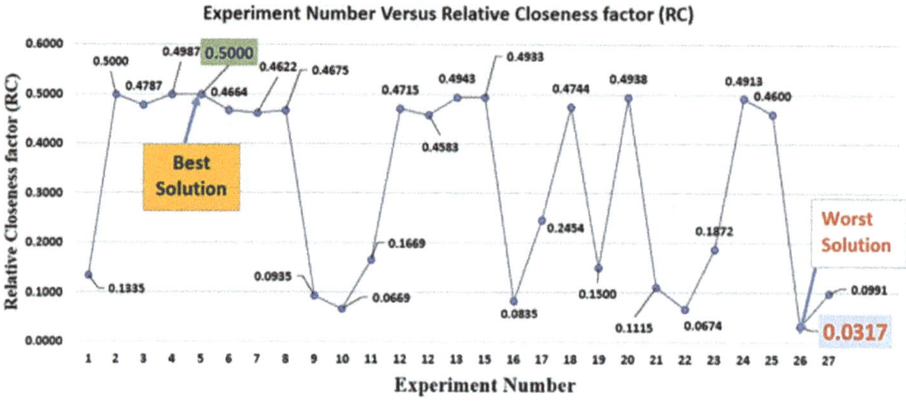

(6) Results and Discussion

A multi-objective problem was converted into a single objective problem to optimize the responses by using the TOPSIS technique. An equal weightage was assigned to wire wear ratio, material removal rate, and surface roughness to identify the best solution by Positive Ideal Solution (PIS) and the worst solution by Negative Ideal Solution (NIS). For maximisation problems, the PIS was maximum and NIS was minimum, and for minimization problems, the PIS was minimum and NIS was maximum. The material removal rate is to be maximized and the wire wear ratio is to be minimized. The closeness coefficients are calculated based on PIS and NIS. The closeness coefficients are ranked in descending order to identify the best input process parameter composition (Boopathi, 2021a). The best-optimized responses of RCF (0.5%), wire wear ratio 0.627210-3, and material removal rate 3.7035 mm3/min resulted in the input parameters gas pressure of 3 bar, flow rate of 10 ml/min, current 3A, pulse width 15s, and pulse interval 45s.The influencing input parameters on the closeness coefficient for the best solution were identified in Figure 6. The closeness coefficients were dominated by current, pulse width, and pulse interval. The confirmation experiments were conducted to validate the TOPSIS results in Table 5.

(7) Prediction of Optimum Results and Confirmation Experiments

The predicted optimal values and confirmation experiment results of air-mist near-dry WEDM are shown in Table 5. These results are verified by conducting validation tests. The best levels of each operating parameter are taken to conduct the experiments. The tests have been replicated two times to avoid experimental errors. The mean value of responses from combination tests agreed well with the predicted results.

Table 5. Best and worst alternative solutions

Exp. No.	P	F	C	PW	PI	RCF	Rank	Description
12	3	15	5	25	45	0.3754	27	**Worst alternative Solutions** MRR: 9.40 mm^3/min; WWR: 0.4333×10^{-3} SR: 4.21 μm (Rough cutting Operation)
25	5	15	3	15	75	0.6553	1	**Best alternative Solutions** MRR: 5.52 mm^3/min; WWR: 0.8345×10^{-3} SR: 1.53 μm (Finish cutting operation)

SUMMARY

In this chapter, the application of TOPSIS techniques to optimize the multiple responses of the micro-machining process is illustrated through a case study. The process parameters of the micro-machining process are automatically adjusted based on the properties of the material, working fluid, tool profile, and machine power by a fuzzy control system. The important steps of the TOPSIS and DOE methods applied in the EDM process have been illustrated using experimental observation. This is very useful to optimization researchers to promote and apply the optimization technique in their research activities. Thus, the TOPSIS technique is very useful to estimate the common process parameters for the two or more conflicting attributes while changing materials, tools, and working mediums.

REFERENCES

Haq, R. H., Marwah, O. M., Rahman, M. N., Haw, H. F., Abdullah, H., & Ahmad, S. (2019). 3D Printer parameters analysis for PCL/PLA filament wire using Design of Experiment (DOE). *IOP Conference Series. Materials Science and Engineering*, *607*(1), 1–10. doi:10.1088/1757-899X/607/1/012001

Bhuyan, R. K., Routara, B. C., & Parida, A. K. (2015). An approach for optimization the process parameter by using TOPSIS Method of Al-24%SiC metal matrix composite during EDM. *Materials Today: Proceedings*, *2*(4–5), 3116–3124. doi:10.1016/j.matpr.2015.07.272

Boopathi, S. (2013). Experimental Study and Multi- Objective Optimization of Near-Dry Wire-Cut Electrical Discharge Machining Process Faculty of Mechanical Engineering. In *Shodhganga@INFLIBNET.* http://hdl.handle.net/10603/16933

Boopathi, S. (2019). Experimental investigation and parameter analysis of LPG refrigeration system using Taguchi method. *SN Applied Sciences*, *1*(8), 892. doi:10.100742452-019-0925-2

Boopathi, S. (2021a). An extensive review on sustainable developments of dry and near-dry electrical discharge machining processes. *Journal of Manufacturing Science and Engineering*, *144*(5), 1–37. doi:10.1115/1.4052527

Boopathi, S. (2021b). An investigation on gas emission concentration and relative emission rate of the near-dry wire-cut electrical discharge machining process. *Environmental Science and Pollution Research International*. doi:10.100711356-021-17658-1 PMID:34837614

Boopathi, S. (2021c). An investigation on gas emission concentration and relative emission rate of the near-dry wire-cut electrical discharge machining process. *Environmental Science and Pollution Research International*, 1–10. doi:10.100711356-021-17658-1 PMID:34837614

Boopathi, S., Balasubramani, V., Kumar, R. S., & Singh, G. R. (2021). The influence of human hair on kenaf and Grewia fiber-based hybrid natural composite material: An experimental study. *Functional Composites and Structures*, *3*(4), 045011. doi:10.1088/2631-6331/ac3afc

Boopathi, S., & Sivakumar, K. (2012). Experimental comparative study of near-dry wire-cut electrical discharge machining (WEDM). *European Journal of Scientific Research*, *75*(4), 472–481.

Boopathi, S., & Sivakumar, K. (2013). Experimental investigation and parameter optimization of near-dry wire-cut electrical discharge machining using multi-objective evolutionary algorithm. *International Journal of Advanced Manufacturing Technology*, *67*(9–12), 2639–2655. doi:10.100700170-012-4680-4

Boopathi, S., & Sivakumar, K. (2016a). Optimal parameter prediction of oxygen-mist near-dry Wire-cut EDM. *International Journal of Manufacturing Technology and Management*, *30*(3–4), 164–178. doi:10.1504/IJMTM.2016.077812

Boopathi, S., Sivakumar, K., & Kalidas, R. (2012). Parametric Study of Dry WEDM Using Taguchi Method. *International Journal of Engineering Research and Development*, *2*(4), 1–6.

Boopathi, S., Thillaivanan, A., Pandian, M., Subbiah, R., & Shanmugam, P. (2021). Friction stir processing of boron carbide reinforced aluminium surface (Al-B4C) composite: Mechanical characteristics analysis. *Materials Today: Proceedings*, *50*, 2430–2435. doi:10.1016/j.matpr.2021.10.261

Boopathi, S., v, H. B., Mageswari, M., & Asif, M. M. (2022). Influences of Boron Carbide Particles on the Wear Rate and Tensile Strength of Aa2014 Surface Composite Fabricated By Friction-Stir Processing. *Materiali in Tehnologije*, *56*(3), 263–270. doi:10.17222/mit.2022.409

Haribalaji, V., Boopathi, S., & Asif, M. M. (2021a). Optimization of friction stir welding process to join dissimilar AA2014 and AA7075 aluminum alloys. *Materials Today: Proceedings*. doi:10.1016/j.matpr.2021.09.499

Haribalaji, V., Boopathi, S., Asif, M. M., Jeyakumar, M., Subbiah, R., & Lewise, K. A. S. (2021). Influences of Friction stir tool parameters for joining two similar AZ61A alloy plates. *Materials Today: Proceedings*, *50*, 2547–2553. doi:10.1016/j.matpr.2021.12.074

Haribalaji, V., Venkatesan, G., Mohammed Asif, M., Pandian, M., Subbiah, R., & Boopathi, S. (2022). Investigation on corrosion and tensile Characteristics: Friction stir welding of AA7075 and AA2014. *Materials Today: Proceedings*. doi:10.1016/j.matpr.2022.04.037

Kannan, E., Trabelsi, Y., Boopathi, S., & Alagesan, S. (2022). Influences of cryogenically treated work material on near-dry wire-cut electrical discharge machining process. *Surface Topography: Metrology and Properties*, *10*(1), 015027. doi:10.1088/2051-672X/ac53e1

Kumar, A., & Rai, R. N. (2020). Grey-Taguchi and TOPSIS-Taguchi-Based Optimisation of Performance Parameters of Spark EDM on Heat-Treated AA7050/5 B4C Composite. *Journal of The Institution of Engineers (India), 101*(1), 71–79. doi:10.1007/s40033-020-00211-8

Mandal, U. K., & Sarkar, B. (2012). Selection of Best Intelligent Manufacturing System (IMS) Under Fuzzy MOORA Conflicting MCDM Environment. *(International Journal of Emerging Technology and Advanced Engineering), 2*(9), 301–310. http://www.ijetae.com/files/Volume2Issue9/IJETAE_0912_52.pdf

Mao, N., Song, M., Pan, D., & Deng, S. (2018). Comparative studies on using RSM and TOPSIS methods to optimize residential air conditioning systems. *Energy, 144*, 98–109. doi:10.1016/j.energy.2017.11.160

Mohanty, C. P., Satpathy, M. P., Mahapatra, S. S., & Singh, M. R. (2018). Optimization of cryo-treated EDM variables using TOPSIS-based TLBO algorithm. *Sadhana - Academy Proceedings in Engineering Sciences, 43*(4), 1–18. doi:10.1007/s12046-018-0829-7

Myilsamy, S., & Sampath, B. (2021). Experimental comparison of near-dry and cryogenically cooled near-dry machining in wire-cut electrical discharge machining processes. *Surface Topography: Metrology and Properties, 9*(3), 035015. doi:10.1088/2051-672X/ac15e0

Nilesh, T. Mohite, P., Shinde, A. P. K., Amar, B., Shubhash, S. T., & Vhagade, R. C. K. (2020). Multi-Response Optimization of Process Parameters of WEDM using TOPSIS Approach. *International Journal of Engineering Research and Technology (IJERT), 9*(3), 182–185. www.ijert.org

Rana, S. C., & Patel, J. N. (2020). Selection of best location for small hydro power project using AHP, WPM and TOPSIS methods. *ISH Journal of Hydraulic Engineering, 26*(2), 173–178. doi:10.1080/09715010.2018.1468827

Sahoo, S. K., Naik, S. S., & Rana, J. (2019). Optimisation of WEDM process parameters during machining of HCHCr steel using TOPSIS method. *International Journal of Process Management and Benchmarking, 9*(2), 216–231. doi:10.1504/IJPMB.2019.099332

Sampath, B., & Haribalaji, V. (2021). Influences of welding parameters on friction stir welding of aluminum and magnesium: A review. *Materials Research Proceedings, 19*, 222–230. doi:10.21741/9781644901618-28

Sampath, B., & Myilsamy, S. (2021). Experimental investigation of a cryogenically cooled oxygen-mist near-dry wire-cut electrical discharge machining process. *Strojniski Vestnik. Jixie Gongcheng Xuebao*, *67*(6), 322–330. doi:10.5545v-jme.2021.7161

Sampath, B., Naveenkumar, N., Sampathkumar, P., Silambarasan, P., Venkadesh, A., & Sakthivel, M. (2021). Experimental comparative study of banana fiber composite with glass fiber composite material using Taguchi method. *Materials Today: Proceedings*, *49*, 1475–1480. doi:10.1016/j.matpr.2021.07.232

Serrano Cardona, L., & Muñoz Mata, E. (2013). Paraninfo Digital. *Early Human Development*, *83*(1), 1–11. doi:10.1016/j.earlhumdev.2006.05.022

Sivaiah, P., & Chakradhar, D. (2017). Multi-objective optimisation of cryogenic turning process using Taguchi-based grey relational analysis. *International Journal of Machining and Machinability of Materials*, *19*(4), 297–312. doi:10.1504/IJMMM.2017.086161

Tansel, Y. I., & Yldrm, S. (2013). MOORA-based Taguchi optimisation for improving product or process quality. *International Journal of Production Research*, *51*(11), 3321–3341. doi:10.1080/00207543.2013.774471

Uthayakumar, M., Babu, K. V., Kumaran, S. T., Kumar, S. S., Jappes, J. T. W., & Rajan, T. P. D. (2019). Study on the machining of Al–SiC functionally graded metal matrix composite using die-sinking EDM. *Particulate Science and Technology*, *37*(1), 103–109. doi:10.1080/02726351.2017.1346020

Wang, C. C., Lin, T. W., & Hu, S. S. (2007). Optimizing the rapid prototyping process by integrating the Taguchi method with the Gray relational analysis. *Rapid Prototyping Journal*, *13*(5), 304–315. doi:10.1108/13552540710824814

Wang, P., Li, Y., Wang, Y. H., & Zhu, Z. Q. (2015). A new method based on TOPSIS and response surface method for MCDM problems with interval numbers. *Mathematical Problems in Engineering*, *2015*, 1–11. doi:10.1155/2015/938535

Yupapin, P., Trabelsi, Y., Nattappan, A., & Boopathi, S. (2022). Performance Improvement of Wire - Cut Electrical Discharge Machining Process Using Cryogenically Treated Super - Conductive State of Monel - K500 Alloy. *Iranian Journal of Science and Technology. Transaction of Mechanical Engineering*, *0123456789*. doi:10.100740997-022-00513-0

Chapter 10
Entomopter Wing Analysis Boon for Mars Mission

Md. Akhtar Khan
Gandhi Institute of Technology and Management, India

Swarnashree K.
Gandhi Institute of Technology and Management, India

ABSTRACT

In this chapter, CFD analysis is performed to understand the aerodynamic characteristics of 3-D Entomopter wing. Entomopter is the MAV which is especially designed for Martian atmosphere exploration. It is a low Reynolds number flyer due to which it can reach to the places where human intervention is not at all possible. The aerodynamic simulations are carried out at different flap angles varying from -20° to 30° in Martian atmosphere. The Entomopter model considered for CFD analysis has a wingspan of 15 cm with varying chord length and Reynolds number ranging from Re=75000 Re=150000 i.e., at a velocity of 2 m/s to 30 m/s. The Entomopter 3-D printed model is designed in CATIA V5 and simulation is carried out using ICEM CFD and ANSYS FLUENT. Entomopters may undertake multimode operations and produce lift by continuously forming and shedding vortices from their flapping wings. So, this research is a viable aspect to understand the future development of MAVs for outer space exploration and can be promising for making impossible things possible.

DOI: 10.4018/978-1-6684-5887-7.ch010

INTRODUCTION

Many researchers in the field of biology, zoology, aerodynamics, and electronics have long been fascinated by the flapping flight of birds and insects (Michelson, 1998; Han et al., 2009; Jaroszewicz, 2013). This flapping flying mechanism provides excellent manueverability and aerodynamic advantages, particularly at low Reynolds numbers (Han et.al, 2009). Hence, miniature flight platforms are the tiny ones which are unmanned and endeavour to invade the flight regime of birds and insects.

A Micro Air Vehicle (MAV) is a flapping unmanned aerial vehicle (UAV) having a size restriction of as small as 15 cm i.e., (approximately 6 inches). It is an insect-sized aircraft that is generally expected to be launched by hand. It uses flapping wings to generate the aerodynamic forces and moments needed for flight, and it maintains stability through a complex wing motion that includes flapping, pitching and plunging, twisting, and lagging motion (Djojodihardjo et al.,2016). Small size and slow flight add to the advantages of a MAVs. Also, a micro air vehicle can enter enclosures and effectively navigate the interiors. Though it is an unmanned aircraft, it can be operated in a similar way to that of a conventional aircraft.

MAVs are generally designed for hobby purposes and remote observation of hazardous environmental conditions (Khan, Khan, & Padhy, 2021).These activities are accomplished by Nano Air Vehicles (NAVs), Ornithopter and Entomopters, of micro air vehicles (MAVs) kind (Gerdes et al.,2012).

An example of a MAVs (micro air vehicles) that demonstrates high degrees of innovation and biological inspiration is the Entomopter,bio-mimicked from Manduca Sexta shown in Figure 1 (a). It is a bioinspired flight that flies using the wing-flapping aerodynamics of an insect. The word is derived from entomo (meaning insect: as in entomology) + pteron (meaning wing) (Michelson, 2004, 2011).

The entomopter is an electromechanical multimode insect capable of flying, crawling, and even swimming, proving it to be an all-time favourite unmanned aerial vehicle. Michelson (2011) and his Georgia Tech Research Institute design team are working on it. This electromechanical device is based on a recent breakthrough known as a Reciprocating Chemical Muscle (RCM), which can generate modest quantities of power for on-board devices using a chemical energy source. In other words, we can say that RCM provides autonomic wing beating for entomopter during its flight.

The entomopter has the extra benefits of indoor operations, high manoeuvrability, slow flying, obstacle avoidance, and multimode locomotion (Michelson, 2004). Furthermore, an entomopter wing can be stationary, rotating, flapping, or lighter-than-air (LTA). Low Reynolds number aerodynamic phenomena, extreme downsizing, high degrees of system integration, and efficient use and re-use of energy are all challenges that the mechanical robot must overcome(Austin, 2011; Michelson, 2004)).The entomopter began as a bio-mimetic design, but rather than attempting to replicate biological kinematics and aerodynamics, improved systems have been devised converting it to be a biologically inspired design to produce a machine that is manufacturable, controllable, and able to generate the power necessary to fly from on-board energy sources.

BIOLOGICAL INSPIRATION

As discussed earlier, birds and insects' flights have always been in limelight for a long time. Many researchers have turned to the Hawk Moth (Manduca Sexta) shown in Figure 1, considering it to be one of the favourite analogs (Michelson, 1998). The mechanisms used to flap the wings are often mechanically coupled, allowing only hovering or steady forward flight. This insect wing-flapping aerodynamics is suitable for flying at a low Reynolds Number regime, with more lift generation.

Figure 1. (a) Hawk Moth (b) Skeletal Structure of Hawk Moth

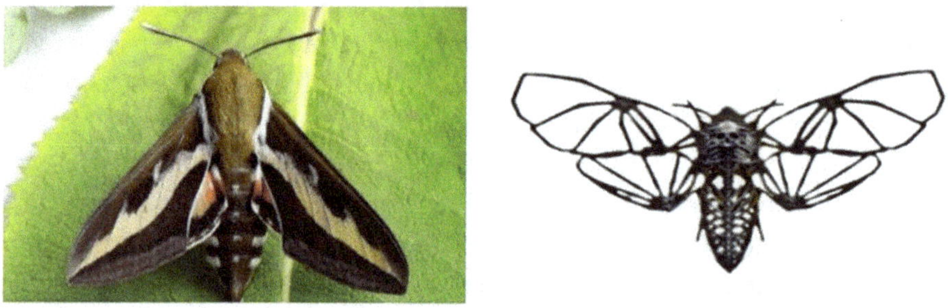

(a) (b)

METHODOLOGY

Entomopter Wing Design

CATIA software was used to design the entomopter wing . The generative design surface is used to represent the entompoter wing, which is made up of 15 sections with variable chord lengths as shown in Figure 2.

Figure 2. CATIA Model of Entomopter Wing Design

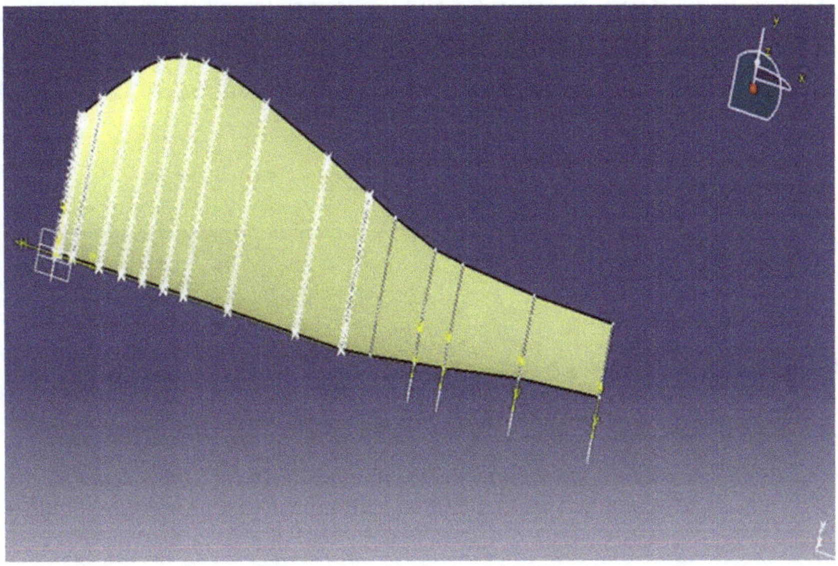

CFD Analysis of the Entomopter Wing

By integrating with ANSYS structural mechanics solutions through the Workbench unified user experience, ANSYS Fluent 2019 simulates the effects of solid motion on fluid flow. It includes pre- and post-processing tools as well as reliable and accurate findings. CFD analysis is performed to find the pressure and velocity vectors. Along with these, we also found the values for lift and drag forces with their dimensionless coefficients. The analysis was performed for angles ranging from -20^0 to 30^0 at a velocity of 15 m/s. The martian atmospheric parameters were considered for the analysis illustrated in table 1.

Table 1. Boundary conditions

Parameters	Boundary conditions
Atmospheric Pressure	610 Pa
Density	0.020 kg/m^3
Viscosity	0.011 kg/ (m.s)
Turbulence Model	k- model

We have considered a low Reynolds Number range of 75,000-1,50,000 to carry out this analysis to get the required results. Aerofoil E 471 was chosen as the cross-section for designing an entomopter wing because the wing design created possesses qualities that are essential for flying in the Martian atmosphere.

Figure 3. E 471 airfoil

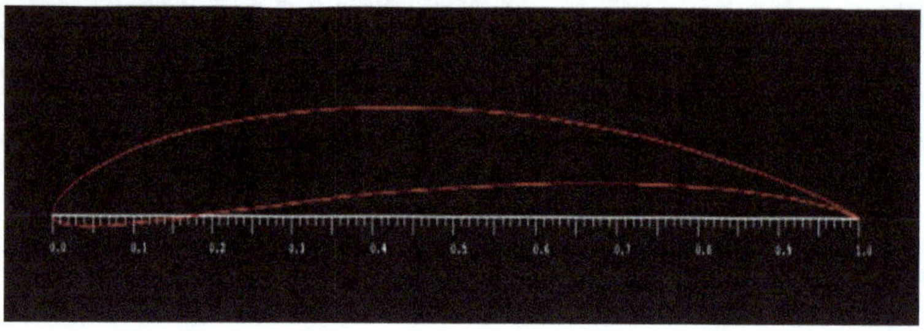

The airfoil has been generated in the X-Foil software with the input of coordinates, refer Figure 3. Later, the airfoil is imported to the ANSYS-19 ICEM CFD software, where the 3D Entomopter Wing is designed and the analysis is performed in the ANSYS FLUENT software shown in Figure 4.

Figure 4. Wing Section (3D view)

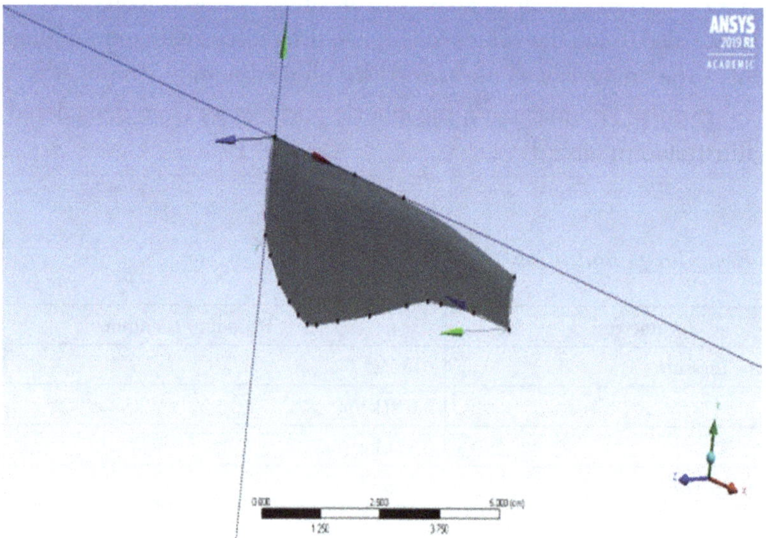

The CFD analysis has been performed on the wing using the above mentioned parameters. Also, the tapered section side of the wing is the inflow of the fluid and, one end of the wing is fixed, whereas, the other end is left free. The boundary condition for the anlaysis is depicted in Table 2.

Table 2. Boundary Conditions for analysis

Parameter	Value
Dynamic Viscosity	0.011 kg(m.s)
Density	0.02 kg/m^3
Nodes for -20^0	240245
Nodes for -10^0	250331
Nodes for 0^0	191965
Nodes for 10^0	189003
Nodes for 20^0	188318
Nodes for 30^0	187168
Domain size	0.5m1m1m
Inlet Velocity	15 m/s

Figure 5. Mesh for 0^0 Flap Angle

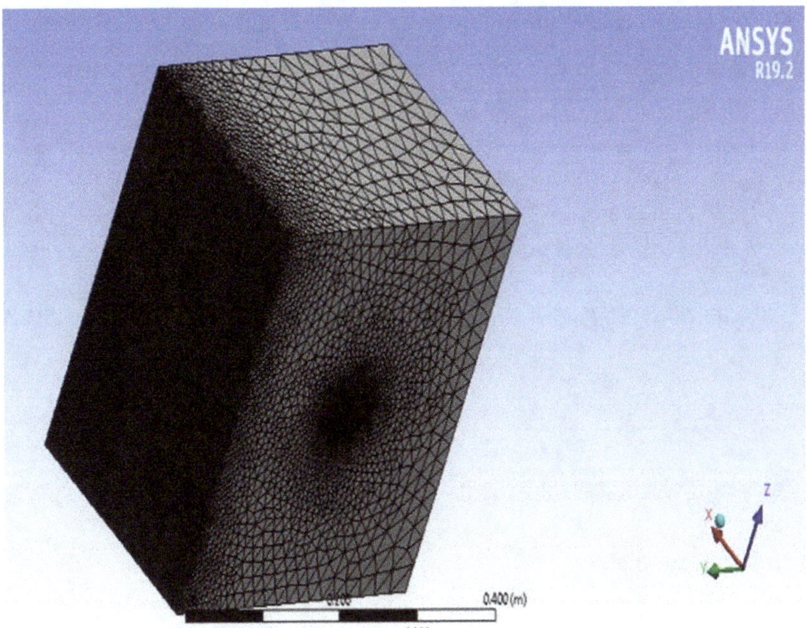

K-Epsilon turbulence model was chosen for simulation. Here, 'K' stands for the turbulent kinetic energy and 'Epsilon' stands for rate of dissipation of turbulent kinetic energy. To perform the meshing, an outer surface in the shape of a cube has been created and the domain is shown in Figure 5. It has to be noted that more number of nodes formed result in greater convergence, more accuracy, high stability and also, more calculation time.

Figure 6. Convergence at different flap angles

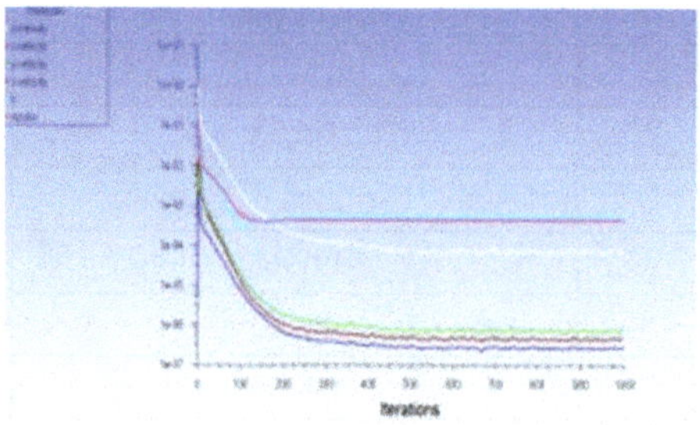

(a) -20° Flap Angle

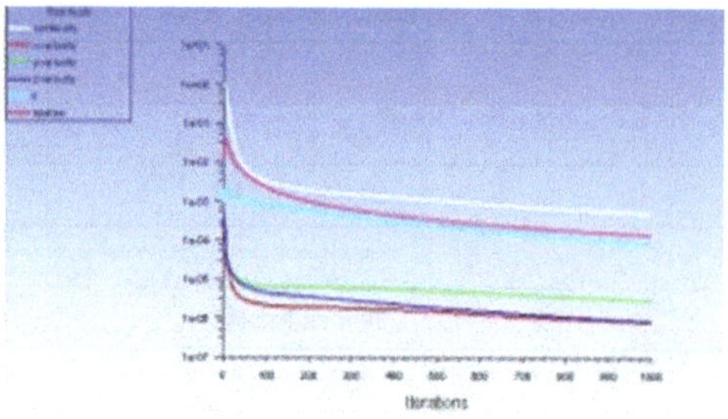

(b) -10° Flap Angle

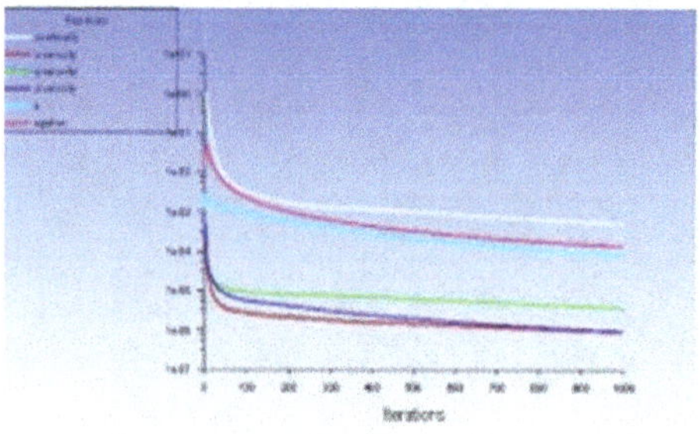

(c) 0° Flap Angle

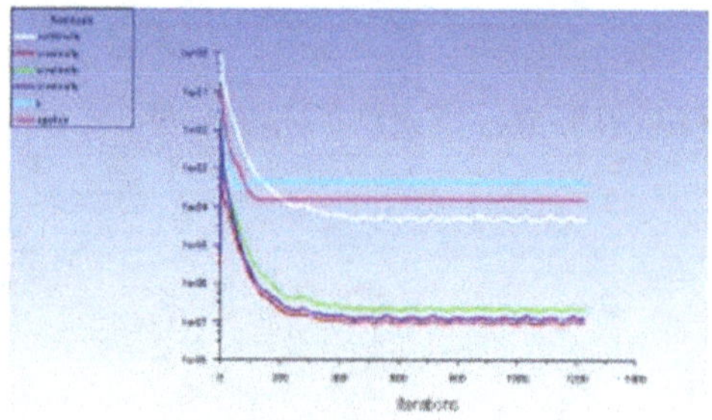

(d) 10° Flap Angle

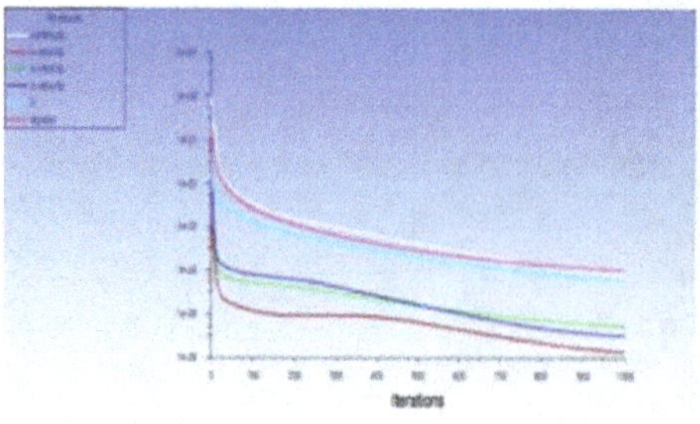

(e) 20° Flap Angle

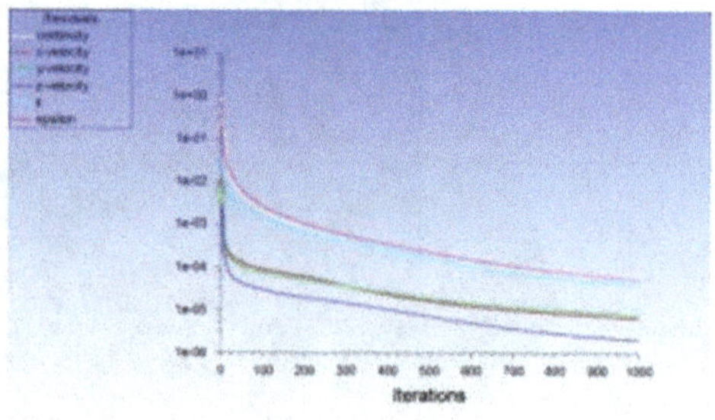

(f) 30° Flap Angle

CFD SIMULATION RESULTS

Residuals (Convergence)

In all the above mentioned cases, the solution is converged. The convergence is considered at an error range of 1 for continuity and convergence at different flap angle is depicted in Figure 6.

Velocity Vectors Plot at Different Flap Angles

From Figure 7, it can be observed that, as the flap angle increases from 0° to 30°, the velocity also increases in the martian atmosphere. But this increase in velocity leads to instability, so a minimum velocity has to be maintained throughout the entire flight to maintain stability and achieve maximum entomopter flight performance.

Figure 7. Velocity vector plot of entomopter wing at different flap angles

(a) -20° Flap Angle

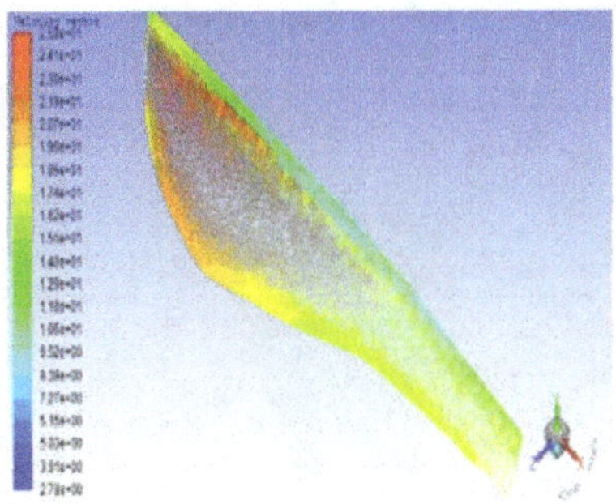

(b) -10° Flap Angle

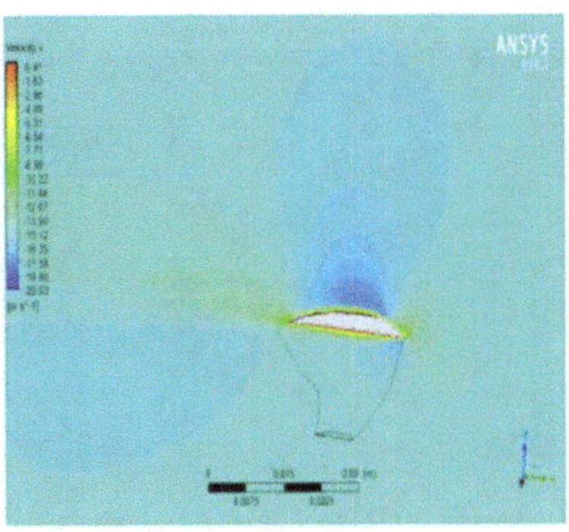

(c) 0° Flap Angle

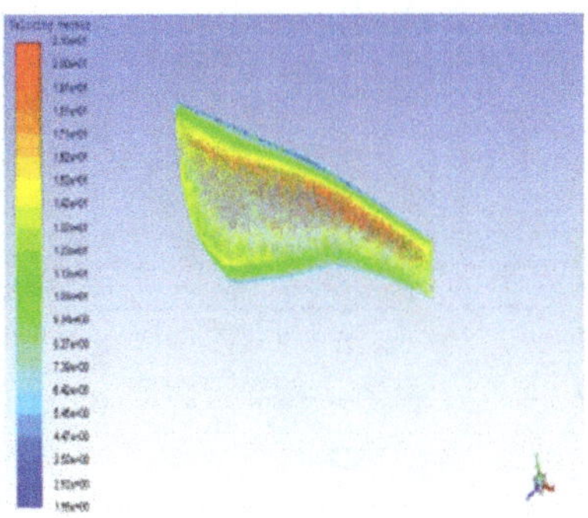

(d) 10° Flap Angle

Static Pressure Contours at Different Flap Angles

From Figure 8, It can be observed that the effect of static pressure on the wing is comparatively less when computed for Martian atmosphere. It experiences the medium to least amount of pressure on the wing surface.

Figure 8. Static pressure contour plot of Entomopter wing at different flap angles

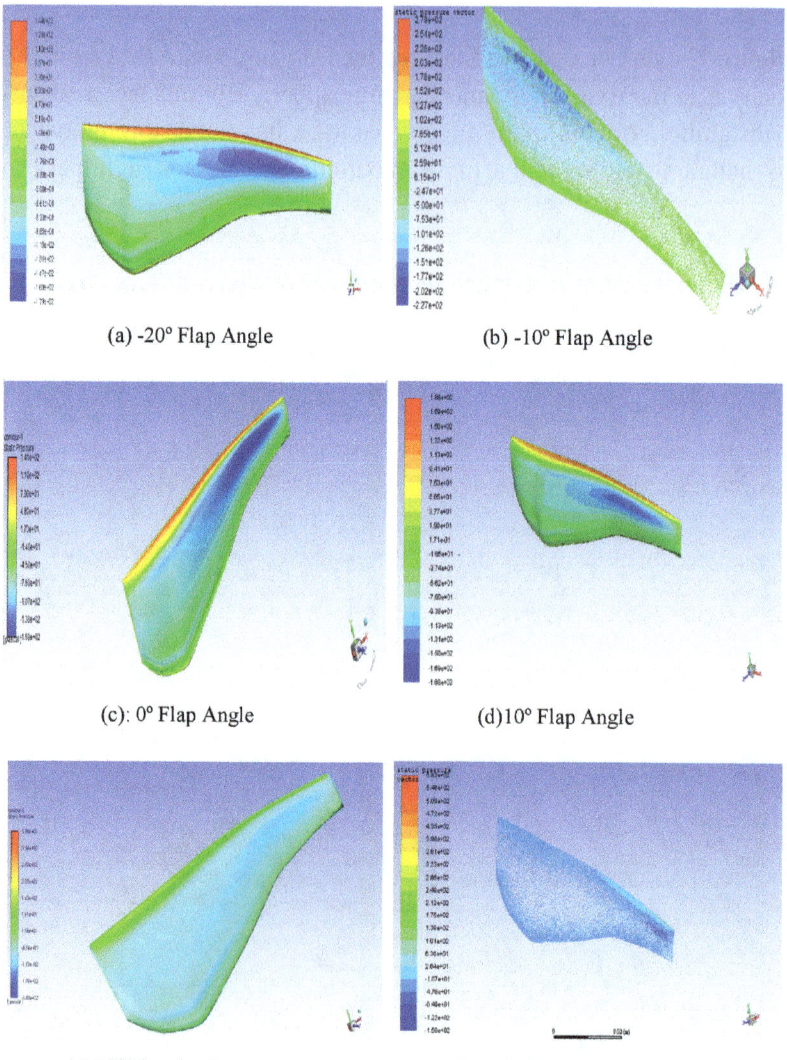

(a) -20° Flap Angle

(b) -10° Flap Angle

(c): 0° Flap Angle

(d)10° Flap Angle

(e) 20° Flap Angle

(f): 30° Flap Angle

Velocity Pathline Plot at Different Flap Angles

At negative angle of attack the velocity on the lower surface is higher which results in lesser lift production. As the angle of attack increases the velocity on the lower surface decreases and pressure increases resulting in higher lift production due to the flapping mechanism. The velocity pathline shows that there is small amount of eddies at the wingtip, refer Figure 9.

Figure 9. Velocity pathline plot of entomopter wing at different flap angle

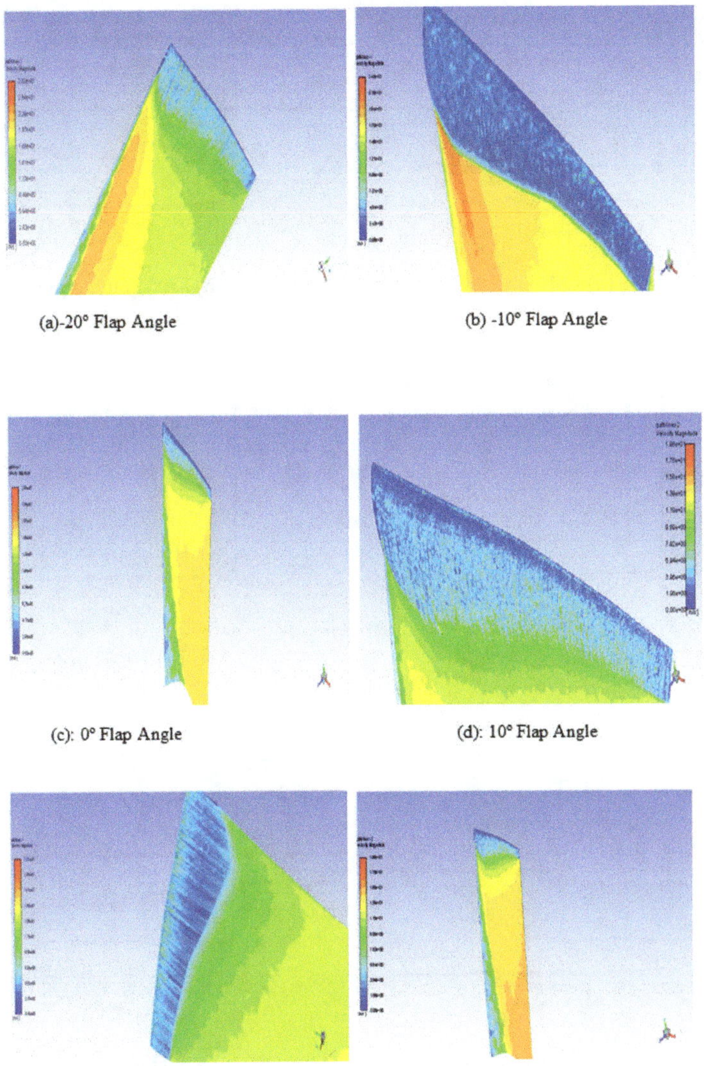

(a)-20° Flap Angle

(b) -10° Flap Angle

(c): 0° Flap Angle

(d): 10° Flap Angle

(e) 20° Flap Angle

(f) 30° Flap Angle

The variation of aerodynamic performance parameters with respect to time iteration is shown from Figure 10 to Figure 13.

The value of the aerodynamic performance parameters are depicted in the Table 3. By observing at the values, we can understand that the gliding ratio (C_L/ C_D) is maximum at -10° as compared to the positive flap angle. It means that the performance is higher during downward stroke of the flight.

Figure 10. Coefficient of Lift (C_L) vs. Iteration of entomopter wing

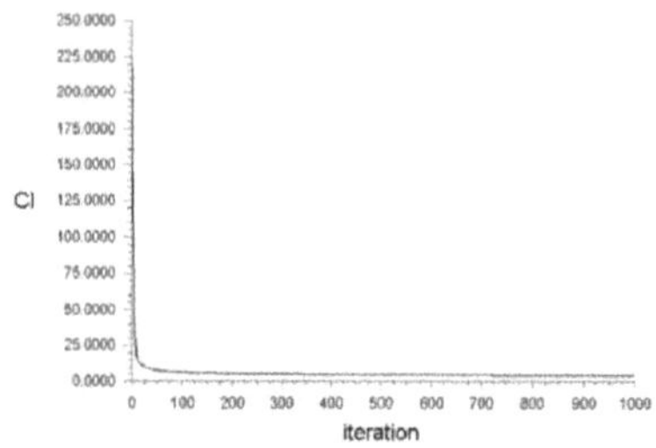

(a) -20° Flap Angle

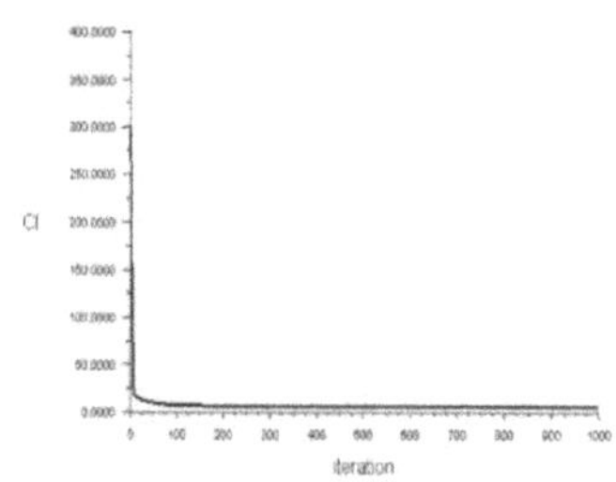

(b) -10° Flap Angle

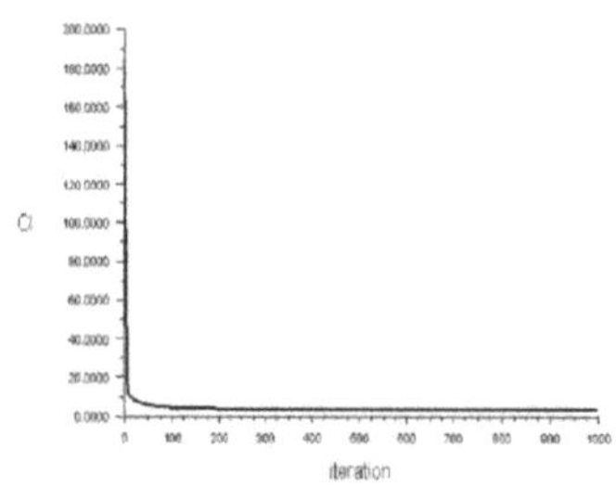

(c) 0° Flap Angle

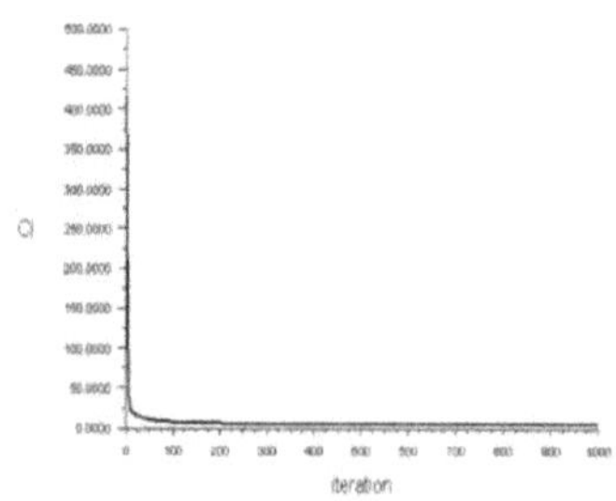

(d) 10° Flap Angle

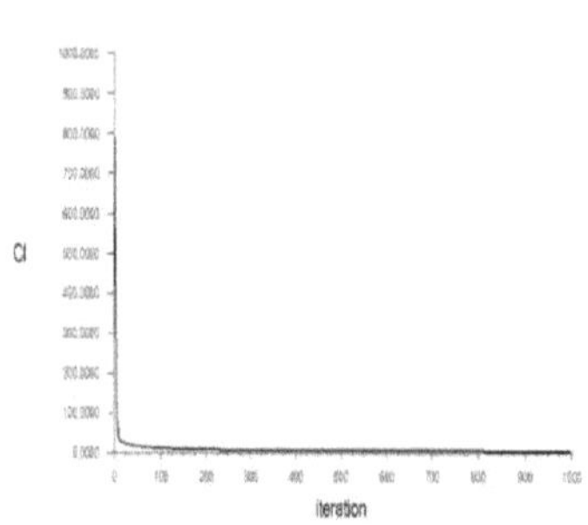

(e) 20° Flap Angle

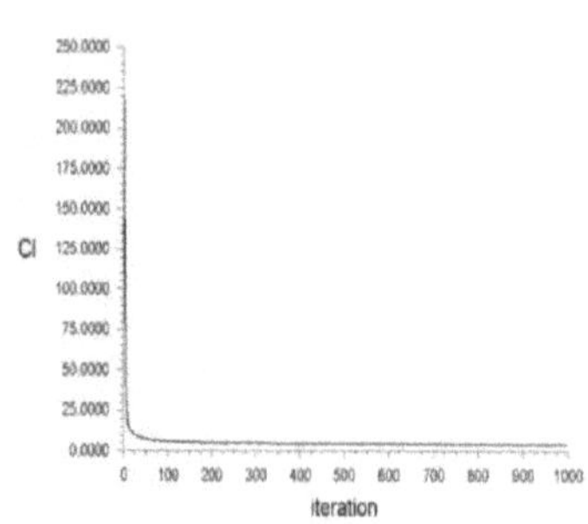

(f) 30° Flap Angle

Figure 11. Coefficient of Drag (C_D) vs. Iteration of entomopter wing

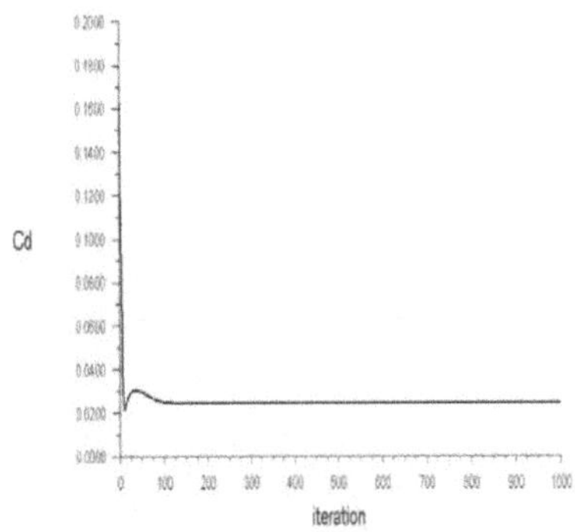

(a) -20° Flap Angle

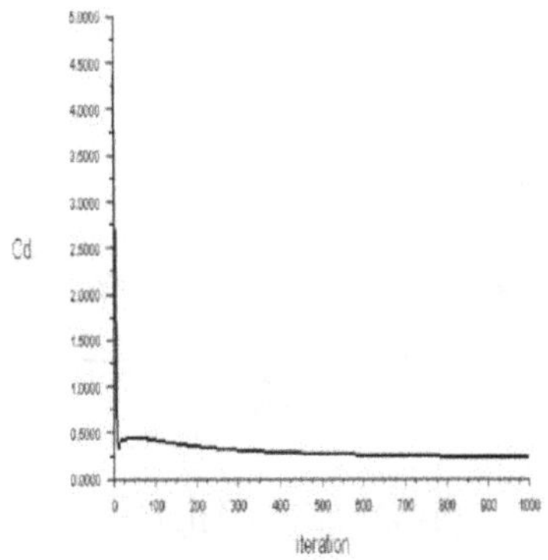

(b) -10° Flap Angle

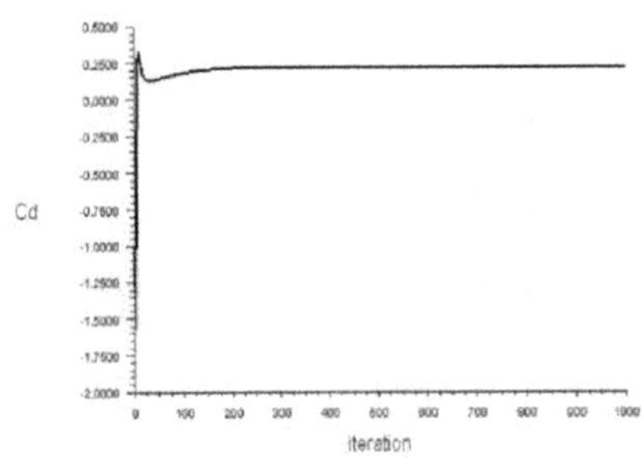

(c) 0° Flap Angle

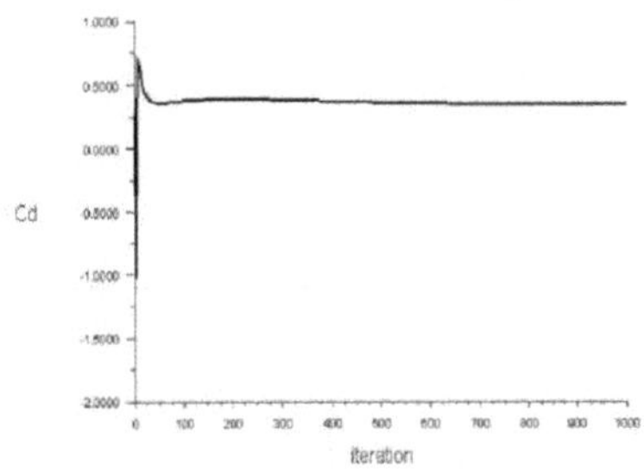

(d) 10° Flap Angle

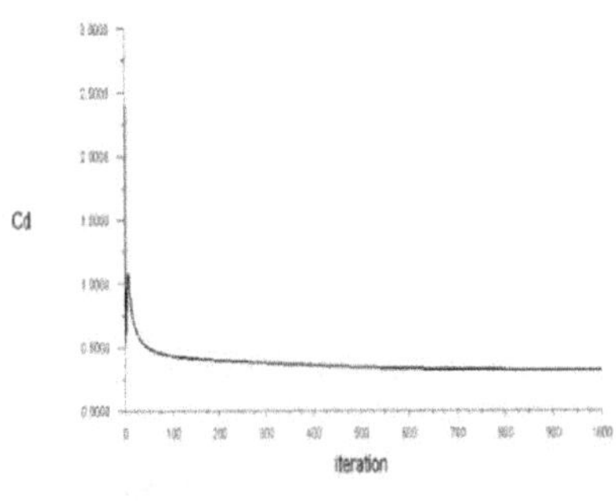

(e): 20° Flap Angle

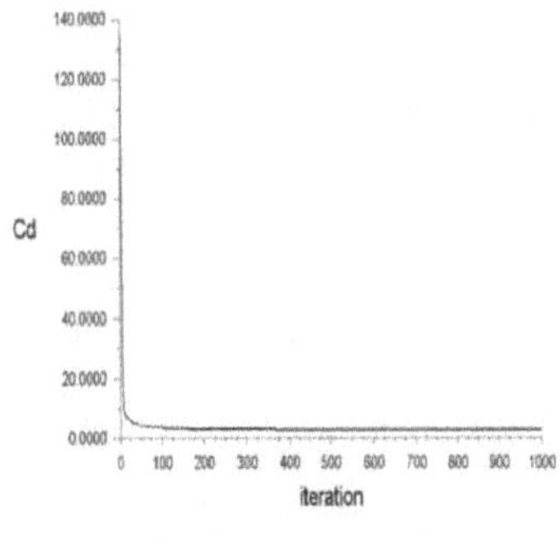

(f): 30° Flap Angle

Figure 12. Lift (L) vs. Iteration of entomopter wing

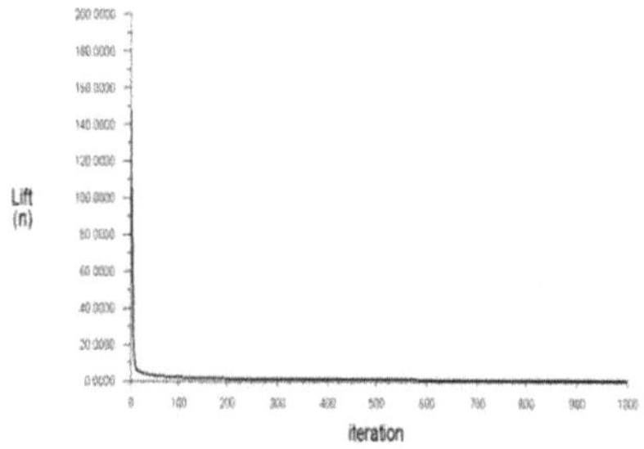

(a) -20° Flap Angle

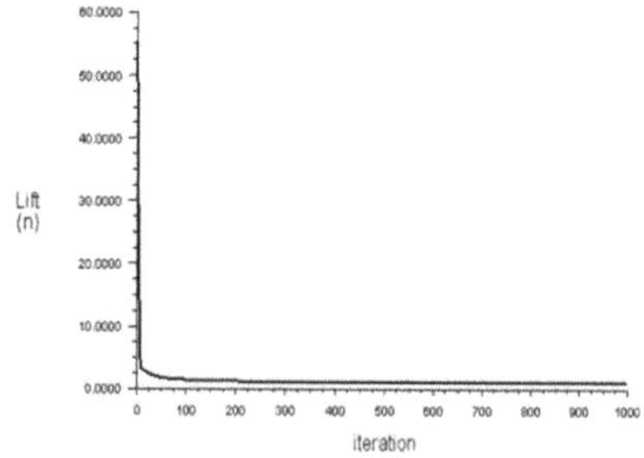

(b) -10° Flap Angle

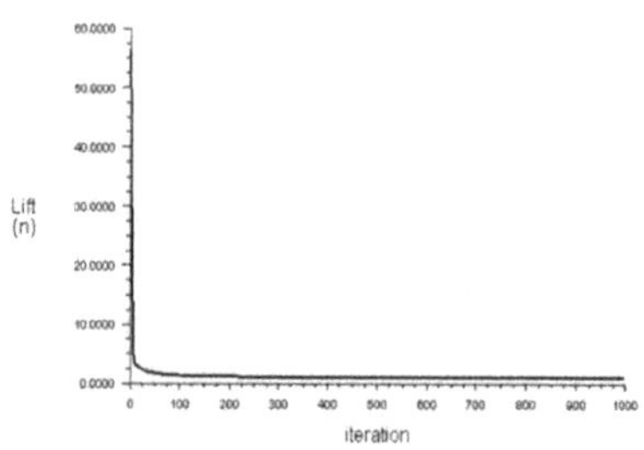

(c) 0° Flap Angle

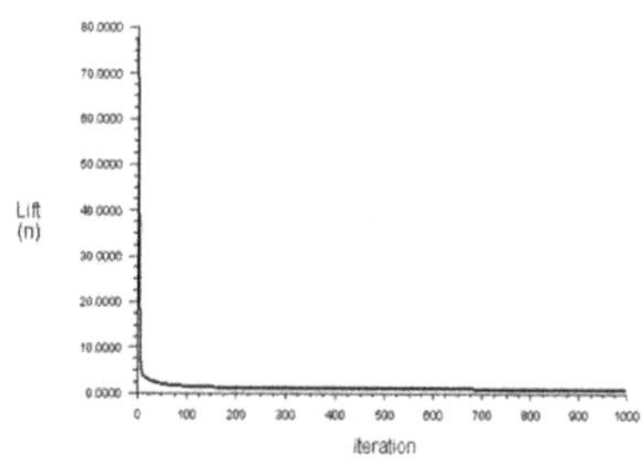

(d) 10° Flap Angle

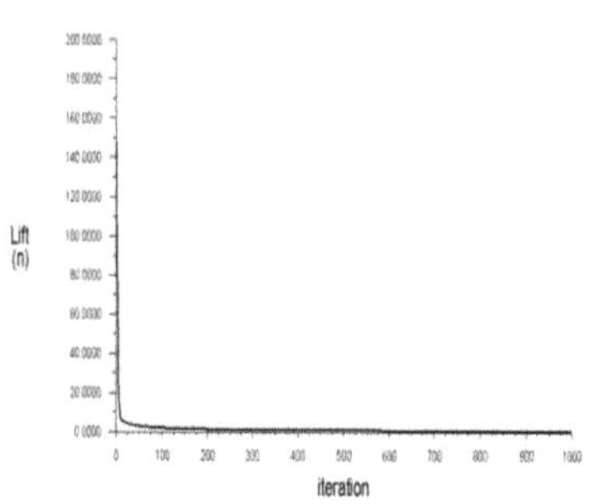

(e) 20° Flap Angle

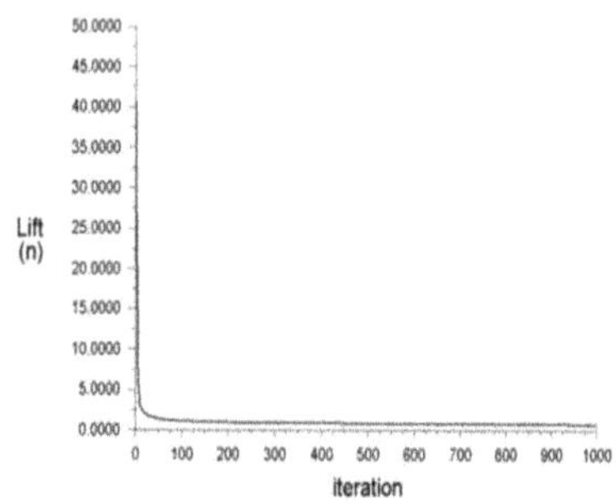

(f) 30° Flap Angle

Figure 13. Drag (D) vs. Iteration of entomopter wing

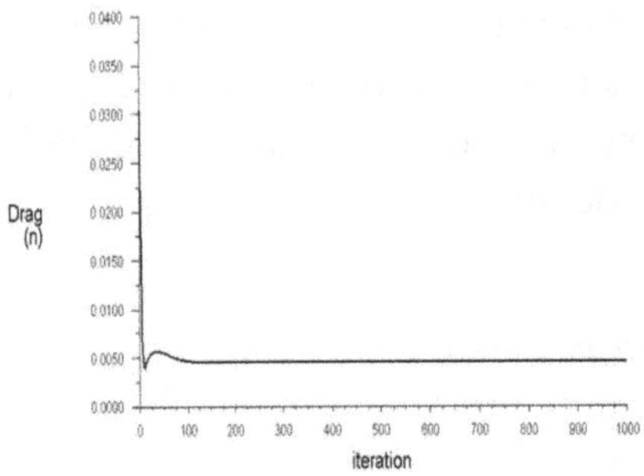

(a) -20° Flap Angle

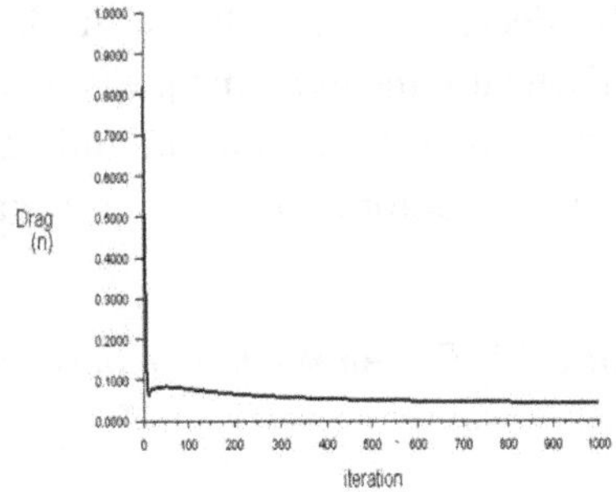

(b) -10° Flap Angle

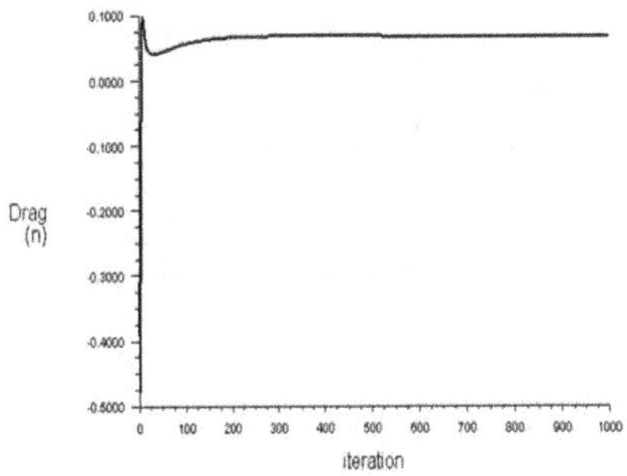

(c) 0° Flap Angle

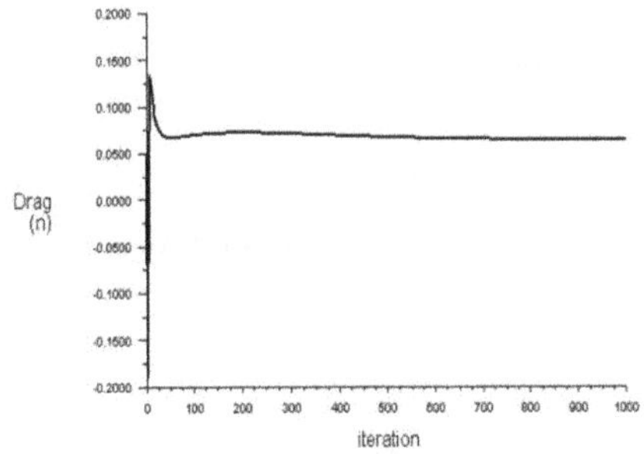

(d)10° Flap Angle

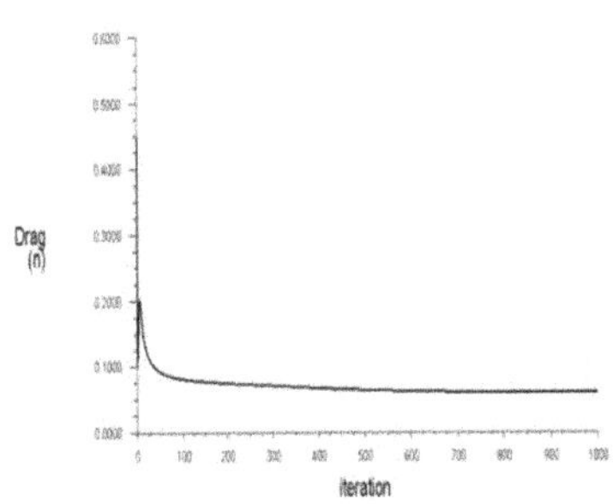

(e) 20° Flap Angle

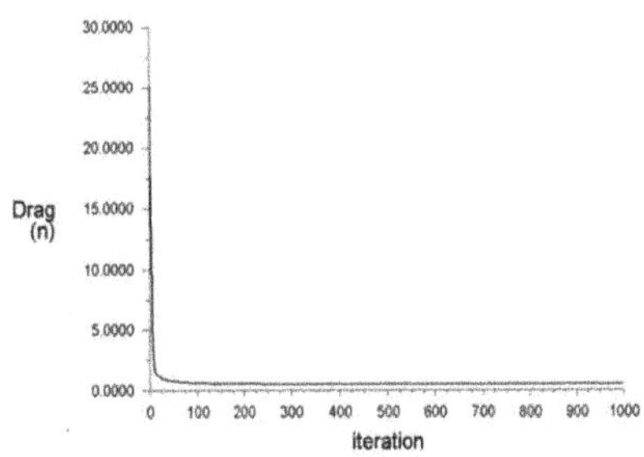

(f) 30° Flap Angle

From the above Figure 14, It can observe that,the entomopter wing produces good amount of lift at low flap angle but with the increase in flap angle, the value of lift decreases.The value of lift/drag is lesser than the value of coefficient of lift/ drag because the dynamic pressure is lesser than the lift.From the figure 15 (c) we can observe that, the value of gliding ratio is maintained nearly equal for all the flap angles except for higher angles, because of instability.

Figure 14. The Aerodynamic performance parameters w.r.t flap angles

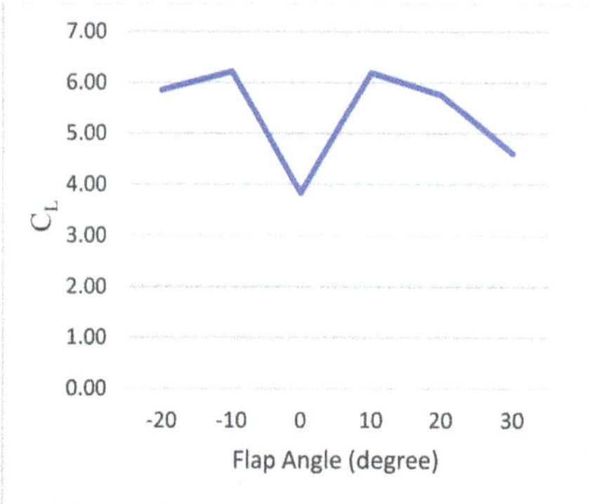

(a) C_L Vs Flap Angle

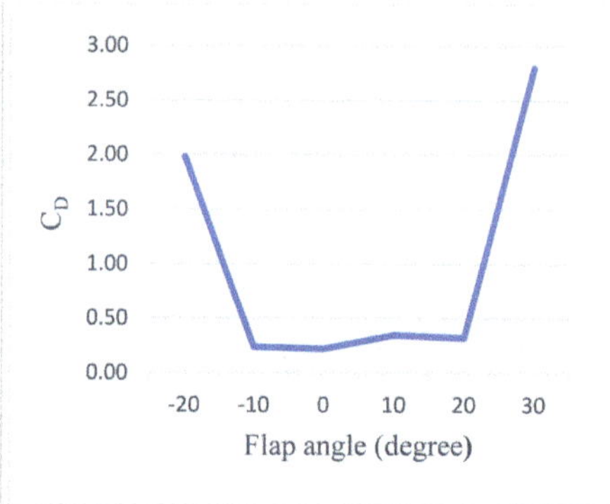

(b) C_D Vs Flap Angle

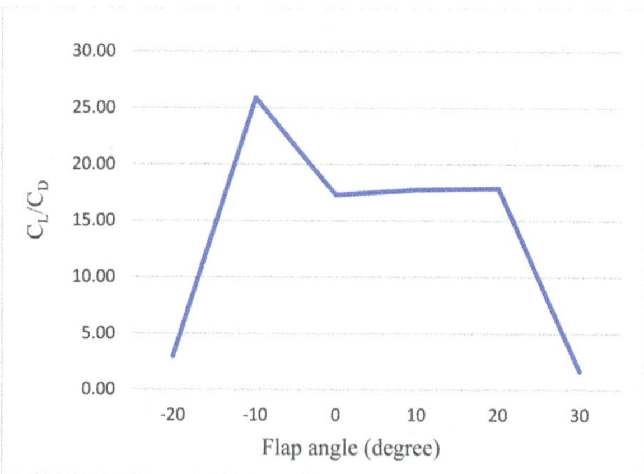

(c) C_L/C_D Vs Flap Angle

Table 3. Aerodynamic performance parameters

Flap Angle	C_L	C_D	Lift (L)	Drag (D)	C_L/C_D
-20°	5.86	1.98	1.08	0.73	2.96
-10°	6.23	0.24	1.16	0.04	25.92
0°	3.84	0.22	1.17	0.07	17.29
10°	6.20	0.35	1.16	0.07	17.76
20°	5.77	0.32	1.08	0.06	17.90
30°	4.62	2.80	0.86	0.52	1.65

CONCLUSION

After performing the CFD analysis we observed uncertain behaviour of the bio-mimicked entomopter wing. it is concluded that the entomopter wing configuration is best suited for the martian atmosphere and more suitable for the low Reynolds number micro air vehicles configuration. The conclusions are listed below:

- E471 aerofoil is found to be stable at higher Flap Angles.
- E471 aerofoil has better aerodynamic performance at low Reynolds number.
- For the wing with E471 aerofoil, the maximum value of Glide ratio is found to be in the range of 15 to 30.
- The drag for this wing is found lesser than the conventional aircraft wing.
- The stall point for this wing is also delayed, beyond 30^0.
- Thus, we can say that, the wing has the efficiency to fly in the range of -10^0 to 20^0 to its fullest capacity.

ACKNOWLEDGMENT

We gratefully acknowledge the support from the GITAM University for providing the Simulation facility for doing this research work more effectively.

REFERENCES

Austin, R. (2011). *Unmanned Aircraft Systems,* (1st ed.). Wiley. https://www.perlego.com/book/1009794/unmanned-aircraft-syste ms-pdf

Bektaş, M., & Guler, M. A. (2018).Steady-state CFD Analysis of 3D Bio-inspired Flapping Wing Models.

Bektaş, M., Kurtuluş, D. F., & Güler, M. A. (2018). Steady-State CFD Analysis of 3D Bio-inspired Flapping Wing Models. *International Symposium on Sustainable Aviation*, Rome, Italy. https://hdl.handle.net/11511/70891

Combes, S. A. (2010). Materials, Structure, and Dynamics of Insect Wings as Bioinspiration for MAVs. In R. Blockley & W. Shyy (Eds.), *Encyclopedia of Aerospace Engineering*. doi:10.1002/9780470686652.eae404

Djojodihardjo, H., Syamim, A., Ramli, S., Wiriadidjaja, S., & Rafie, A. M. (2016). Kinematic And Aerodynamic Modeling Of Bi- And Quad-wing Flapping Wing *Micro-air-vehicle, Advances in Aerospace Science and Technology 1*(1),83-101.

Gerdes, J., Gupta, S., & Wilkerson, S. (2012). A Review of Bird-inspired Flapping Wing Miniature Air Vehicle Designs, Khan, M. A., & Padhy, C. (2021). Influence of aerodynamic parameters on dragonfly inspired corrugated aerofoil. In K. B. Rushi, R. Sivaraj, & J. Prakash (Eds.), *Advances in Fluid Dynamics*. Lecture Notes in Mechanical EngineeringSpringer. doi:10.1007/978-981-15-4308-1_23

Han, J. H., Lee, J. S., Kim, D. K. (2009). Bio-inspired Flapping UAV design: A University Perspective. *Proc. SPIE.7295,Health Monitoring of Structural and Biological Systems.* doi:. doi:10.1117/12.815337

Jaroszewicz, A. (2013). Modeling And Simulation of Flapping Wing Entomopter In Martian Atmosphere, Springer, GeoPlanet. *Earth and Planetary Sciences, 8*, 143–162. doi:10.1007/978-3-642-34020-8_12

Michelson, R. C. (2004). Novel approaches to miniature flight platforms. *Proceedings of the Institution of Mechanical Engineers. Part G, Journal of Aerospace Engineering, 218*(6), 363–373. doi:10.1243/0954410042794911

Michelson, R. C. (2011), Slow flight in the lower Mars Atmosphere in support of NASA science missions.

Michelson, R.C., & Reece, S. (1998). Update on flapping wing micro air vehicle research ongoing work to Develop a Flapping Wing, Crawling "Entomopter".

Chapter 11
Fabrication and Stabilization of a Low-Cost Rotary-Inverted Pendulum Setup (STRIPS 1.0): Using Full-State Feedback Algorithm

Tufail Ahmad Bhat
https://orcid.org/0000-0002-6296-7140
Islamic University of Science and Technology, India

Majid Hameed Koul
Department of Mechanical Engineering, Islamic University of Science and Technology, India

ABSTRACT

This paper presents the fabrication and Stabilization of a low-cost in-house developed Rotary Inverted Pendulum Setup (STRIPS 1.0), commonly known as a Furuta Pendulum. The developed setup consists of a base frame on which an arm is mounted to allow rotary motion in the horizontal plane and a pendulum link subsequently mounted on the other end of the arm to allow rotation in the vertical plane. Essentially the setup possesses 2- DOF (Degrees of Freedom); however, since the actuation is provided at the arm axis only, the setup behaves as an under-actuated one. Owing to the inherently unstable nature of these under-actuated systems, full-state feedback and an LQR (Linear Quadratic Regulator) controller are designed and implemented, which maintain the pendulum link in the upright position even if disturbed about its linear region of control. In particular, a set of gains are chosen for a particular performance criterion and implemented using a MATLAB interface with the micro-controller on the developed setup, allowing a restricted arm movement.

DOI: 10.4018/978-1-6684-5887-7.ch011

INTRODUCTION

One of the primary, however fascinating, problems in the field of control theory is the inverted pendulum control. It is a challenge that can be readily envisioned since everyone at one time has attempted to balance a broomstick on their hand. While the broomstick is a relatively simplistic demonstration, it highlights the inherent instability of the inverted pendulum system. The problem of balancing a broomstick on the hand can be modeled as a rotary inverted pendulum in which the human arm is mimicked by a rotating arm, actuated by an electric motor, and the broomstick modeled as a pendulum connected to the arm using the revolute joint, with control torque given to the arm link in order to stabilize the pendulum in an upright position. Inverted pednulum has two equilibrium states: vertical downward stable equilibrium and vertical upright unstable equilibrium. The downward position is achieved by gravity; therefore, no feedback approach is required; but in order to achieve the unstable upright position, an external feedback approach is implemented; one such technique utilized in the system is LQR.

The inverted pendulum is inherently an under-actuated system with more degrees of freedom than the number of actuations. In this system, a single actuator at the arm indirectly impacts the motion of the pendulum also, owing to the multibody nature of the arm and pendulum. This indirect impact of the arm motion on the pendulum and the absence of an actuator at the pendulum axis makes the system intrinsically unstable at various operating points. A good control strategy would allow one to control the pendulum in the upright position without requiring an additional actuator. Such systems thereby widely find applications ranging from balancing humanoid robots, where the robot itself can be considered an inverted pendulum, to the Segway human transporter, where a vehicle moves in the direction that the rider leans.

However, because of the system's inherent instability and ability to operate in a non-linear range, the inverted pendulum problem is widely researched and considered a classical benchmark problem in control systems engineering. Figure 1 shows a commercial setup of a rotary inverted pendulum used to teach control systems and experiment with new strategies in the laboratories of UG (Undergraduate) and PG (Postgraduate) courses on Control Systems. Although the setup is highly optimized and useful, the primary reason for fabricating the in-house setup (STRIPS 1.0) was the affordability and that the commercial setups act as a black box to the learner. Several other solutions also exist; however, they are available only in the laboratories of various research institutions. In addition to our interest in fabrication, the primarily focused on the controller design, simulation, and implementation. Before the such attempt, the authors present a review of various inverted pendulum systems and associated control techniques that have been investigated in the past.

Figure 1. Rotary Inverted Pendulum (Qube-Servo 2)

BACKGROUND

Many identical setups have been designed and modified by experts throughout the years,such as Quanser's (Quanser, 2021) QNET 2.0, QUBE - Servo 2, and Rotary Servo Base Unit inverted pendulum setups, Edukit's (Edukit, 2021) Rotary Inverted Pendulum Control System offers the capability to create both conventional control systems and state-space modern control systems and Energized lab (EnergizedLab,2021) developed Prayog-I Rotary Inverted Pendulum experiment setup controlled by double PID (Proportional Derivative Integral) controller to teach the concepts of the control system. Muskinja (2006) proposed a new fuzzy swinging algorithm on a cart inverted pendulum (CIP) in which the cart would move horizontally on the rail in order to balance the pendulum upright. Owing to the fixed length, however, the cart movement was constrained. In order to compensate for this, Katsuhisa Furuta (2002) and his colleagues invented the rotary inverted pendulum (RIP) in 1992 at Tokyo Institute of Technology where the cart was replaced by a rotary arm, which was named "Furuta pendulum". Chen, Yung-Feng (2014) proposed an adaptive controller to balance a rotary inverted pendulum with time-varying uncertainties. K Selcuk, Z Bingul, (2014) used real-time fuzzy logic control of inverted pendulum and conducted experiments to verify the robustness of the controller for disturbances. Akhtaruzzaman, Md (2010) covered two modern control techniques: Full-State Feedback (FSF) and Linear Quadratic Regulator (LQR). They concluded that LQR controller was faster and more robust than the FSF controller. Hua, Jingwen (2019) observed that the model-free backstepping (MFBS) control technique performed better than the LQR control for the RIP control problem. M. Mansouri and M. Teshnehlab (2011) proposed a hybrid control structure, using both sliding mode controllers and fuzzy logic controllers simultaneously, and found the controller stabilizes the system significantly faster than single sliding mode control addition, and the controller also maintained its robustness against uncertainties.

In this work, a low-cost RIP (Rotary Inverted Pendulum) setup was fabricated in-house and controlled using MATLAB's full-state feedback algorithm. At first, a mathematical model of the system is formulated using Lagrangian dynamics. The mathematical model of the system is linearized using Taylor's approximation in the upright position to apply the linear control theory. Subsequently, the modeling, simulation, and controller design are presented. The implementation of the controller on the physical system and the experimental results are discussed towards the end.

Figure 2. Line diagram of the moving parts

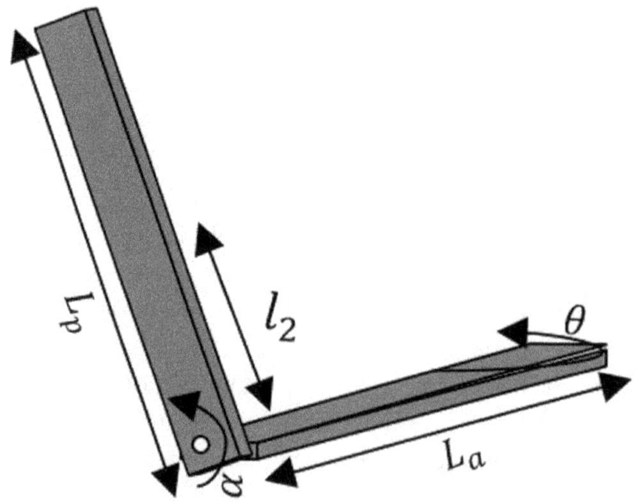

Mathematical Model

Different people have adopted different approaches and methods to solve these problems. First, a mathematical model of the problem is formulated, which is done in different ways. Nevertheless, two methods have been mainly used: The Newtonian and the Langrangian. Although the Newtonian approach is simple, it becomes very complex for multi-link body dynamics problems because it usually deals with vectors. In contrast, as the Langrangian approach contains more rigorous mathematics, it proves a compelling method for deriving a mathematical model of complex systems. To derive the equations of motion of the system, the Langrangian approach is used. These equations are used to derive the system's transfer functions and state-space equations.

The model was built using the assumptions and simplifications mentioned below:

1. Friction is not taken into account.
2. Backlash effects are assumed to be negligible.
3. The base of the setup is static.

The dynamic model of the inverted rotary pendulum was derived by using the Lagrangian, based on energy expressions with respect to variable independent generalized coordinates (Öksüz, 2018). The velocity of the pendulum lumped mass has two components shown Figure 3.

$$\vartheta_x = L_a\dot{\theta} - l_2\dot{\alpha}\cos\alpha$$

$$\vartheta_y = -l_2\dot{\alpha}\sin\alpha$$

Figure 3. Velocity analysis of pendulum link

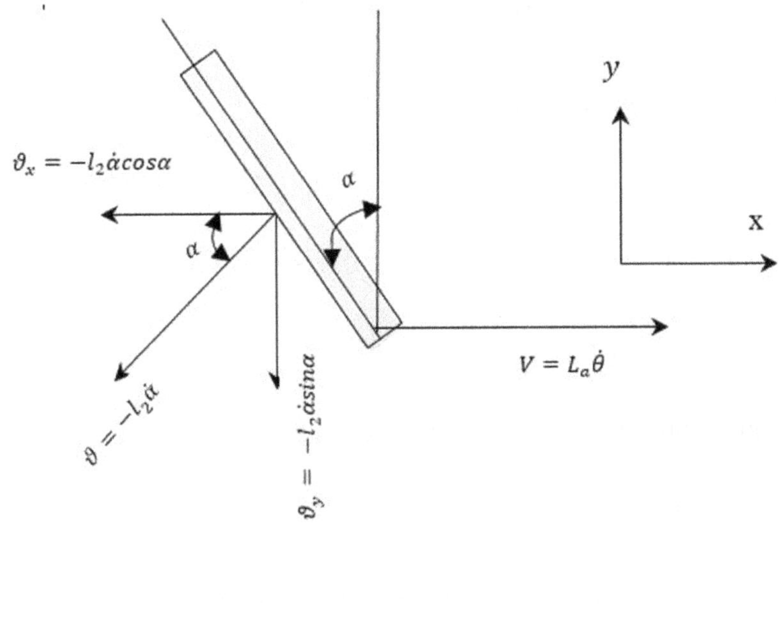

Now the total kinetic energy of the system is,

$$\mathrm{T} = \frac{1}{2}\,\mathrm{J}_{\mathrm{eq}}\dot{\theta}^2 + \frac{1}{2}\,\mathrm{J}_{\mathrm{p}}\dot{\alpha}^2 + \frac{1}{2}m\left(\vartheta_x^2 + \vartheta_y^2\right)$$

$$\mathrm{T} = \frac{1}{2}\left(J_{eq} + ml_2^2\right)\dot{\theta}^2 + \frac{2}{3}ml_2^2\dot{\alpha}^2 - ml_2L_a\cos\alpha\dot{\alpha}\dot{\theta}$$

Potential energy in the system due to gravity.

$$\mathrm{V} = mgl_2\cos\alpha$$

where T is the total kinetic energy and V is the total potential energy of the system.

$$L = \frac{1}{2}\left(J_{eq} + mL_a^2\right)\dot{\theta}^2 + \frac{2}{3}ml_2^2\dot{\alpha}^2 - ml_2L_a\cos\alpha\dot{\alpha}\theta - mgl_2\cos\alpha$$

Now according to Langrangian equation for the arm link

$$\frac{\partial}{\partial \mathrm{t}}\left(\frac{\partial \mathrm{L}}{\partial \dot{\theta}}\right) - \frac{\partial \mathrm{L}}{\partial \theta} = \tau - \mathrm{C}_{\mathrm{a}}\dot{\theta}$$

$$\left(J_{eq} + mL_a^2\right)\ddot{\theta} - ml_2L_a\cos\alpha\ddot{\alpha} + ml_2L_a\sin\alpha\dot{\alpha}^2 + C_a\dot{\theta} = \tau$$

Langrangian equation for the pendulum link

$$\frac{\partial}{\partial \mathrm{t}}\left(\frac{\partial L}{\partial \dot{\alpha}}\right) - \frac{\partial \mathrm{L}}{\partial \alpha} = 0$$

$$\frac{4}{3}ml_2^2\ddot{\alpha} - ml_2L_a\cos\alpha\ddot{\theta} - mgl_2\sin\alpha = 0$$

The above equations can be written as,

$$\left(z_1\right)\ddot{\theta} - z_2\cos\alpha\ddot{\alpha} + z_2\sin\alpha\dot{\alpha}^2 + z_5\dot{\theta} = \tau \qquad (1)$$

τ describes the applied torque produced by the motor at the base of the rotary arm.

$$\tau = \frac{K_t V_a}{R_a} - \frac{K_t K_v \dot{\theta}}{R_a}$$

where K_v is the motor back-emf constant, K_t is the torque constant, V_a is the input voltage to the motor and R_a is the armature resistance.

Where,

$$z_1 = J_{eq} + mL_a^2$$

$$z_2 = ml_2L_a$$

$$z_3 = \frac{4}{3}ml_2^2$$

$$z_4 = mgl_2$$

$$z_5 = C_a$$

$$z_3\ddot{\alpha} - z_2\cos\alpha\ddot{\theta} - z_4\sin\alpha = 0 \tag{2}$$

To simulate the system and tune the controller, the following non-linear equations are used to create the Simulink model. The model of the rotary inverted pendulum is shown below in Figure 4. To obtain the state space representation, the four states are chosen.

Figure 4. Dynamic model of rotary inverted pendulum

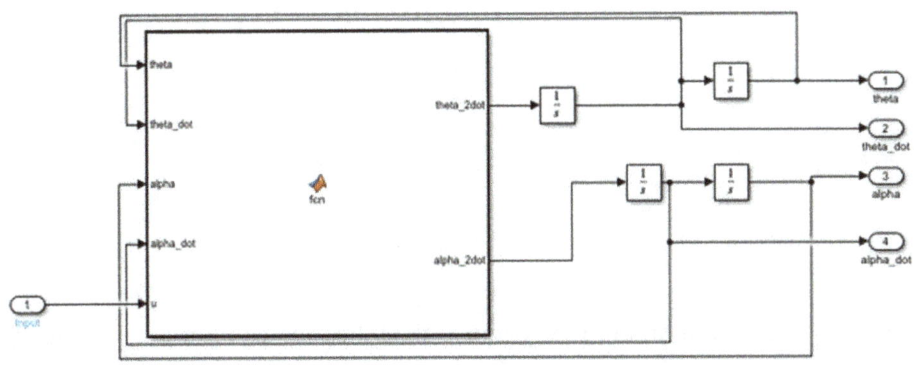

$$x_1 = \theta$$

$$x_2 = \dot{\theta}$$

$$x_3 = \alpha$$

$$x_4 = \dot{\alpha}$$

The corresponding state space representation:

$$\dot{x} = Ax + Bu$$

Where x and u are state and control vector respectively.

To design an LQR controller for the system, linearize the system at the following equilibrium point in using Taylor's approximation and find its state-space representation.

$$\dot{x} = \mathrm{f}\left(x, u\right)$$

where x is the state vector of the system, u is the vector of inputs, and f (·, ·) is a nonlinear vector function that represents the dynamics of the system.

$$\ddot{\theta} = f_1\left(\theta, \dot{\theta}, \alpha, \dot{\alpha}, u\right)$$

$$\ddot{\alpha} = f_2\left(\theta, \dot{\theta}, \alpha, \dot{\alpha}, u\right)$$

$$\dot{x}_1 = x_2$$

$$\dot{x}_2 = f_1\left(\theta, \dot{\theta}, \alpha, \dot{\alpha}, u\right)$$

$$\dot{x}_3 = x_4$$

$$\dot{x}_4 = f_2\left(\theta, \dot{\theta}, \alpha, \dot{\alpha}, u\right)$$

Linearize the dynamics of the system at the following equilibrium point using Taylor's approximation:

$$x_e = \begin{bmatrix} 0 \\ 0 \\ 0 \\ 0 \end{bmatrix}, u_e = 0$$

$$A = \begin{bmatrix} 0 & 1 & 0 & 0 \\ \frac{\partial f_1}{\partial x_1} & \frac{\partial f_1}{\partial x_2} & \frac{\partial f_1}{\partial x_3} & \frac{\partial f_1}{\partial x_4} \\ 0 & 0 & 0 & 1 \\ \frac{\partial f_2}{\partial x_1} & \frac{\partial f_2}{\partial x_2} & \frac{\partial f_2}{\partial x_3} & \frac{\partial f_2}{\partial x_4} \end{bmatrix} B = \begin{bmatrix} 0 \\ \frac{\partial f_1}{\partial u} \\ 0 \\ \frac{\partial f_2}{\partial u} \end{bmatrix}$$

Table 1. Parameters of the model

Symbol	Description of Parameter	Value
L_a	Length of arm link	0.145m
L_p	Length of pendulum link	0.195m
l_2	Length of pendulum centre of mass from pivot	0.0975m
m	Mass of pendulum link	0.1096 kg
C_a	Viscous coefficient of arm link	0.001 Nm.s
g	Gravity	9.81 m/s^2
J_{eq}	Moment of inertia of the arm link	0.0001333 kg/m^2

So, the linearized matrix becomes

$$A = \begin{bmatrix} 0 & 1 & 0 & 0 \\ 0 & -1.4 & 164.83 & 0 \\ 0 & 0 & 0 & 1 \\ 0 & -1.573 & 259.3 & 0 \end{bmatrix} B = \begin{bmatrix} 0 \\ 1409.7 \\ 0 \\ 1572.3 \end{bmatrix}$$

With the help of MATLAB (lqr) command, the LQR for the rotating inverted pendulum is designed according to the weight matrices. By selecting Q and R.

$$Q = \begin{bmatrix} 1 & 0 & 0 & 0 \\ 0 & 1 & 0 & 0 \\ 0 & 0 & 1 & 0 \\ 0 & 0 & 0 & 1 \end{bmatrix} \qquad R = [1]$$

Therefore, the optimal gain values for the system are

K= [-1.000 -1.2588 19.8554 2.4797]

The above gain matrix K is used to stabilize the Simulink simulation of the rotary inverted pendulum, shown in Figure 5. To verify the effectiveness of the system, further additional fine-tuning is needed for weight matrices Q and R. Since the system is linear, it is assumed that the pendulum link and arm link start at 0.1 rad and 0.2 rad, respectively, from their initial positions, i.e., the vertical upright unstable equilibrium position of the pendulum and initial position of the arm anywhere in the horizontal plane.

Figure 5. Simulink block diagram of the rotary inverted pendulum

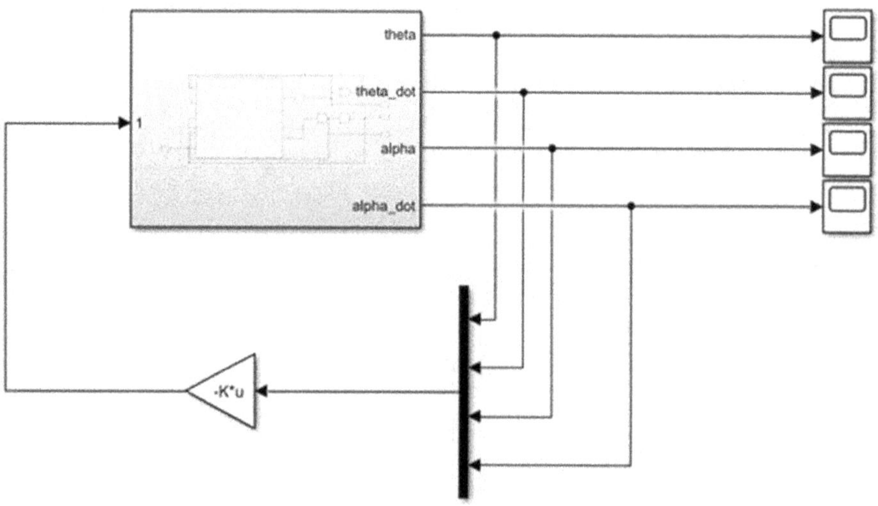

Figure 6. Angular position of the Arm vs. time for simulation

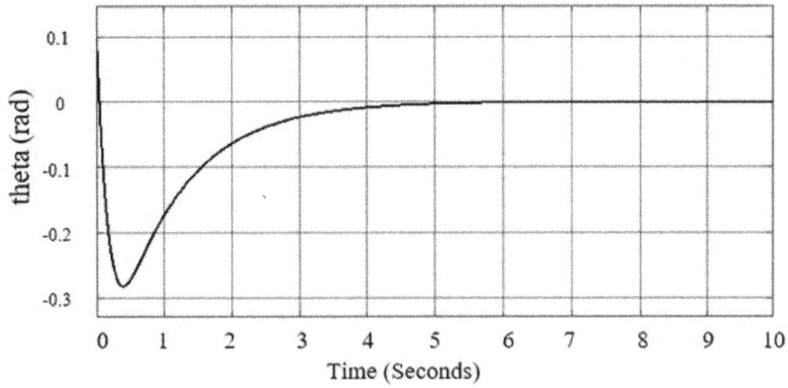

Figure 7. Angular position of the Pendulum vs. time for simulation

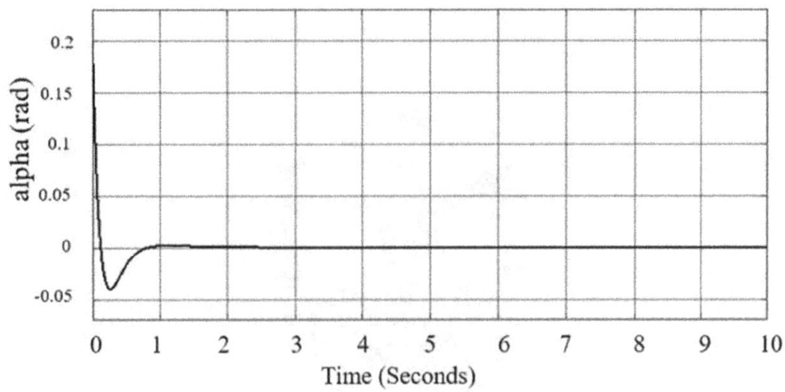

Figure 8. Angular velocity of the Pendulum vs. time for simulation

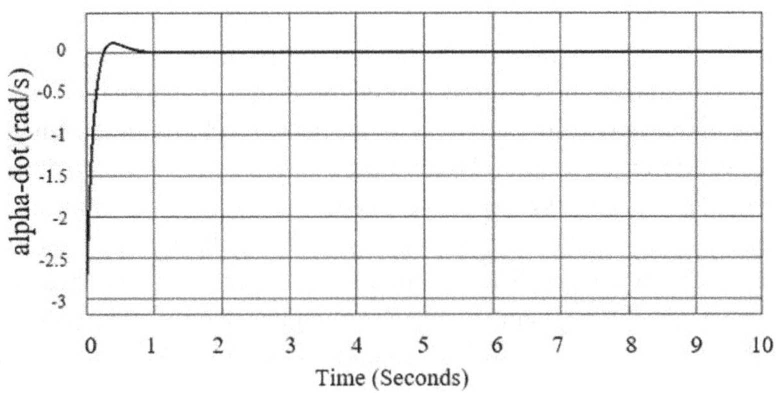

Figure 9. Angular velocity of the Arm vs. time for simulation

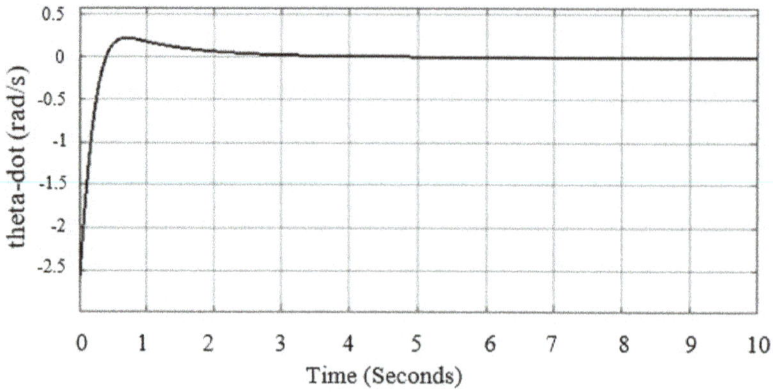

Figure 10. Final CAD model of the rotary inverted pendulum

EXPERIMENTAL RESULTS

In a real-time implementation, the Arduino board controller is utilized because it is an open-source electronics platform; a diverse set of software backs these boards. Simulink is employed to program the system. All the blocks for interacting with the Arduino came from the official Arduino Hardware Support Package. In addition, the Simulink model of the system was created to tune and test the controller before testing on the actual hardware to address any problems that could develop without danger of hardware damage. The controller was first designed and then simulated and discretized, and the obtained gains were implemented in the existing system with further fine-tuning. The optimal control law $u = -Kx$ is utilized, where the gain matrix, k, is composed of the following feedback gains derived from the optimal process controller design. The sensor utilized in this project is an incremental rotary encoder. According the rotational movement, it creates an electrical signal, either analog or digital. It is not powered by a motor and is simply a free-spinning axle. The sensor has a significantly greater precise accuracy due to its 2400 counts per rotation. The motor utilized has a 12 V DC geared motor with torque of 4.5 kg-cm featuring an additional quadrature type rotary encoder providing a resolution of 1560 CPR (Counts per revolution). A rotary incremental encoder is coupled to the pendulum link at the end of the arm link, and a high torque DC motor with an encoder is connected at the arm's pivot point. The data from the two encoders give the 4-state variables θ, $\dot{\theta}$, α, and $\dot{\alpha}$ are obtained. However, since the velocities are not obtained directly from the sensor, there is a noise problem in two state variables while passing through the discrete derivative. Therefore, angular velocities are passed to the filter to suppress the noise, as shown in Figure 11.

Figure 11. Filtering of the state variables

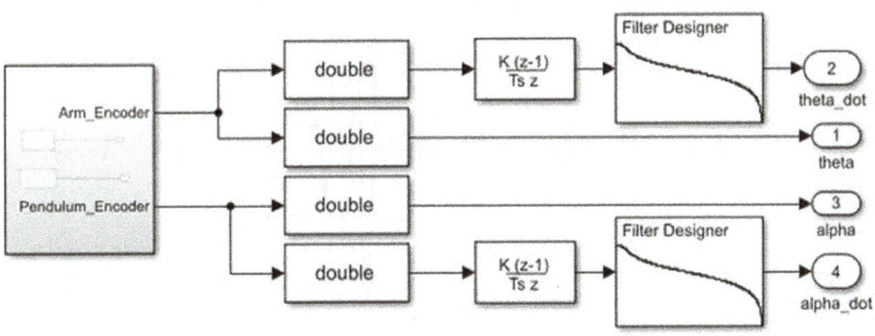

Figure 12. LQR controller block diagram for the experimental setup

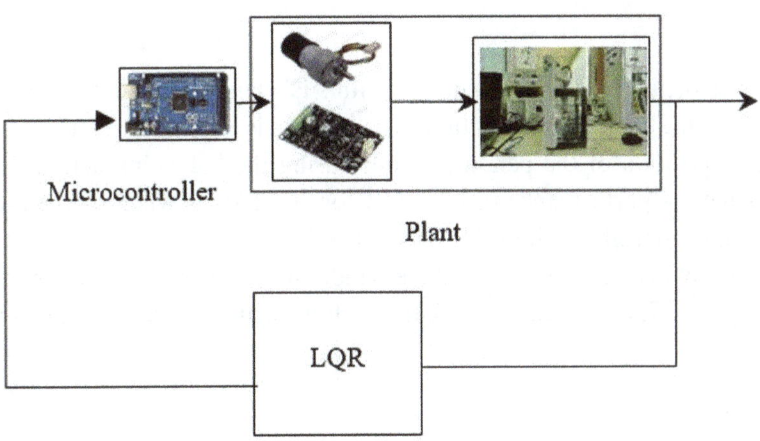

The following data is obtained during stabilization and illustrate the good performance of the LQR controller.

Figure 13. Angular position of the arm in real-time

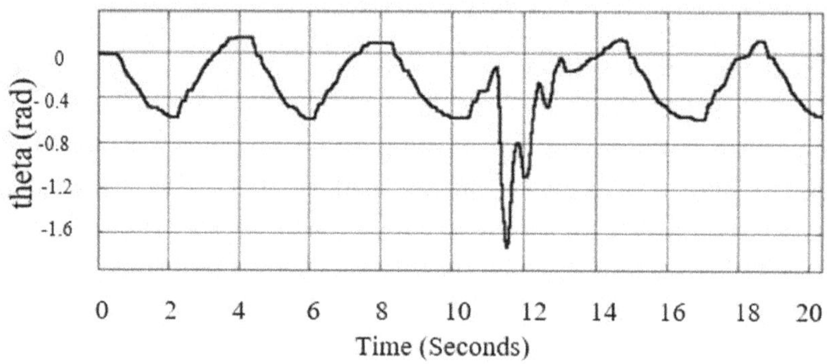

Figure 14. Angular position of the pendulum in real-time

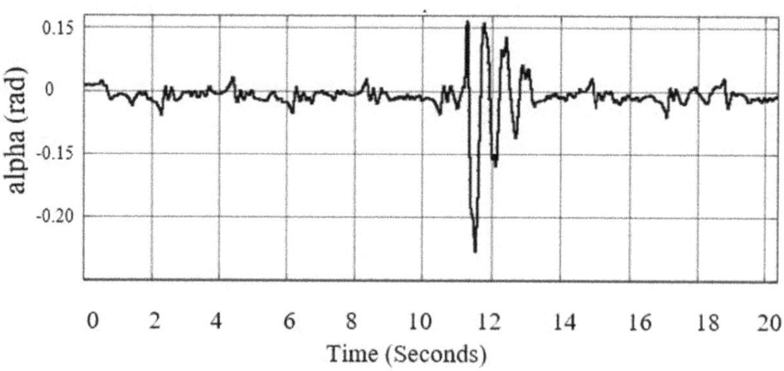

The trial begins when the pendulum link is manually placed in its upright position unstable equilibrium position as the system is linear. The Figure 14 depicts the angular position of the pendulum link during the balancing action. It shows the experimental performance of the LQR for 20 seconds when the pendulum is stabilized at the upright position. To check its robustness, an external disturbance is provided following 11th second, and the pendulum recovered from a range of less than 20 degrees. The pendulum oscillates slightly (within a small range) around the upright position, the balancing controller has negligible overshoot, and the steady-state error is as expected, it is linked to backlash present in the DC geared motor, couplings, experiment base vibrations, and so on

Figure 15. Completed pendulum Assembly (Controller switched on)

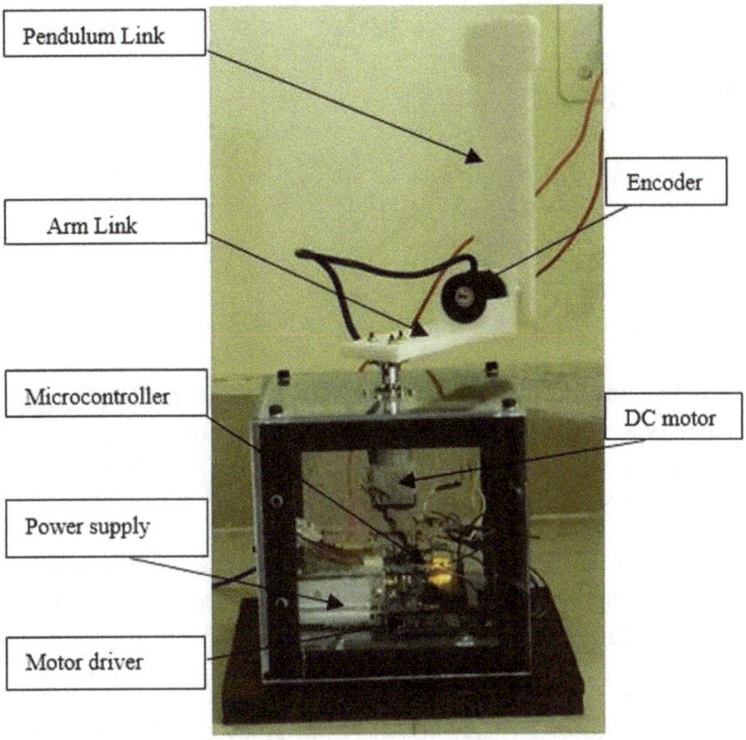

CONCLUSION

The work intends to develop a cheap control system rotary inverted pendulum. It would be helpful for undergraduate/graduate students majoring in Mechanical, Electrical Engineering, or Mechatronics to give a practical experience in experimentation setup for control courses. The novelty of the STRIPS 1.0 system is the open-source approach, which accomplishes repeatability, robustness, and low cost. The design of the STRIPS 1.0 gives access to an individual to customize or replicate it and assemble their copies of the setup. Such a modularized basic control setup benefits students in better understanding the control theory. The project aims to make it accessible to students in the future to grasp the fundamentals and advanced aspects of control systems. Also, the overall cost of the complete design, development, and implementation of the setup was pushed under 150 USD, indicating an excellent chance for developing a control test platform for educational purposes.

In terms of robustness, the system recuperates to equilibrium quickly and with minimal overshoot. In fact, despite having a lot of disturbance applied to it, the pendulum never fell during the trial. As a result, the controller significantly outperforms the expectations.

ACKNOWLEDGMENT

This work was supported by the Design Innovation Centre (DIC) of the Islamic University of Science and Technology, J&K.

REFERENCES

Ahangar-Asr, H., Mansouri, M., Teshnehlab, M., & Pazoki, A. R. (2011, September). A hybrid strategy for the control of Rotary Inverted Pendulum. In *2011 International Conference on Electrical and Control Engineering* (pp. 5656-5659). 10.1109/ICECENG.2011.6057971

Akhtaruzzaman, M., & Shafie, A. A. (2010, August*). Modeling and control of a rotary inverted pendulum using various methods, comparative assessment and result analysis*. In *2010 IEEE International Conference on Mechatronics and Automation,* (pp. 1342-1347). IEEE. 10.1109/ICMA.2010.5589450

Al-Mahturi, A., Santoso, F., Garratt, M. A., & Anavatti, S. G. (2019, June). An intelligent control of an inverted pendulum based on an adaptive interval type-2 fuzzy inference system. In *2019 IEEE International Conference on Fuzzy Systems (FUZZ-IEEE)* (pp. 1-6). IEEE. 10.1109/FUZZ-IEEE.2019.8858948

Anvar, S. M. M., Hassanzadeh, I., & Alizadeh, G. (2010, October). Design and implementation of sliding mode-state feedback control for stabilization of Rotary Inverted Pendulum. In *ICCAS 2010* (pp. 1952–1957). IEEE. doi:10.1109/ICCAS.2010.5670108

Chen, Y. F., & Huang, A. C. (2014). Adaptive control of rotary inverted pendulum system with time-varying uncertainties. *Nonlinear Dynamics, 76*(1), 95–102. doi:10.100711071-013-1112-4

Edukit. (2021). *Rotary Inverted Pendulum* https://sites.google.com/view/ucla-st-motor-control/home

EnergizedLab. (2021). *Rotary Inverted Pendulum* https://energizedlab.com/

Fantoni, I., Lozano, R., & Lozano, R. (2002). *Non-linear control for underactuated mechanical systems*. Springer Science & Business Media. doi:10.1007/978-1-4471-0177-2

Huang, J., Zhang, T., Fan, Y., & Sun, J. Q. (2019). Control of rotary inverted pendulum using model-free backstepping technique. *IEEE Access: Practical Innovations, Open Solutions*, *7*, 96965–96973. doi:10.1109/ACCESS.2019.2930220

Kim, S., & Kwon, S. (2017). Nonlinear optimal control design for underactuated two-wheeled inverted pendulum mobile platform. *IEEE/ASME Transactions on Mechatronics*, *22*(6), 2803–2808. doi:10.1109/TMECH.2017.2767085

Kizir, S., Bingul, Z., & Oysu, C. (2010). *Fuzzy control of a real time inverted pendulum system. Journal of Intelligent & Fuzzy Systems*, *21*(1, 2), 121-133.

Krishen, J., & Becerra, V. M. (2006, October). *Efficient fuzzy control of a rotary inverted pendulum based on LQR mapping*. In IEEE Conference on Computer Aided Control System Design, IEEE International Conference on Control Applications, IEEE International Symposium on Intelligent Control, (pp. 2701-2706). IEEE. 10.1109/ISIC.2006.285548

Kuo, T. C., Huang, Y. J., & Hong, B. W. (2009, July). Adaptive PID with sliding mode control for the rotary inverted pendulum system. In *2009 IEEE/ASME International Conference on Advanced Intelligent Mechatronics* (pp. 1804-1809). IEEE. 10.1109/AIM.2009.5229784

Kwon, T., & Hodgins, J. K. (2010, July). *Control Systems for Human Running using an Inverted Pendulum Model and a Reference Motion Capture Sequence*. In *Symposium on Computer Animation,* (pp. 129-138).

Li, W., Ding, H., & Cheng, K. (2012, September). *An investigation on the design and performance assessment of double-PID and LQR controllers for the inverted pendulum*. In *Proceedings of 2012 UKACC international conference on control,* (pp. 190-196). IEEE. 10.1109/CONTROL.2012.6334628

Loram, I. D., & Lakie, M. (2002). Human balancing of an inverted pendulum: Position control by small, ballistic-like, throw and catch movements. *The Journal of Physiology*, *540*(3), 1111–1124. doi:10.1113/jphysiol.2001.013077 PMID:11986396

Muskinja, N., & Tovornik, B. (2006). Swinging up and stabilization of a real inverted pendulum. [IEEE.]. *IEEE Transactions on Industrial Electronics*, *53*(2), 631–639. doi:10.1109/TIE.2006.870667

Nise, N. S. (2020). *Control systems engineering*. John Wiley & Sons.

Ogata, K. (2010). *Modern control engineering* (Vol. 5). Prentice hall.

Öksüz, M., Önal, M. B., Halicioğlu, R., & Dülger, L. C. (2018). Alternative controller design for rotary inverted pendulum. *Tehnički glasnik, 12*(3), 139-145.

Park, M. S., & Chwa, D. (2009). Swing-up and stabilization control of inverted-pendulum systems via coupled sliding-mode control method. *IEEE Transactions on Industrial Electronics*, *56*(9), 3541–3555. doi:10.1109/TIE.2009.2012452

Prasad, L. B., Tyagi, B., & Gupta, H. O. (2012, May). *Modelling and simulation for optimal control of nonlinear inverted pendulum dynamical system using PID controller and LQR*. In *2012 Sixth Asia Modelling Symposium,* (pp. 138-143). IEEE. 10.1109/AMS.2012.21

Quanser. (2021). *Rotary Inverted Pendulum* https://www.quanser.com/products/rotary-inverted-pendulum/

Roman, M., Bobasu, E., & Sendrescu, D. (2008, May). *Modelling of the rotary inverted pendulum system.* In 2008 IEEE International Conference on Automation, Quality and Testing [). IEEE.]. *Robotics*, *2*, 141–146.

Saifan Rafiq. (2017). *Design, Fabrication and Control of 2-DOF Rotary Inverted Pendulu*m [Bachelors' dissertation]. National Institute of Technology, Srinagar, Kashmir, India.

Solihin, M. I., & Akmeliawati, R. (2010). Particle swam optimization for stabilizing controller of a self-erecting linear inverted pendulum. *International Journal of Electrical and Electronic Systems Research*, *3*, 410–415.

Tang, T. F., Chong, S. H., & Pang, K. K. (2020). *Stabilisation of a rotary inverted pendulum system with double-PID and LQR control:* Experimental verification. *International Journal of Automation and Control*, *14*(1), 18–33. doi:10.1504/IJAAC.2020.103799

Yue, M., An, C., Du, Y., & Sun, J. (2016). Indirect adaptive fuzzy control for a nonholonomic/underactuated wheeled inverted pendulum vehicle based on a data-driven trajectory planner. *Fuzzy Sets and Systems*, *290*, 158–177. doi:10.1016/j.fss.2015.08.013

Chapter 12
Hydroponics, Aeroponics, and Aquaponics Technologies in Modern Agricultural Cultivation

Praveen Kumar S.
Muthayammal Engineering College, India

Boopathi Sampath
https://orcid.org/0000-0002-2065-6539
Muthayammal Engineering College, India

Sanjeev Kumar R.
Swarnandhra College of Engineering and Technology, India

Bachina Harish Babu
VNR Vignana Jyothi Institute of Engineering and Technology, India

Ahalya N.
MS Ramaiah Institute of Technology, India

ABSTRACT

In the modern agriculture cultivation field, hydroponics, aeroponics, and aquaponics technologies are currently playing an important role in growing vegetables, fruits, and crops without using soil. These have been used for high productivity, minimum usage of water resources, and compatibility of farming in every place. In this chapter, the authors cover the implementation procedures of hydroponics, aeroponics, and aquaponics technologies in the real world; various inert and artificial mediums, possible types of plants; cultivating methods; technological advancements and

DOI: 10.4018/978-1-6684-5887-7.ch012

application of the internet of things (IOT); and pros and cons for implementations. The various cultivation methods are compared to conventional methods (soil-based cultivation) based on their technology advancements, environmental impacts, yields, and costs.

INTRODUCTION

Agriculture has been a sacred occupation since the dawn of time. Agriculture has been used to feed the living. In agriculture, "producing with seed" (soil can be plowed, seed sown, nutrients can be applied to crops and harvested) is an art. Since the beginning of the world, agriculture has depended on the soil. Crops are easier to grow as the nutrients are high in the soil.

"Plow life will live"

That is what Tiruvallur has mentioned in Thirukkural. Those who ploughed and fed the living were considered equal to the deity.

Nowadays, the population of this world is increasing rapidly. As a result, the need for food is critical. The government has faced the challenge of producing food in large quantities with a lack of suitable land and sufficient water without stressing the environment. Traditional farming methods cannot produce large quantities in the same way that artificial fertilizers are used on the land. The land does not grow enough healthy food. So, a new farming technique is introduced, namely "soil-less farming." With this technique, there is no need for soil. This is an eco-friendly method. The basic needs of growing plants are soil, air, water, and sunlight.

The farming methods are divided into two types, namely (Lakhiar et al. 2020):

- **Open system:** In an open system, the supply of nutrient solutions cannot be recycled.
- **Closed system:** In a closed system, the excess nutrient solution can be recycled again and again, providing nutrients to the plants.

CONVENTIONAL FARMING METHOD

Conventional farming is the method of growing plants in soil. The crops are in an open-air medium, so irrigation is needed, as is the use of artificial fertilizers. The conventional method utilizes more water and has large land requirements, soil degradation, and a high concentration of nutrients (Sathyanarayana et al. 2022). Higher productivity is needed in agriculture, due to the rapid growth of the world population.

SOILLESS FARMING METHOD

In a soilless farming culture, there is no need for soil for the growth of plants. It can grow with the help of substrates, nutrient solutions, air, and water. The farm provides essential micronutrients and macronutrients for the development and growth of plants. The new modern technique in the agriculture sector is that water can be used efficiently. It produces a high yield in a short period. The advanced technologies in agriculture are hydroponics, aeroponics, and aquaponics. Growing in a hydroponics system consumes 10-20 times less water compared to the traditional method of farming. The yield of soilless farming in the hydroponics system is increased by 5% compared to the open system. The soilless culture has come into existence since ancient times. NASA has recently developed a controlled ecological life support system (Lakhiar et al. 2020).

IMPORTANT PARAMETERS IN SOILLESS FARMING

Nutrient Solution

Soilless farming uses nutrient solutions and substrates. The solutions of an element are made up of elements like nitrogen, boron, zinc, copper, molybdenum, manganese, phosphorus, iron, calcium, and magnesium (Pickering et al. 2002). But the required quantities differed for each plant. The solution helps the plant's growth and development.

pH Level

The pH level of the nutrient solution is very essential for the growth of the plant. The range of the nutrient solution in soilless farming is between 5.8 and 6.5[5]. The monitoring and controlling of the pH are important. If it is not regulated, it creates an effect on the plants.

Electrical Conductivity

The electrical conductivity of the soilless culture is significant. The electrical conductivity (Ec) is measured in dS/m and represents the strength of the nutrient solution. In hydroponic systems, the range of electrical conductivity is between 1.5 and 2.5 dS/m(Rouphael and Colla 2005). The imbalance of the Ec creates a barrier to the plant's growth.

MODERN CULTIVATION METHODS

Hydroponics System

The term hydroponics is derived from the Latin word hydro, meaning "water". Ponics means "labor". It is followed by the Italians. In this type of farming, water can be used as the main resource for the growth of plants.

Wick System

In this system, the plants are placed in pots. Each pot has a small hole at the bottom [7]. The pots are filled with perlite, vermiculite, and sterile mediums like gravel, sand, and coconut coir with a wick of cotton or nylon in a growing tray. The wicks absorb the nutrient solution from the reservoir and deliver it to the roots in a pot through the capillary method. The oxygen is created by the air stone. The wick system is suitable for growing small plants, herbs, and other species. It is the simplest method in a

Figure 1. Schematic Arrangements of Wick System

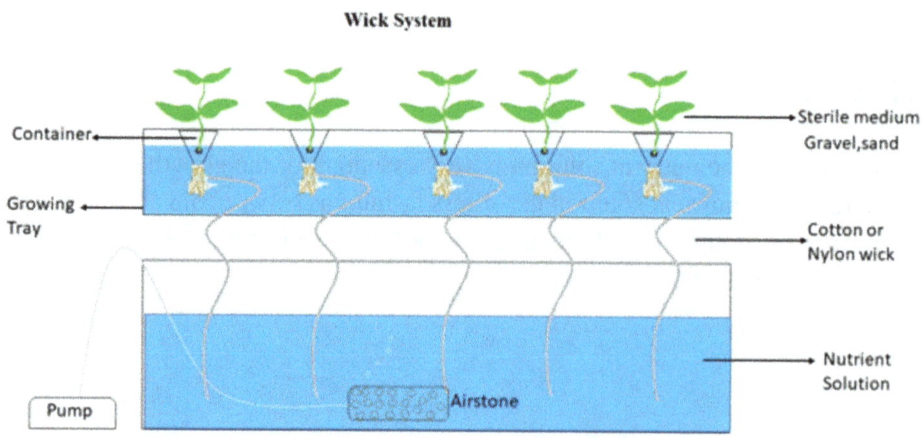

hydroponic system. The lettuces are fast growing, as are herbs in the wick system. These include herbs such as rosemary.

Ebb and Flow System

In this ebb and flow system, vine crops are suitable. The growing medium is rocks, gravel, or granular Rockwool. It works on the principle of flood and drains. Nutrient solution and water from a reservoir are flooded through a water pump to the grow bed until it reaches a certain level and stays there for a certain period so that it provides nutrients and moisture to plants. Besides, it is possible to grow different kinds of crops, but the problems of root rot, algae, and mold are very common (Nielsen et al. 2006)

Figure 2. Schematic Arrangements of Ebb and Flow System

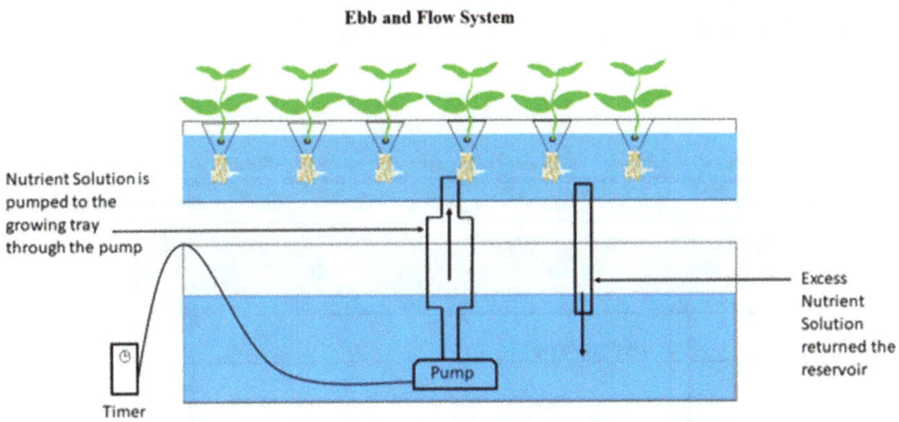

The Growing Plants in Ebb and Flow Systems are:

- Strawberries
- Tomatoes
- Cucumbers
- Peppers
- Lettuce
- Spinach
- Radishes
- Beans
- Basil

- Oregano
- Mint
- Chinese money plant
- Peace lilies

Drip System

In the drip system in the hydroponics method, the drip lines are used for growing the plants. It is widely used for both home and commercial growers. The plants are placed in a growing tray with a suitable growing medium. The nutrient solution can be stored in the reservoir. The drip lines are attached to the trays. The nutrient solution is supplied to the base of each plant root through the drip lines with the help of a pump (Rouphael and Colla 2005). The overflow of the solution is stored in the reservoir. The pump is connected to the timer. It controls the entire supply system.

Figure 3. Schematic Arrangements of Drip System

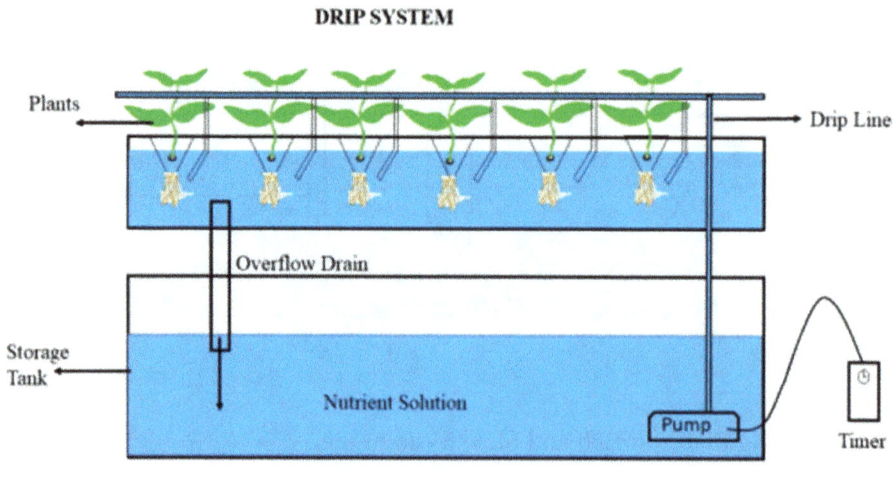

The growing plants in the Drip system are:

- Lettuce
- Leeks
- Onions
- Melons
- Peas
- Tomatoes

- Radishes
- Cucumbers
- Strawberries
- Zucchini
- Pumpkins

4.Deep Water Culture

In this deep-water culture method, the plant roots are immersed in the nutrient solution for growing the plants. Styrofoam or net pots are used to make the platform that holds the plants. The roots of plants are suspended in nutrient-enriched solutions or water where they grow quickly into a large mass. The air is provided directly to the roots by an air stone. The hydroponics bucket system is a classic example of this system. It is mandatory to monitor the oxygen and nutrient concentrations, salinity, and pH values (Domingues et al. 2012). This system works, especially for cucumbers and tomatoes. The Styrofoam platform is suitable for leafy crops like lettuce.

Figure 4. Schematic Arrangements of Deep-Water Culture

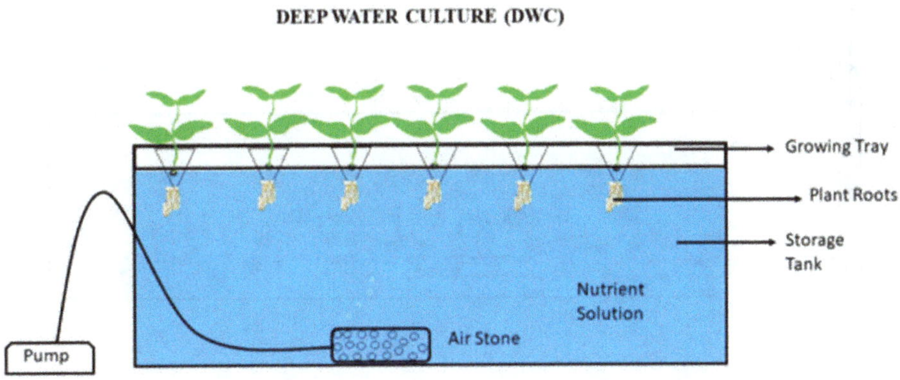

The plants to grow in a deep-water culture system are:

- Lettuce.
- Spinach.
- Strawberries.
- Bell Peppers.
- Herbs

5.Nutrient Flim Technique (NFT)

The Nutrient Flim Technique is a method of growing plants in water with a tapered setup. It is similar to the Ebb and Flow method (Domingues et al. 2012). The plants are placed in a growing tray. The nutrient solution is supplied from the reservoir through the pipe with the help of the pump. The excess nutrient solution is stored in the reservoir through the pipe that is placed at the bottom of the tray. The circulation of the solution happens regularly at a time. It is controlled by the timer. This technique is suitable for achieving high yields. The suitable growing media for the Nutrient Flim Technique are coconut husks (Coco coir), expanded clay pellets, perlite, growl tones, and gravel.

Figure 5. Schematic Arrangements of Nutrient Film Technique

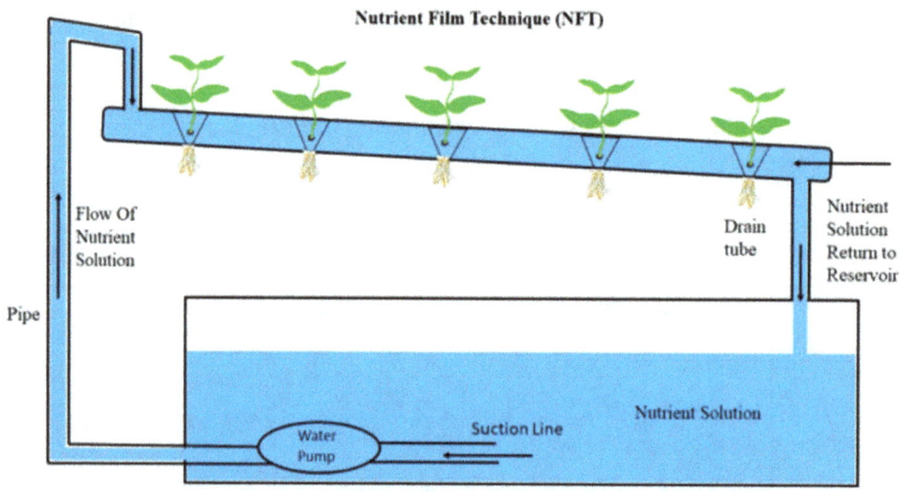

The Growing Lettuces

- Romaine Lettuce
- Red Lettuce
- Bok Choy
- Kale
- Broccoli
- Raab

The Growing Herbs

- Basil
- Thyme
- Cilantro
- Parsley
- Sage
- Thyme
- Oregano
- Mint

The Growing Fruits And Vegetables

- Mini Peppers
- Beefsteak Tomatoes
- Cherry Tomatoes
- Chili Peppers
- Mini Cucumbers
- Zucchini
- Strawberries
- Green Beans

Figure 6. Schematic Arrangements of Nutrient Cycle in Aquaponics System

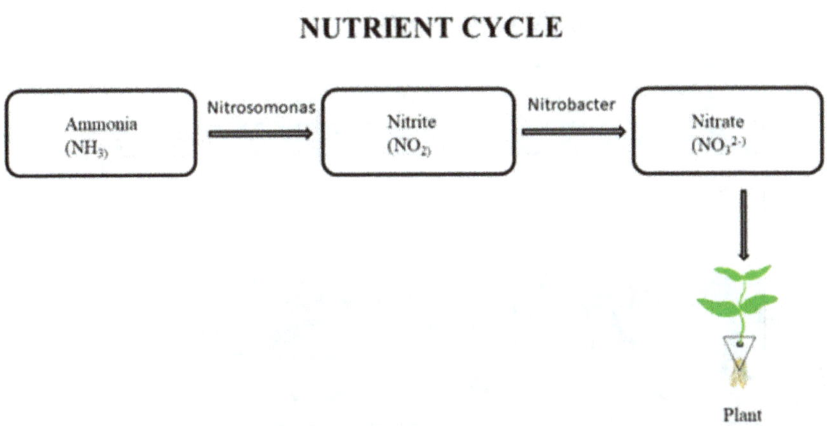

AQUAPONICS SYSTEM

The term "aquaponics" is a combination of aquaculture and hydroponics. Aquaculture is the "controlled raising of fish, shrimp, algae, and other seafood," while Ponics is the "growing in a soilless medium." Aquaponics is a method of growing plants in the water while also providing a habitat for fish [(Verdonck et al. 1984). "Aqua" means water and "Ponics" means work or growing. It is a type of closed cycle.

In the aquaponics system, the plants are growing in the nitrogen cycle. It is the beginning step of this system. The fish can start to eat the food. The ammonia will be released from the respiratory system. Ammonia is a waste product of it. It accumulates in high concentrations; it becomes toxic to the fish. On the wet surface of the aquaponics system, nitrifying bacteria live. It is known as Nitrosomonas. These are naturally occurring and extremely beneficial because they convert ammonia into nitrite. Unfortunately, nitrite is even more toxic to fish than ammonia. Luckily, there is another class of bacteria called Nitrobacter, which converts the nitrites to nitrates. And this is fantastic because nitrates are a great nutrient source for plants. So, the nitrogen cycle is an extremely important process that converts toxic waste products into less toxic nutrients that plants can use. Now that the nutrient solution is supplied to the plants, they absorb the enriched nutrients through the roots. In that way, the water is also cleaned. So, the water can be reused. This whole farming process happens in a closed water cycle (Calone and Orsini 2022).

Figure 7. Schematic Arrangements of Aquaponics System

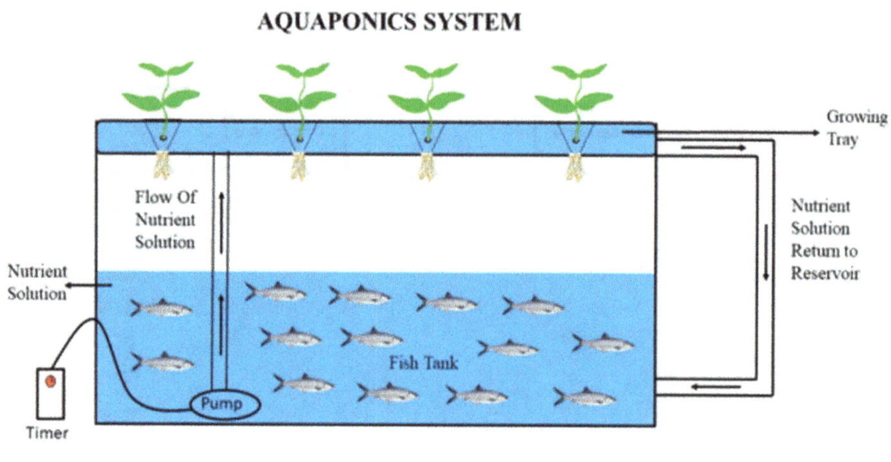

The Small System Plants grown in Aquaponics systems are:

- Kale
- Spinach
- Swiss Chard
- Arugula
- Chives
- Pak Choi
- Wheatgrass
- Radish Sprouts
- Basil
- Watercress
- Leafy Lettuce

The Large System Plants grown in Aquaponics systems are:

- Peppers
- Cucumbers
- Cauliflower
- Tomatoes
- Cabbage
- Peas
- Broccoli
- Squash
- Beans

AEROPONICS SYSTEM

Aeroponics is a new farming method of growing plants in a semi-closed system (Sathyanarayana et al. 2022). This farming method, is done by using air and a limited amount of water in a controlled environment. This technique is economical in the use of fertilizers and it saves water and nutrients compared to other soilless systems. The aeroponic system consists of the pump, temperature sensor, humidity sensor, light intensity sensor, timer, mist nozzles, reservoir, nutrient solutions, and growing tray. The growing tray helps to hold the net cups which are attached to the container. The nutrient solution is sprayed onto the roots through the nozzles with the help of a submerged pump in the delivery lines. The timer controls the flow of the nutrient solutions. The temperature sensor is used to measure the temperature inside the system. The humidity sensor is used to measure the humidity level inside the environment.

Figure 8. Schematic Arrangements of Aeroponics System

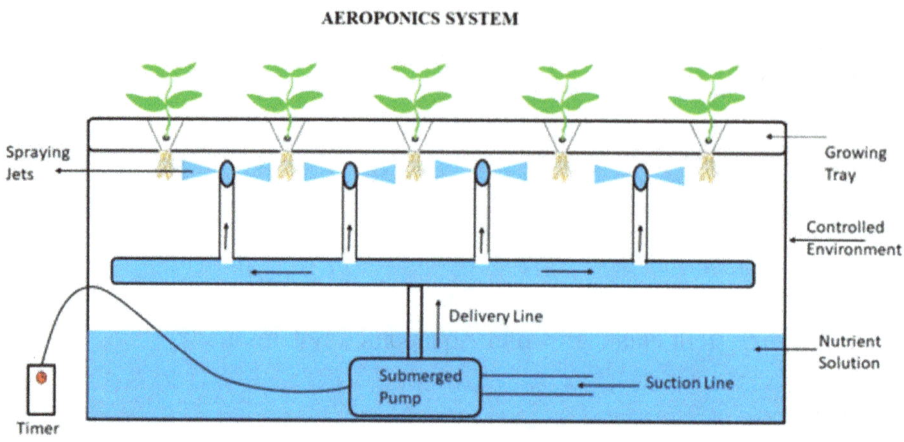

The growing herbs grown in the Aeroponics system are:

- Chives
- Oregano
- Basil
- Sage
- Rosemary

The leafy green vegetables grown in the Aeroponics system are:

- Romaine Lettuce
- Butterhead Lettuce
- Red Leaf Lettuce
- Tuscan Kale

The Fruits and Vegetables grown in the Aeroponics system are:

- Beets
- Broccoli
- Cabbage

- Carrots
- Cauliflower
- Corn
- Cucumber
- Eggplant
- Grapes
- Melons
- Onions
- Peas
- Peppers
- Potatoes
- Radish
- Raspberry
- Strawberry
- Sweet Potato
- Tomatoes
- Watermelon

SUBSTRATES

The substrates are mainly used for soilless farming. It provides anchorage for the roots, helping the plant retain plenty of moisture and ensure proper aeration. The substrates are composed of a mixture of organic, inorganic, and synthetic origins. The organic origin can be derived from nature (e.g., coco coir, straw); the inorganic origin can be derived from human activities such as agricultural by-products and industrial and urban activities. The industrial origin is obtained from synthesis processes (e.g., polystyrene) (Rouphael and Colla 2005).

Peat

Peat is the most important material in organic farming for the preparation of substrates. It can be used alone or mixed with other substrates. The peat is derived from the bryophytes, cyperaceous, and other residues. The brown peat and blond peat have good structural stability, the availability of nutrients is very low, and the acidic pH, whilst they are mainly different from their structure.

Coconut Fibre

Coconut fiber is also called coco coir. The extraction is obtained from the fibrous husks of coconuts. The pH level is high and the environmental impact level is lower than peat. These can be used before some processes can be done on them, such as composting, dehydrating, compressing, and rehydrating (Fornes et al. 2003).

Sand

Sand is a natural inorganic material that forms from various weathering materials. The sand consists of 98.0-99.5% silica (Maucieri et al. 2019). The size of the particles of the sand ranges between 0.05- 2.0 mm. Fine sands are more suitable for hydroponic systems. The coarse sands are used to increase the drainage capacity of the substrate.

Volcanic Tuffs

The volcanic tuffs can be derived from volcanic eruptions with the size of the particles ranging between 2 and 10 mm in diameter. The bulk density ranges between 850 and 1100 kg m-3, and the water retention capacity ranges between 15% and 25% by volume (Plants 1988).

Vermiculite

Vermiculite includes hydrous phyllosilicates of magnesium, iron, and aluminum. The exfoliated vermiculite is used in the horticultural industry because it has a high buffer competence. It can be used alone. But it is preferred to mix it with perlite or peat. The vermiculite helps to aerate the soil, retaining water and nutrients. It is useful at the time of seed sowing and propagation.

Perlite

Perlite is most suitable for horticulture because it helps to retain water and create an airy compost for seedlings. It is mainly used in aerating compost. It is marketed in different sizes. The perlite contains aluminum silicates. It occurs in nature of volcanic origin and comprises 75% SiO_2 and 13% Al_2O_3 (Verdonck et al. 1984).

Expanded Clay

The expanded clay is made by heating clay powder to 700 degrees Celsius. It can be used in 10-35% volume mixtures with organic materials to provide supplementary ventilation and drainage (Rahmad et al. 2019). The pH value of the expanded clay is not suitable for soilless culture.

Stone Wool

The stone wool originates from the fusion of magnesium, silicate, aluminum, carbon coke, and calcium at a temperature range of between 1500° C and 2000° C(Abad et al. 2002). The chemically inert stone wool is added to the substrate. It is an excellent anchorage for the plant roots and it improves aeration. The stone wool can be used as a sowing substrate all by itself.

Zeolites

The zeolites include aluminum silicate. It can absorb gaseous elements. It has all the higher micro and macro elements. The substrates create more absorption and it slowly releases the K+ and NH4+ ions. It is toxic to plants because they cannot absorb Cl- and Na+.

Expanded Polystyrene

The expanded polystyrene improves the porosity and drainage of the plant. It is formed into granules with a diameter of 4–10 mm. The ability of porosity is very high, but the water retention capacity is extremely low.

Polyurethane Foam

Polyurethane foam is used for soilless cultivation. It has a low-density material with a porous structure that allows the level of water absorption to be equal to 70% of its volume(Espinal and Matulić 2019). It does not contain the useful nutrients that are available to the plants. The foam is available on the market in the form of granules or blocks, like stone wool.

Internet of Things (IoT)

The Internet of Things, or IoT, is defined as the process of connecting physical devices through the internet for sharing. collecting and transferring data. Anything that can be connected through the internet falls under IoT technology. So, the IoT unit is a device that collects information or data by using the sensors and actuators from its surroundings and sends this captured information to the internet for processing of that data. IoT devices use machine learning and artificial intelligence to automate processes and systems, including medical equipment, industrial manufacturing, agriculture, etc.

Figure 9. Schematic Arrangements of IoT

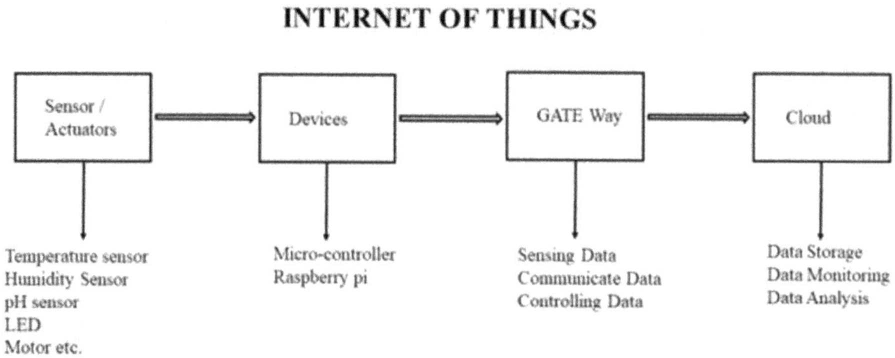

Application Of IoT in the Real World

- Agriculture and Pest Control
- Environment
- Smart Home Applications
- Health Care
- Smart Cities
- Safe Driving
- Waste Management
- Tackling Industrial Issues
- Supply Chain Management

Sensor

The sensors can be used to sense the data of the system and transfer it into the cloud with the help of the devices. Sensors such as temperature sensors, humidity sensors, pH sensors, DO sensors, and so on..

Devices

The devices are connected to the sensors and the cloud. The devices are Raspberry PI and micro-controllers like Arduino that can receive data from the sensor and transfer it to the GATEway.

GATE way

In a GATE way, it is used for communicating data between the systems. It is connected to the cloud. It receives and transfers the files and documents to the user.

Cloud

A cloud can be used to store the data of the system. It can be used to monitor, control, and analyse the data.

Figure 10. Schematic Arrangements of IoT in Soilless Farming

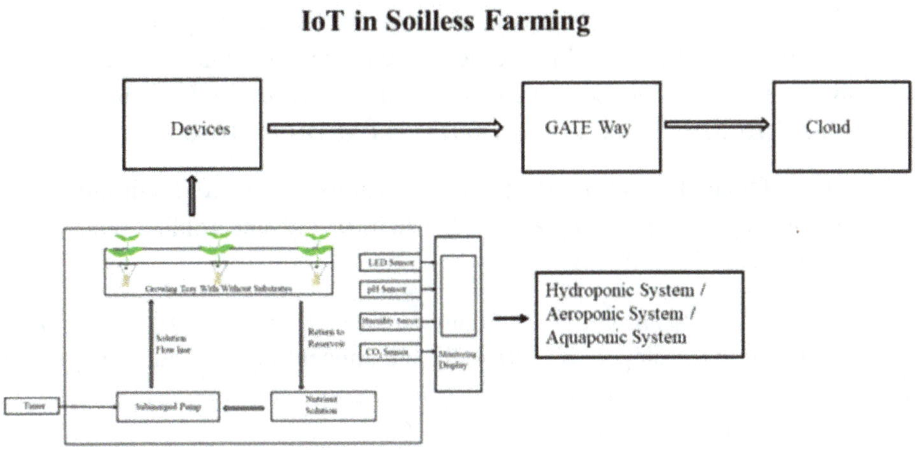

IoT in Soilless Farming

An IoT system can be used in soilless farming methods like hydroponics, aquaponics, and aeroponics. The IoT system can be used for controlling, monitoring, and analyzing data in a closed environment. Soilless farming is done in a closed environment or system. These systems consist of sensors, monitoring devices, and flow control devices. The concept of the IoT in soilless farming, in a closed system, is controlling, deploying, and monitoring data through the Internet with the help of devices. This data is stored in the cloud. The data can be accessed by the user. Micro-controllers are used for managing the automation, monitoring, and control of the entire system. It helps farmers solve the problem of producing food in larger quantities and higher quality with fewer resources and in greener ways.

Summary

The soilless farming method is suitable in all places, which conserves less water and increases the rate of productivity compared to the traditional method. However, it is not possible to grow all the plants in these systems. Both the traditional method and the conventional method have advantages and disadvantages in their techniques. Modern cultivation is replacing the demand for land and water. In this case, because the farmers use soil-less farming, they gain more productivity in less time, and their profit will increase.

REFERENCES

Abad, M., Noguera, P., Puchades, R., Maquieira, A., & Noguera, V. (2002). Physico-chemical and chemical properties of some coconut coir dusts for use as a peat substitute for containerised ornamental plants. *Bioresource Technology*, *82*(3), 241–245. doi:10.1016/S0960-8524(01)00189-4 PMID:11991072

Calone, R., & Orsini, F. (2022). Aquaponics: A Promising Tool for Environmentally Friendly Farming. *Front. Young Minds*, *10*, 707801. Advance online publication. doi:10.3389/frym.2022.707801

Domingues, D. S., Takahashi, H. W., Camara, C. A. P., & Nixdorf, S. L. (2012). Automated system developed to control pH and concentration of nutrient solution evaluated in hydroponic lettuce production. *Computers and Electronics in Agriculture*, *84*, 53–61. doi:10.1016/j.compag.2012.02.006

Espinal, C. A., & Matulić, D. (2019). Recirculating Aquaculture Technologies. *Aquaponics Food Prod. Syst.*, *01*, 35–76. doi:10.1007/978-3-030-15943-6_3

Fornes, F., Belda, R. M., Abad, M., Noguera, P., Puchades, R., Maquieira, A., & Noguera, V. (2003). The microstructure of coconut coir dusts for use as alternatives to peat in soilless growing media. *Australian Journal of Experimental Agriculture*, *43*(9), 1171–1179. doi:10.1071/EA02128

Lakhiar, I. A., Gao, J., Syed, T. N., Ali Chandio, F., Tunio, M. H., Ahmad, F., & Ali Solangi, K. (2020). Overview of the aeroponic agriculture – An emerging technology for global food security. *International Journal of Agricultural and Biological Engineering*, *13*(1), 1–10. doi:10.25165/j.ijabe.20201301.5156

Maucieri, C., Nicoletto, C., van Os, E., Anseeuw, D., Van Havermaet, R., & Junge, R. (2019) Hydroponic Technologies. In Aquaponics Food Production Systems. doi:10.1007/978-3-030-15943-6_4

Nielsen, C. J., Ferrin, D. M., & Stanghellini, M. E. (2006). Efficacy of biosurfactants in the management of Phytophthora capsici on pepper in recirculating hydroponic systems. *Canadian Journal of Plant Pathology*, *28*(3), 450–460. doi:10.1080/07060660609507319

Pickering, H. W., Menzies, N. W., & Hunter, M. N. (2002). Zeolite/rock phosphate - A novel slow release phosphorus fertiliser for potted plant production. *Scientia Horticulturae*, *94*(3–4), 333–343. doi:10.1016/S0304-4238(02)00006-7

Plants, M. C. (1988). *Media and mixes container-grown for plants*. British Library.

Rahmad, I. F., Nababan, E. B., Tanti, L., Triandi, B., Ekadiansyah, E., & Fragastia, V. A. (2019). Application of the Alcohol Sensor MQ-303A to Detect Alcohol Levels on Car Driver. *2019 7th Int. Conf. Cyber IT Serv. Manag. CITSM 2019*, 247–255. 10.1109/CITSM47753.2019.8965395

Rouphael, Y., & Colla, G. (2005). Growth, yield, fruit quality and nutrient uptake of hydroponically cultivated zucchini squash as affected by irrigation systems and growing seasons. *Scientia Horticulturae*, *105*(2), 177–195. doi:10.1016/j.scienta.2005.01.025

Sathyanarayana, S. R., Gangadhar, W. V., Badrinath, M. G., Ravindra, R. M., & Shriramrao, A. U. (2022). Hydroponics: An Intensified Agriculture Practice to Improve Food Production. In Reviews in Agricultural Science (Vol. 10, pp. 101–114). doi:10.7831/ras.10.0_101

Verdonck, O., Penninck, R., & De Boodt, M. (1984) The Physical Properties of Different Horticultural Substrates. In Acta Horticulturae (Issue 150, pp. 155–160). doi:10.17660/ActaHortic.1984.150.16

Chapter 13
A Review on Predictive Models Designed From Artificial Intelligence Techniques in the Wastewater Treatment Process

Varun Geetha Mohan
https://orcid.org/0000-0002-5153-3216
Universiti Malaysia Pahang, Malaysia

Mohamed Ariff Ameedeen
Universiti Malaysia Pahang, Malaysia

Bincy Lathakumary Vijayan
Universiti Malaysia Pahang, Malaysia

Al-Fahim Mubarak Ali
Universiti Malaysia Pahang, Malaysia

ABSTRACT

Modeling and optimization of pollutant removal processes are the best solutions to increase the efficiency of wastewater treatment. The relationship between input and output parameters in wastewater treatment processes (WWTP) are complicated. Artificial intelligence (AI) models are generally more flexible when compared with statistical models while modeling complex datasets with nonlinearity and missing data. Studies on AI-based WWTP are increasing day by day. Therefore, it is crucial to review the AI techniques available which are implemented for WWTP. Such a review helps classifying the techniques that are invented and helps to identify challenges as well as gaps for future studies. Lastly, it can sort out the best AI technique to design predictive models for WWTPs.

DOI: 10.4018/978-1-6684-5887-7.ch013

INTRODUCTION

Artificial Intelligence (AI) has been considered as an invincible tool in the experimental designs that can generate the optimal variables for the modeling and optimization in wastewater industry (Zhao et al., 2020). Pollutant removal processes are the ultimate procedure for every wastewater industry (Ghaedi & Vafaei, 2017; Oller, Malato, & Sánchez-Pérez, 2011; Wilson, 2017). These industries always trying to improve the efficiency of pollutant removal processes without increasing the cost (Hasheminasab, Gholipour, Kharrazi, & Streimikiene, 2018; Rahimpour, shojaeimehr, & Khadivi, 2013). This leads to the AI modelling and optimization of pollutant removal processes in wastewater industry. As we know, wastewater treatment is a tough procedure, and these procedures are facing huge environmental management challenges (Ahmed, Al-Dhafeeri, & Mydlarczyk, 2019; Benedetti, Dirckx, Bixio, Thoeye, & Vanrolleghem, 2008; Leitão, van Haandel, Zeeman, & Lettinga, 2006). AI offers reliable solutions to solve practical problems which are commonly in many engineering disciplines including wastewater treatment. There are some technologies existing for wastewater treatment are Advanced oxidation processes (T. Liu et al., 2022; Yang et al., 2022), Coagulation and flocculation (Jabbar, Barzinjy, & Hamad, 2022; Kurniawan et al., 2022), Membrane filtration (Hube et al., 2020; Xiang, Min, Tang, Sillanpää, & Zhao, 2022), Adsorption process (Rajendran et al., 2022), biological process (Chowdhury, Viraraghavan, & Srinivasan, 2010), and Integrated method (N. A. Khan, Khan, Tiwari, Zubair, & Naushad, 2021; Lin et al., 2022). Photoreforming (Nwosu et al., 2021) is the latest study in WWTP, and which is in its initial stage. Therefore, AI modelling of WWTP using photoreforming parameters is the ultimately new research going on and this is the limitation of the current study. Photoreforming can be considered as a promising approach to produce renewable as well as sustainable chemicals and fuels. The increasing number of research focus to selectively obtain precious chemicals as well as high-quality fuels with the help of photoreforming. A potential worldwide crisis with the quick development of industry in worldwide has arisen due to the issues of continuously intensifying environmental pollutions and health disorders. Uncertainly, it is required to develop a standard and efficient technique to face these current situations (Puga, 2016; Sagir et al., 2022). This leads to the AI modelling on wastewater treatment processes (WWTP) for the optimization of pollutant removal processes and its general outline has shown in the Figure 1.

AI has been applying worldwide in various aspects of engineering and technology (Fan, Hu, Cao, Ruan, & Wei, 2018) such as autonomous driving (Di & Shi, 2021; M. A. Khan et al., 2022), big data (Anthopoulos & Kazantzi, 2022), pattern recognition (Chen, 2013), intelligent search engine (Bekiroglu, Duru, Gulay, Su, & Lagoa, 2018), image understanding (Matsuyama & Hwang, 2013), automatic programming

(Cohen & Feigenbaum, 2014), robotics (Emaminejad & Akhavian, 2022), and human-computer games (S. Liu, Cao, Wang, Chen, & Liu, 2021). To achieve these milestones, AI technology deals with the design of computer systems and programs that are capable of mimic the human features (Fan et al., 2018; Naveen, Naidu, Rao, & Radha, 2019). Accordingly, AI models are generally very efficient when comparing with models developed by statistical techniques in the case of modeling complex datasets that having missing data as well as nonlinearities (Karlaftis & Vlahogianni, 2011). And there are lot of AI techniques available now and the problem is to find out which one is the best for making predictive models in Wastewater treatment processes (WWTP). Therefore, the main aim of this review work is to sort out the latest AI techniques suitable to design predictive models for WWTP.

The remaining content of the paper is organized as follows: In Section 2, we introduce AI models which can implement for WWTP and showing the impact of these models based on literature study. The discussion and the findings of the research questions are discussed in Section 3 and finally, section 4 concludes the study along with future study of the research.

Figure 1. AI modelling on wastewater treatment processes

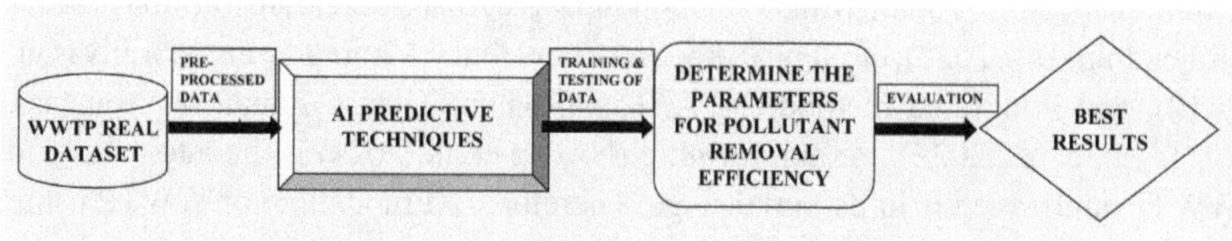

SURVEY ON AI APPLICATIONS FOR WASTEWATER INDUSTRY

In this section, introducing machine learning techniques using for AI modelling for WWTP and show what impact these models given for WWTP. Then go with the status of study on WWTP modeling.

Artificial Neural Networks

The Artificial Neural Networks (ANN) model having some considerable constraints such as estimate bugs, over lapping errors, and optimize configuration of network, was extensively hired for projecting the performance of wastewater treatments due to its simplicity (Al-Mutairi, Kartam, Koushki, & Al-Mutairi, 2004; Areerachakul, 2012; Bahaa, Ayman, Hussein, & Ashraf, 2012). Also, there are few research have been reported on the combination of optimization algorithms with ANN (Pai et al., 2009). Therefore, it is decisive to enhance the development of network configuration for modelling wastewater treatment using an transformative computation methods (Smalheiser, 2000; Yetilmezsoy, 2006). ANN is an auspicious tool that can use for simulation, modelling, and prediction in wastewater industry (Hamed, Khalafallah, & Hassanien, 2004; Tezel, Yel, & Sinan, 2010; Vyas, Modhera, Vyas, & Sharma, 2011). Moreover, ANN can be used to manage big data in environmental studies like urban geography as well as perform topographical analysis (Coelho, Giasson, Campos, de Oliveira e Silva, & Costa, 2021; Feng, Wang, Zhang, & Du, 2021; Grekousis, 2019).

Applications of ANN for WWTP by using Backpropagation Techniques

The most widely used neural network structure is the multilayer feed forward neural network (MFNN). In that, Backpropagation (BP) algorithm (Cruz-López, Boyer, & El-Baz, 2017) along with ANN called BPNN as one of the reliable learning algorithms due to its simplicity and its efficiency as well as its enormous power to calculate large amount of data (Cruz-López et al., 2017; Leskovec, Rajaraman, & Ullman, 2020; Vlachas et al., 2020) which can be hired to grasp these MFNNs with differentiable transfer functions (Varun Geetha Mohan, Ameedeen, & Azad, 2022; Schmidhuber, 2015; Smalheiser, 2000). There are some studies of applications of BPNN in WWTP which are more flexible while compare with usual methods which have been discussed in the Table 1.

Few Studies of Ensembled Neural Network Techniques AI Modelling Implemented for WWTP

Prediction can be considered as a kind of data analysis that can be used to predict future data trends by extract models. This type of data analysis helps to provide better understanding of the large dataset. Also, several studies have been performed in prediction since its birth. Here, some categories of prediction techniques which are ensembled with ANN are presented in Figure 2.

Various AI prediction techniques ensembled with ANN such as decision tree, logistic regression and naïve bayes techniques are being followed in wastewater treatment system and some studies are discussed in the Table 2.

It has become clear from the study that AI-based prediction models, such as ANN can be successfully implemented as integral technologies and best AI technique for real applications of wastewater industry.

Table 1. Applications of BPNN for WWTP

Applications	Findings	References
Modelling of Chemical Oxygen Demand (COD) removal	The design of the BPNN (activation functions, number of neurons in the hidden layer, and BP algorithm) used for this study giving the least mean square error (MSE).	(Elmolla, Chaudhuri, & Eltoukhy, 2010)
Modelling of discoloration process using feed forward neural network.	The neural network was trained with sample dataset and presenting best correlation coefficient (R^2) which is less than 0.98 showing that this model has the high-level accurate prediction efficiency.	(Guimarães et al., 2008)
ANN prediction model for the removal of humic substances	As a result of ANN modelling, the values obtained for R^2, standard deviation ratio (SDR), mean absolute error (MAE), and root mean square error (RMSE) are 0.995, 0.065, 4.057, and 5.496, accordingly.	(Oguz, Tortum, & Keskinler, 2008)
Prediction of Titanium Oxide (TiO2) photocatalytic efficiency for nitrogen oxides removal	Overall optimization error was in the average under 5% and R^2 were gained from the database with a level of 98.57%.	(Toma et al., 2004)
Degradation of polyvinyl alcohol in aqueous solution	The explanation of the trial results depended on the fixing of neural network models for predicting solid phase arrangement and Dissolved Organic Carbon Concentration (DOC) fixation along the response time.	(Giroto, Guardani, Teixeira, & Nascimento, 2006)
ANN prediction on azo dye decolorization	The model gave better expectations of R^2 and the MSE were 0.996 and 0.004 respectively.	(Aleboyeh, Kasiri, Olya, & Aleboyeh, 2008)
Colour Removal of Textile Wastewater	The ANN models have a better efficiency to correlate the relationship among the parameters and achieved the values for R^2 in the range of 0.90-0.97 which were showing better results than linear regression models.	(Yu, Chen, Cheng, & Hsieh, 2009)
Modelling of the WWTP contaminated with methyl *tert*-butyl ether (MTBE)	The network was tested, and the MSE is 0.0004 and while comparing actual and predicted results, the R^2 is 0.998.	(Salari, Daneshvar, Aghazadeh, & Khataee, 2005)
A BPNN modelling to determine the relationship between sewage samples and its Biochemical Oxygen Demand (BOD) values.	The samples collected from different locations of a wastewater plant can be successfully classify by proposed BPNN model.	(Onkal-Engin, Demir, & Engin, 2005)
The BPNN modelling for anaerobic tapered fluidized bed reactor.	The prediction results are showing that the values are closer to the experimental results.	(Parthiban & Parthiban, 2012)
Prediction of ion removal efficiencies from aqueous solution by BPNN modelling.	The proposed model was able to predict adsorption efficiency with a R^2 of approximately 0.94.	(Yetilmezsoy & Demirel, 2008)

Figure 2. Some AI based prediction techniques ensembled with ANN used in WWTP

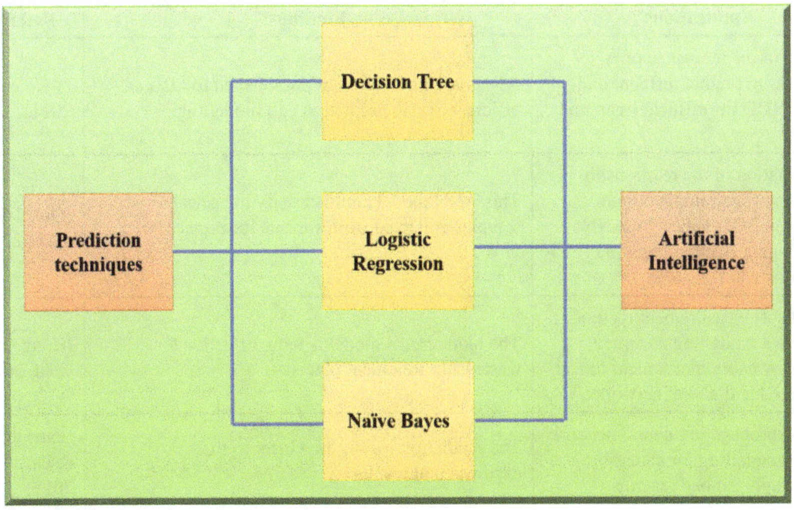

Table 2. Random studies of ensembled ANN techniques involved in WWTP for AI modelling

Applications	Findings	References
Neural network modelling on a contact stabilization process for an effluent biological product	General Regression Neural Network (GRNN) model can be used as a suitable technique for the modelling process.	(Al-Mutairi et al., 2004)
Evaluation of effluent COD by a conceptual neural-fuzzy model for a large-scale anaerobic wastewater treatment plant.	The correlation coefficient (R^2) value 0.894 is considered as a good fit between estimated and measured values of output variable COD.	(Altunay Perendeci, Arslan, Tanyolaç, & Çelebi, 2007)
Comparison of the predictive ability of the adaptive neuro fuzzy inference system (ANFIS) and ANN model for finding the Biochemical Oxygen Demand (BOD).	The experimental results show that the ANN model predicts the effluent BOD better than the ANFIS model.	(Areerachakul, 2012)
To examine the ANN potential among different water quality parameters.	ANN models are comparatively more reliable.	(Bahaa et al., 2012)
To apply an ANN black-box modelling approach to generate a process model for a real wastewater plant.	The plant operation parameters can be captured with a better accuracy by ANN models.	(Mjalli, Al-Asheh, & Alfadala, 2007)
Two models have been developed based on ANNs to predict the effluent concentrations of BOD and Suspended Solids (SS).	ANN models provided good estimates which covered a range of data for training and testing purposes.	(Hamed et al., 2004)

continues on following page

Table 2. Continued

Applications	Findings	References
Analysis of the relevance of ANN techniques to predict influent and effluent BOD for effluent treatment process.	ANN based models were established to offer an efficient tool in prediction and modelling.	(Vyas et al., 2011)
Determination of the relationship between sewage sample odours and related BOD values by ANN trained with a Backpropagation (BP) algorithm.	This ANN model could classify the sewage samples collected from various locations of a WWTP.	(Onkal-Engin et al., 2005)
A novel soft measurement method to resolve a modelling accuracy problem by using mechanical model and hierarchical neural networks.	The hierarchical model is more effective for wastewater treatment plants.	(Cong, Yu, & Chai, 2009)
The Backpropagation neural network (BPNN) modelling for anaerobic tapered fluidized bed reactor.	The prediction results are closer to the experimental results.	(Parthiban & Parthiban, 2012)
To predict key features from a biological wastewater treatment plant in an industrial park by analysis of three types of ANFIS and ANN models	ANFIS is better than ANN in terms of effluent prediction.	(Pai et al., 2009)
To predict COD removal efficiency of Up-Flow Anaerobic Sludge Blanket (UASB) reactors by ANN-based modelling.	ANN model predicted precise and values with a satisfactory R^2 of approximately 0.83 for nine different process parameters.	(Yetilmezsoy, 2006)
To develop an ANN model to predict the efficiencies of ion removal from aqueous solution.	The proposed ANN model which is BP training combined with principal component analysis (PCA) which could predict the adsorption efficiency with a R^2 of approximately 0.94.	(Yetilmezsoy & Demirel, 2008)
To predict the COD concentrations of wastewater from the effluents of a Sugar industry.	ANFIS expresses high prediction accuracy than the statistical model.	(Altunay Perendeci, Arslan, Çelebi, & Tanyolaç, 2008)
Modelling of anaerobic digestion system of primary sludge of the WWTP.	A highly complex system could be easily modelled with a highly nonlinear structured ANFIS model.	(Cakmakci, 2007)
To optimize coagulant dosage that is used for turbidity removal in a water treatment plant by an ANFIS-based model	Obtained a better performance than ANFIS for works with ANN.	(Chun, Kwak, & Ryu, 1999)
The prediction of carbon and nitrogen removal in the aerobic biological treatment in large scale WWTP by employing ANFIS-based modelling.	The ANFIS modelling approach could be an effective advanced technique for performance prediction and control of treatment processes.	(Civelekoglu, Perendeci, Yigit, & Kitis, 2007)

continues on following page

Table 2. Continued

Applications	Findings	References
Estimation of COD concentration by using different input variables developed by three ANFIS techniques.	The three different ANFIS techniques provided similar accuracy in estimating COD.	(Kisi & Ay, 2013)
A conceptual ANFIS-based model to estimate the effluent COD for a real-scale anaerobic WWTP.	The altered ANFIS model could show the behaviour of the considered treatment system.	(Altınay Perendeci, Arslan, Tanyolaç, & Çelebi, 2009)
ANN and ANFIS models have been used to model poly-aluminium chloride for dosing of the surface water.	The self-predicting model of ANFIS is better than the ANN model for poly-aluminium chloride dosage forecasting.	(Wu & Lo, 2008)
An ANN model has been introduced and applied it to a case study of a domestic WWTP.	The two stages developed model can indicate the performance of the treatment system adequately.	(Tezel et al., 2010)
BPNN predictive model for petroleum wastewater treatment dataset	The mean square error (MSE) of 0.0025 obtained from the 80 datasets.	(V. G. Mohan, Ali, Vijayan, Azad, & Ameedeen, 2022)

Regression and Classification Learning

Francis Galton used the term "regression" to describe a biological phenomenon in 19th century for his work and later Udny Yule and Karl Pearson developed his work into statistical context. Common regression analysis may include of immense or a wide range of values which involves inputs of numerical data. Legendre published the primitive mode of regression in 1805 called least squares and further development of the theory of least squares was published later by Gauss in 1821 which include the Gauss-Markov theorem (Freedman, 2009; Siew & Nordin, 2012).

The aim of regression techniques is for predicting as much as accurate for certain real-valued properties of items based on their properties. While go through classifiers, the main aim is to classify automatically that can induce a model from data, such that given a description of a particular it can generate as output a class label. There are many important concepts for regression and classification to outline in ML viz hypothesis, error functions, optimization, and cost analysis (Castillo-Botón et al., 2022; McCaffrey, 2020).

The regression techniques which are using for AI modelling called regression learning (Maulud & Abdulazeez, 2020) and such models are called regression models (Johnson et al., 2018; Lloyd, Duvenaud, Grosse, Tenenbaum, & Ghahramani, 2014). These kinds of models are analytical models for evaluating that how the output is related to the variables which have an impact on the output (Mata, 2011). The mathematic equation of regression learning is given below:

$$y^{\wedge} = a_1 . x_1 + a_2 . x_2 + \ldots + a_n . x_n \quad (1)$$

$$y = y^{\wedge} + e \quad (2)$$

where, y and $y^{\wedge}$ shows the actual output and the predicted output of regression model respectively, then x_1 to x_n indicate the influence parameters as well as a_1 to a_n represent the coefficients for the corresponding influence parameters, and finally e will be the associated error term.

In linear regression, the main goal is the minimization of the root mean square error (RMSE) by calculating the parameter estimates (Garre, Ruiz, & Hontoria, 2020). The analysis of linear regression examines the relation between several independent variables and a dependent variable and the relationship of these variables are assumed to be linear (Montgomery, Peck, & Vining, 2021; Uyanık & Güler, 2013). The following shows a multiple linear regression equation:

$$Y = \alpha + \beta 1X1 + \beta 2X2 + \ldots + \beta k\, Xk \quad (3)$$

where α is the model intercept are the regression coefficients, and β_k is the number of independent variables.

Similarly, the relationship between the dependent and the independent variables are assumed to be non-linear for the analysis of non-linear regression and can estimate a model using random relationship between dependent as well as independent variables (Archontoulis & Miguez, 2015; Bates & Chambers, 2017).

Application of Classification and Regression Learning in WWTP

Support Vector Machine (SVM) is a classification technique which is popular for pattern recognition which is well applied for most of the studies (Boser, Guyon, & Vapnik, 1992; Cortes & Vapnik, 1995). The working of SVM is based on the concept of a non-linear mapping which an input vector is mapping into a high dimensional feature space (Cortez, 2012). The other studies of SVM regarding classification as well as including some mathematical formulations and proofs have been explained in (Burges, 1998; Cortes & Vapnik, 1995; Vapnik, 1995). Later, SVM starts applied in regression problems are known as support vector regression (SVR) (Ribeiro, Sanfins, & Belo, 2013). Also, SVR is an extensively suitable machine learning technique for difficult regression problems especially for WWTP simulations (M. S. Zaghloul, R. A. Hamza, O. T. Iorhemen, & J. H. Tay, 2020).

A composite model coordinating latent variables of kernel partial least squares with relevance vector machine (KPLS-RVM) has introduced for improving the prediction efficiency of typical soft sensors while go through industrial processes. Two case studies have been evaluated mainly for the performance of the proposed model – Subway Indoor Air Quality data and WWTP. The results show the superiority of KPLS-RVM in prediction performance over the other counterparts including least squares support vector machine (LSSVM), PLS-LSSVM, PLS-RVM, KPLS-LSSVM. The coefficient of determination value of KPLS-RVM has been improved by approximately 7.30-19.65% in comparison with the other methods for the prediction of effluent chemical oxygen demand (COD) in WWTP data (H. Liu, Yang, Huang, & Yoo, 2020). Then, predictive models were proposed to show the potential of AI for aerobic granular sludge (AGS) reactors based on influent characteristics and operational conditions, provided insight into the selection of the appropriate algorithms - adaptive neuro-fuzzy inference system (ANFIS), and SVR (M. S. Zaghloul et al., 2020). The other studies about WWTP implemented by Regression learning has discussed in the following Table 3.

Table 3. Studies of applications of Regression and Classification learning for WWTP

Applications	Findings	References
Integrated time series models from a sewage treatment plant (STP) in Malaysia.	The results shown that the integrated model made accurate predictions than the SVM and the Non-linear Autoregressive models.	(Ansari, Othman, Abunama, & El-Shafie, 2018)
A nowcasting system to predict fecal pollutants efficiency at recreational beaches in Korea.	The performance of the Artificial Neural Network (ANN) model prediction was substantially higher than that of SVR model.	(Park et al., 2018)
A data-driven modelling of aerobic granular sludge reactors.	The outcomes have been shown that SVR showed more accuracy than ANFIS, in the case of prediction errors. Anyhow, by looking through generalization ability ANFIS provided much better performance than SVR.	(Mohamed Sherif Zaghloul, Rania Ahmed Hamza, Oliver Terna Iorhemen, & Joo Hwa Tay, 2020)
An activated sludge process modelling using SVMs, cascade neural networks and boosted trees.	It is very useful operationally while only a limited access to the wastewater quality measurements is available.	(Szeląg, Barbusiński, & Studziński, 2018)

continues on following page

Table 3. Continued

Applications	Findings	References
Showed some machine learning techniques for the predicting of wastewater quality indicators.	SVR and Regression Trees are the two techniques have been used in the research showed a comparable performance in the case of reliability, robustness, and high generalization ability.	(Granata, Papirio, Esposito, Gargano, & de Marinis, 2017)
Implementation of a nonlinear system modelling using radial basis function (RBF) networks for industrial application and an Improved ErrCor (IErrCor) algorithm is proposed.	Finally, the simulation results show the outperformance and potential of IErrCor RBF network to analyze with SVM, ELM, and ErrCor algorithms.	(Meng, Rozycki, Qiao, & Wilamowski, 2018)
A novel prediction model of a biofilm system in domestic wastewater.	RMSE of different ratios of wastewater obtained were 1.27, 1.26, and 5.94 mg/L respectively and the study suggests that Stacked denoising auto-encoders (SDAE) deep learning network has the better outcomes compared with other AI models.	(Shi & Xu, 2018)
Prediction of pollutant concentration level in a WWTP by establishing two ML models – ANNs and SVMs.	Results indicated that both models can be adequately do prediction and however, in the case of sensitive analysis concluded that the ANN model is the superior one.	(Guo et al., 2015)
An implementation of an unsupervised ML technique for recognizing potential defects in WWTPs.	The quality of detection algorithm can be affected in the presence of highly noisy parameters. Therefore, it is essential for developing a robust technique that can tackle noisy data.	(Harrou, Dairi, Sun, & Senouci, 2018)

From the above limited study regarding SVM, it showed a higher prediction accuracy in the training stage as well as in validation stage. But in the case of noisy data or heterogenous data, the study suggested to go with the efficient and robust techniques for the best predictions in WWTPs.

DISCUSSION AND FINDINGS

As per the articles reviewed, ANN with multilayer perceptron which leads to BPNN, a supervised neural network is the technique mostly used AI modelling of WWTP due to its simplicity and accuracy. These models can learn complex model functions as well as can accord with a great deal of information and this is the greatest advantage of ANN over other techniques especially going through non-linear processes without having to estimate the form of the relationship between training datasets of input and output variables. Much of the work goes with BPNN from the initial stage and still this technique is using by ensemble with other methods. Such type of models found to be simpler in topology and quicker in computation. This difference may be due to the type of training data as well as training algorithm used to design the ANN models. A few studies have recommended a more convenient way of determining the parameters of models such as epochs, number of hidden layers, and number of neurons instead of using trial and error directly in terms of ANN. The most common training algorithm among all the literature is the Levenberg-Marquardt (LM) BP algorithm which provides accuracy and better performance. Moreover, other training algorithms such as batch BP, BFGS quasi-Newton BP and Bayesian should be examined, and few studies have shown the success of these techniques in the literature. The activation function, architecture of the network, and other important parameters are reliant on the nature of process and data used to train the ANN model. Accordingly, the aspect of the model demand to be tested for the implementation of ANN on different processes.

CONCLUSION

To conclude, methodology of the ANN shows to be a successful alternative to the conventional techniques. The ANN models can be used to simulate, control, and optimize the operations of WWTP. Mostly, ANN is limited to interpolation between the range of the data used for training the network. Also, in a few cases, ANN has the ability for forecasting and extrapolating as well. This survey further concentrates the challenges in modelling using ANN especially in the selection of optimization methods and training algorithms. Along with, some inconsistencies and contradictions have also been detected. It might be due to ANN being a black-box model with different types of training data and algorithms.

The main aim of the survey was to prove the importance of neural network techniques to design predictive models for WWTPs. ANN with multilayer perceptron approach with BP algorithm become a supervised neural network called BPNN is the most used predictive technique for AI modelling of WWTP. Most of the experimental research done with BPNN and now this technique is implementing by ensemble with other AI techniques. This review is just a beginning in our studies for analyzing that how crucial is ANN modelling for major areas of wastewater sector. Next time, the focus is to improve the interpretability of ANNs by preparing an analysis for explaining why an ANN model is more robust than other models.

ACKNOWLEDGMENT

The authors would like to thank the Ministry of Education Malaysia and Universiti Malaysia Pahang (UMP) for providing us with the Trans-Disciplinary Research

Grant (TRGS/1/2018/UMP/02/2/2) under Grant No. RDU191802-2 for us to execute this research.

REFERENCES

Ahmed, M. E., Al-Dhafeeri, A., & Mydlarczyk, A. (2019). Predominance of attached versus suspended growth in a mixed-growth, continuous-flow biological reactor treating primary-treated petrochemical wastewater. *Arabian Journal for Science and Engineering*, *44*(5), 4111–4117. doi:10.100713369-018-3315-y

Al-Mutairi, N., Kartam, N., Koushki, P., & Al-Mutairi, M. (2004). Modeling and predicting biological performance of contact stabilization process using artificial neural networks. *Journal of Computing in Civil Engineering*, *18*(4), 341–349. doi:10.1061/(ASCE)0887-3801(2004)18:4(341)

Aleboyeh, A., Kasiri, M., Olya, M., & Aleboyeh, H. (2008). Prediction of azo dye decolorization by UV/H2O2 using artificial neural networks. *Dyes and Pigments*, *77*(2), 288–294. doi:10.1016/j.dyepig.2007.05.014

Ansari, M., Othman, F., Abunama, T., & El-Shafie, A. (2018). Analysing the accuracy of machine learning techniques to develop an integrated influent time series model: Case study of a sewage treatment plant, Malaysia. *Environmental Science and Pollution Research International*, *25*(12), 12139–12149. doi:10.100711356-018-1438-z PMID:29455350

Anthopoulos, L., & Kazantzi, V. (2022). Urban energy efficiency assessment models from an AI and big data perspective: Tools for policy makers. *Sustainable Cities and Society*, *76*, 103492. doi:10.1016/j.scs.2021.103492

Archontoulis, S. V., & Miguez, F. E. (2015). Nonlinear regression models and applications in agricultural research. *Agronomy Journal*, *107*(2), 786–798. doi:10.2134/agronj2012.0506

Areerachakul, S. (2012). Comparison of ANFIS and ANN for estimation of biochemical oxygen demand parameter in surface water. *International Journal of Chemical and Biological Engineering*, *6*, 286–290.

Bahaa, M. K., Ayman, G. A., Hussein, K., & Ashraf, E.-S. (2012). Application of artificial neural networks for the prediction of water quality variables in the Nile Delta. *Journal of Water Resource and Protection*.

Bates, D. M., & Chambers, J. M. (2017). *Nonlinear models Statistical models in S*. Routledge.

Bekiroglu, K., Duru, O., Gulay, E., Su, R., & Lagoa, C. (2018). Predictive analytics of crude oil prices by utilizing the intelligent model search engine. *Applied Energy*, *228*, 2387–2397. doi:10.1016/j.apenergy.2018.07.071

Benedetti, L., Dirckx, G., Bixio, D., Thoeye, C., & Vanrolleghem, P. A. (2008). Environmental and economic performance assessment of the integrated urban wastewater system. *Journal of Environmental Management*, *88*(4), 1262–1272. doi:10.1016/j.jenvman.2007.06.020 PMID:17766032

Boser, B. E., Guyon, I. M., & Vapnik, V. N. (1992). Training algorithm for optimal margin classifiers. *Proceedings of the Fifth Annual ACM Workshop on Computational Learning Theory*. 10.1145/130385.130401

Burges, C. J. (1998). A tutorial on support vector machines for pattern recognition. *Data Mining and Knowledge Discovery*, *2*(2), 121–167. doi:10.1023/A:1009715923555

Cakmakci, M. (2007). Adaptive neuro-fuzzy modelling of anaerobic digestion of primary sedimentation sludge. *Bioprocess and Biosystems Engineering*, *30*(5), 349–357. doi:10.100700449-007-0131-2 PMID:17593401

Castillo-Botón, C., Casillas-Pérez, D., Casanova-Mateo, C., Ghimire, S., Cerro-Prada, E., Gutierrez, P. A., Deo, R. C., & Salcedo-Sanz, S. (2022). Machine learning regression and classification methods for fog events prediction. *Atmospheric Research*, *272*, 106157. doi:10.1016/j.atmosres.2022.106157

Chen, C.-h. (2013). *Pattern recognition and artificial intelligence*. Elsevier.

Chowdhury, P., Viraraghavan, T., & Srinivasan, A. (2010). Biological treatment processes for fish processing wastewater – A review. *Bioresource Technology*, *101*(2), 439–449. doi:10.1016/j.biortech.2009.08.065 PMID:19751972

Chun, M.-G., Kwak, K.-C., & Ryu, J.-W. (1999). *Application of ANFIS for coagulant dosing process in a water purification plant.* Paper presented at the FUZZ-IEEE'99. 1999 IEEE International Fuzzy Systems. Conference Proceedings (Cat. No. 99CH36315).

Civelekoglu, G., Perendeci, A., Yigit, N. O., & Kitis, M. (2007). Modeling Carbon and Nitrogen Removal in an Industrial Wastewater Treatment Plant Using an Adaptive Network-Based Fuzzy Inference System. *CLEAN–Soil, Air, Water*, *35*(6), 617–625.

Coelho, F. F., Giasson, E., Campos, A. R., de Oliveira e Silva, R. G. P., & Costa, J. J. F. (2021). Geographic object-based image analysis and artificial neural networks for digital soil mapping. *Catena*, *206*, 105568. doi:10.1016/j.catena.2021.105568

Cohen, P. R., & Feigenbaum, E. A. (2014). *The Handbook of Artificial Intelligence* (Vol. 3). Butterworth-Heinemann.

Cong, Q., Yu, W., & Chai, T. (2009). *Hierarchical neural network model for water quality prediction in wastewater treatment plants. In Advances in Computational Intelligence*. Springer.

Cortes, C., & Vapnik, V. (1995). Support-Vector Networks. *Machine Learning*, *20*(3), 273–297. doi:10.1007/BF00994018

Cortez, P. (2012). *Data Mining with Multilayer Perceptrons and Support Vector Machines* (Vol. 24). Intelligent Systems Reference Library. doi:10.1007/978-3-642-23241-1_2

Cruz-López, J. A., Boyer, V., & El-Baz, D. (2017). *Training many neural networks in parallel via back-propagation.* Paper presented at the 2017 IEEE International Parallel and Distributed Processing Symposium Workshops (IPDPSW). 10.1109/IPDPSW.2017.72

Di, X., & Shi, R. (2021). A survey on autonomous vehicle control in the era of mixed-autonomy: From physics-based to AI-guided driving policy learning. *Transportation Research Part C, Emerging Technologies*, *125*, 103008. doi:10.1016/j.trc.2021.103008

Elmolla, E. S., Chaudhuri, M., & Eltoukhy, M. M. (2010). The use of artificial neural network (ANN) for modeling of COD removal from antibiotic aqueous solution by the Fenton process. *Journal of Hazardous Materials*, *179*(1-3), 127–134. doi:10.1016/j.jhazmat.2010.02.068 PMID:20307930

Emaminejad, N., & Akhavian, R. (2022). Trustworthy AI and robotics: Implications for the AEC industry. *Automation in Construction*, *139*, 104298. doi:10.1016/j.autcon.2022.104298

Fan, M., Hu, J., Cao, R., Ruan, W., & Wei, X. (2018). A review on experimental design for pollutants removal in water treatment with the aid of artificial intelligence. *Chemosphere*, *200*, 330–343. doi:10.1016/j.chemosphere.2018.02.111 PMID:29494914

Feng, L., Wang, Y., Zhang, Z., & Du, Q. (2021). Geographically and temporally weighted neural network for winter wheat yield prediction. *Remote Sensing of Environment*, *262*, 112514. doi:10.1016/j.rse.2021.112514

Freedman, D. A. (2009). *Statistical models: Theory and practice.* Cambridge University Press.

Garre, A., Ruiz, M. C., & Hontoria, E. (2020). Application of Machine Learning to support production planning of a food industry in the context of waste generation under uncertainty. *Operations Research Perspectives*, 100147.

Ghaedi, A. M., & Vafaei, A. (2017). Applications of artificial neural networks for adsorption removal of dyes from aqueous solution: A review. *Advances in Colloid and Interface Science*, *245*, 20–39. doi:10.1016/j.cis.2017.04.015 PMID:28473053

Giroto, J. A., Guardani, R., Teixeira, A. C. S., & Nascimento, C. A. O. (2006). Study on the photo-Fenton degradation of polyvinyl alcohol in aqueous solution. *Chemical Engineering and Processing: Process Intensification*, *45*(7), 523–532. doi:10.1016/j.cep.2005.12.001

Granata, F., Papirio, S., Esposito, G., Gargano, R., & de Marinis, G. (2017). Machine learning algorithms for the forecasting of wastewater quality indicators. *Water (Switzerland)*, *9*(2), 105. Advance online publication. doi:10.3390/w9020105

Grekousis, G. (2019). Artificial neural networks and deep learning in urban geography: A systematic review and meta-analysis. *Computers, Environment and Urban Systems*, *74*, 244–256. doi:10.1016/j.compenvurbsys.2018.10.008

Guimarães, O. L. C., dos Reis Chagas, M. H., Villela Filho, D. N., Siqueira, A. F., Izário Filho, H. J., de Aquino, H. O. Q., & Silva, M. B. (2008). Discoloration process modeling by neural network. *Chemical Engineering Journal*, *140*(1-3), 71–76. doi:10.1016/j.cej.2007.09.021

Guo, H., Jeong, K., Lim, J., Jo, J., Kim, Y. M., Park, J., Kim, J. H., & Cho, K. H. (2015). Prediction of effluent concentration in a wastewater treatment plant using machine learning models. *Journal of Environmental Sciences (China)*, *32*, 90–101. doi:10.1016/j.jes.2015.01.007 PMID:26040735

Hamed, M. M., Khalafallah, M. G., & Hassanien, E. A. (2004). Prediction of wastewater treatment plant performance using artificial neural networks. *Environmental Modelling & Software*, *19*(10), 919–928. doi:10.1016/j.envsoft.2003.10.005

Harrou, F., Dairi, A., Sun, Y., & Senouci, M. (2018). Statistical monitoring of a wastewater treatment plant: A case study. *Journal of Environmental Management*, *223*, 807–814. doi:10.1016/j.jenvman.2018.06.087 PMID:29986328

Hasheminasab, H., Gholipour, Y., Kharrazi, M., & Streimikiene, D. (2018). A novel metric of sustainability for petroleum refinery projects. *Journal of Cleaner Production*, *171*, 1215–1224. doi:10.1016/j.jclepro.2017.09.223

Hube, S., Eskafi, M., Hrafnkelsdóttir, K. F., Bjarnadóttir, B., Bjarnadóttir, M. Á., Axelsdóttir, S., & Wu, B. (2020). Direct membrane filtration for wastewater treatment and resource recovery: A review. *The Science of the Total Environment, 710*, 136375. doi:10.1016/j.scitotenv.2019.136375 PMID:31923693

Jabbar, K. Q., Barzinjy, A. A., & Hamad, S. M. (2022). Iron oxide nanoparticles: Preparation methods, functions, adsorption and coagulation/flocculation in wastewater treatment. *Environmental Nanotechnology, Monitoring & Management, 17*, 100661. doi:10.1016/j.enmm.2022.100661

Johnson, K. W., Torres Soto, J., Glicksberg, B. S., Shameer, K., Miotto, R., Ali, M., Ashley, E., & Dudley, J. T. (2018). Artificial intelligence in cardiology. *Journal of the American College of Cardiology, 71*(23), 2668–2679. doi:10.1016/j.jacc.2018.03.521 PMID:29880128

Karlaftis, M. G., & Vlahogianni, E. I. (2011). Statistical methods versus neural networks in transportation research: Differences, similarities and some insights. *Transportation Research Part C, Emerging Technologies, 19*(3), 387–399. doi:10.1016/j.trc.2010.10.004

Khan, M. A., El Sayed, H., Malik, S., Zia, M. T., Alkaabi, N., & Khan, J. (2022). A journey towards fully autonomous driving - fueled by a smart communication system. *Vehicular Communications, 100476*, 100476. Advance online publication. doi:10.1016/j.vehcom.2022.100476

Khan, N. A., Khan, A. H., Tiwari, P., Zubair, M., & Naushad, M. (2021). New insights into the integrated application of Fenton-based oxidation processes for the treatment of pharmaceutical wastewater. *Journal of Water Process Engineering, 44*, 102440. doi:10.1016/j.jwpe.2021.102440

Kisi, O., & Ay, M. (2013). *Modelling COD concentration by using three different ANFIS techniques*. Academic Press.

Kurniawan, S. B., Imron, M. F., Chik, C. E. N. C. E., Owodunni, A. A., Ahmad, A., Alnawajha, M. M., Rahim, N. F. M., Said, N. S. M., Abdullah, S. R. S., Kasan, N. A., Ismail, S., Othman, A. R., & Hasan, H. A. (2022). What compound inside biocoagulants/bioflocculants is contributing the most to the coagulation and flocculation processes? *The Science of the Total Environment, 806*, 150902. doi:10.1016/j.scitotenv.2021.150902 PMID:34653447

Leitão, R. C., van Haandel, A. C., Zeeman, G., & Lettinga, G. (2006). The effects of operational and environmental variations on anaerobic wastewater treatment systems: A review. *Bioresource Technology, 97*(9), 1105–1118. doi:10.1016/j.biortech.2004.12.007 PMID:16551532

Leskovec, J., Rajaraman, A., & Ullman, J. D. (2020). *Mining of massive data sets*. Cambridge University Press. doi:10.1017/9781108684163

Lin, R., Li, Y., Yong, T., Cao, W., Wu, J., & Shen, Y. (2022). Synergistic effects of oxidation, coagulation and adsorption in the integrated fenton-based process for wastewater treatment: A review. *Journal of Environmental Management*, *306*, 114460. doi:10.1016/j.jenvman.2022.114460 PMID:35026715

Liu, H., Yang, C., Huang, M., & Yoo, C. (2020). Soft sensor modeling of industrial process data using kernel latent variables-based relevance vector machine. *Applied Soft Computing*, *90*, 106149. Advance online publication. doi:10.1016/j.asoc.2020.106149

Liu, S., Cao, J., Wang, Y., Chen, W., & Liu, Y. (2021). Self-play reinforcement learning with comprehensive critic in computer games. *Neurocomputing*, *449*, 207–213. doi:10.1016/j.neucom.2021.04.006

Liu, T., Yao, B., Luo, Z., Li, W., Li, C., Ye, Z., Gong, X., Yang, J., & Zhou, Y. (2022). Applications and influencing factors of the biochar-persulfate based advanced oxidation processes for the remediation of groundwater and soil contaminated with organic compounds. *The Science of the Total Environment*, *836*, 155421. doi:10.1016/j.scitotenv.2022.155421 PMID:35472360

Lloyd, J., Duvenaud, D., Grosse, R., Tenenbaum, J., & Ghahramani, Z. (2014). Automatic construction and natural-language description of nonparametric regression models. *Proceedings of the AAAI Conference on Artificial Intelligence*. 10.1609/aaai.v28i1.8904

Mata, J. (2011). Interpretation of concrete dam behaviour with artificial neural network and multiple linear regression models. *Engineering Structures*, *33*(3), 903–910. doi:10.1016/j.engstruct.2010.12.011

Matsuyama, T., & Hwang, V. S.-S. (2013). *SIGMA: A knowledge-based aerial image understanding system*. Springer Science & Business Media.

Maulud, D., & Abdulazeez, A. M. (2020). A review on linear regression comprehensive in machine learning. *Journal of Applied Science and Technology Trends*, *1*(4), 140–147. doi:10.38094/jastt1457

McCaffrey, P. (2020). Introduction to machine learning: Regression, classification, and important concepts. In P. McCaffrey (Ed.), *An Introduction to Healthcare Informatics* (pp. 191–210). Academic Press. doi:10.1016/B978-0-12-814915-7.00014-4

Meng, X., Rozycki, P., Qiao, J. F., & Wilamowski, B. M. (2018). Nonlinear System Modeling Using RBF Networks for Industrial Application. *IEEE Transactions on Industrial Informatics, 14*(3), 931–940. doi:10.1109/TII.2017.2734686

Mjalli, F. S., Al-Asheh, S., & Alfadala, H. (2007). Use of artificial neural network black-box modeling for the prediction of wastewater treatment plants performance. *Journal of Environmental Management, 83*(3), 329–338. doi:10.1016/j.jenvman.2006.03.004 PMID:16806660

Mohan, V. G., Ali, A. F. M., Vijayan, B. L., Azad, S., & Ameedeen, M. A. B. (2022). *A Supervised Neural Network-based predictive model for petrochemical wastewater treatment dataset.* Paper presented at the 2022 First International Conference on Electrical, Electronics, Information and Communication Technologies (ICEEICT). 10.1109/ICEEICT53079.2022.9768566

Mohan, V. G., Ameedeen, M. A., & Azad, S. (2022). *A Supervised Learning Neural Network Approach for the Prediction of Supercapacitive Energy Storage Materials.* doi:10.1007/978-981-33-4597-3_76

Montgomery, D. C., Peck, E. A., & Vining, G. G. (2021). *Introduction to linear regression analysis.* John Wiley & Sons.

Naveen, G., Naidu, M. A., Rao, B. T., & Radha, K. (2019). *A Comparative Study on Artificial Intelligence and Expert Systems.* Academic Press.

Nwosu, U., Wang, A., Palma, B., Zhao, H., Khan, M. A., Kibria, M., & Hu, J. (2021). Selective biomass photoreforming for valuable chemicals and fuels: A critical review. *Renewable & Sustainable Energy Reviews, 148*, 111266. doi:10.1016/j.rser.2021.111266

Oguz, E., Tortum, A., & Keskinler, B. (2008). Determination of the apparent rate constants of the degradation of humic substances by ozonation and modeling of the removal of humic substances from the aqueous solutions with neural network. *Journal of Hazardous Materials, 157*(2), 455–463. doi:10.1016/j.jhazmat.2008.01.018 PMID:18289778

Oller, I., Malato, S., & Sánchez-Pérez, J. (2011). Combination of advanced oxidation processes and biological treatments for wastewater decontamination—A review. *The Science of the Total Environment, 409*(20), 4141–4166. doi:10.1016/j.scitotenv.2010.08.061 PMID:20956012

Onkal-Engin, G., Demir, I., & Engin, S. N. (2005). Determination of the relationship between sewage odour and BOD by neural networks. *Environmental Modelling & Software, 20*(7), 843–850. doi:10.1016/j.envsoft.2004.04.012

Pai, T.-Y., Wang, S., Chiang, C., Su, H., Yu, L., Sung, P., Lin, C. Y., & Hu, H. (2009). Improving neural network prediction of effluent from biological wastewater treatment plant of industrial park using fuzzy learning approach. *Bioprocess and Biosystems Engineering*, *32*(6), 781–790. doi:10.100700449-009-0304-2 PMID:19253022

Park, Y., Kim, M., Pachepsky, Y., Choi, S. H., Cho, J. G., Jeon, J., & Cho, K. H. (2018). Development of a nowcasting system using machine learning approaches to predict fecal contamination levels at recreational beaches in Korea. *Journal of Environmental Quality*, *47*(5), 1094–1102. doi:10.2134/jeq2017.11.0425 PMID:30272778

Parthiban, R., & Parthiban, L. (2012). Back propagation Neural network modeling approach in the anaerobic digestion of wastewater treatment. *International Journal of Environmental Sciences*, *2*(4), 1944.

Perendeci, A., Arslan, S., Çelebi, S. S., & Tanyolaç, A. (2008). Prediction of effluent quality of an anaerobic treatment plant under unsteady state through ANFIS modeling with on-line input variables. *Chemical Engineering Journal*, *145*(1), 78–85. doi:10.1016/j.cej.2008.03.008

Perendeci, A., Arslan, S., Tanyolaç, A., & Çelebi, S. S. (2007). Evaluation of input variables in adaptive-network-based fuzzy inference system modeling for an anaerobic wastewater treatment plant under unsteady state. *Journal of Environmental Engineering*, *133*(7), 765–771. doi:10.1061/(ASCE)0733-9372(2007)133:7(765)

Perendeci, A., Arslan, S., Tanyolaç, A., & Çelebi, S. S. (2009). Effects of phase vector and history extension on prediction power of adaptive-network based fuzzy inference system (ANFIS) model for a real scale anaerobic wastewater treatment plant operating under unsteady state. *Bioresource Technology*, *100*(20), 4579–4587. doi:10.1016/j.biortech.2009.04.049 PMID:19525108

Puga, A. V. (2016). Photocatalytic production of hydrogen from biomass-derived feedstocks. *Coordination Chemistry Reviews*, *315*, 1–66. doi:10.1016/j.ccr.2015.12.009

Rahimpour, F., Shojaeimehr, T., & Khadivi, M. A. (2013). A modeling study by response surface methodology (RSM) and artificial neural network (ANN) on Cu2+ adsorption optimization using light expended clay aggregate (LECA). *Journal of Industrial and Engineering Chemistry*, *20*. Advance online publication. doi:10.1016/j.jiec.2013.06.017

Rajendran, S., Priya, A. K., Senthil Kumar, P., Hoang, T. K. A., Sekar, K., Chong, K. Y., Khoo, K. S., Ng, H. S., & Show, P. L. (2022). A critical and recent developments on adsorption technique for removal of heavy metals from wastewater-A review. *Chemosphere*, *303*, 135146. doi:10.1016/j.chemosphere.2022.135146 PMID:35636612

Ribeiro, D., Sanfins, A., & Belo, O. (2013) Wastewater treatment plant performance prediction with support vector machines. Lecture Notes in Computer Science, 7987, 99-111.

Sagir, M., Tahir, M. B., Waheed, U., Akram, J., Ullah, S., & Waseem, A. (2022). Photocatalysts for Hydrogen Production. In A.-G. Olabi (Ed.), *Encyclopedia of Smart Materials* (pp. 371–377). Elsevier. doi:10.1016/B978-0-12-815732-9.00039-5

Salari, D., Daneshvar, N., Aghazadeh, F., & Khataee, A. (2005). Application of artificial neural networks for modeling of the treatment of wastewater contaminated with methyl tert-butyl ether (MTBE) by UV/H2O2 process. *Journal of Hazardous Materials*, *125*(1-3), 205–210. doi:10.1016/j.jhazmat.2005.05.030 PMID:15996818

Schmidhuber, J. (2015). Deep learning in neural networks: An overview. *Neural Networks*, *61*, 85–117. doi:10.1016/j.neunet.2014.09.003 PMID:25462637

Shi, S., & Xu, G. (2018). Novel performance prediction model of a biofilm system treating domestic wastewater based on stacked denoising auto-encoders deep learning network. *Chemical Engineering Journal*, *347*, 280–290. doi:10.1016/j.cej.2018.04.087

Siew, H. L., & Nordin, M. J. (2012). *Regression techniques for the prediction of stock price trend.* Paper presented at the 2012 International Conference on Statistics in Science, Business and Engineering (ICSSBE). 10.1109/ICSSBE.2012.6396535

Smalheiser, N. R. (2000). Walter pitts. *Perspectives in Biology and Medicine*, *43*(2), 217–226. doi:10.1353/pbm.2000.0009 PMID:10804586

Szeląg, B., Barbusiński, K., & Studziński, J. (2018). Activated sludge process modelling using selected machine learning techniques. *Desalination and Water Treatment*, *117*, 78–87. doi:10.5004/dwt.2018.22095

Tezel, G., Yel, E., & Sinan, R. K. (2010). *Artificial neural network (Ann) model for domestic wastewater treatment plant control.* Paper presented at the The fourth scientific conference on water observation and information system for decision support (BALWOIS), Ohrid, Macedonia.

Toma, F.-L., Guessasma, S., Klein, D., Montavon, G., Bertrand, G., & Coddet, C. (2004). Neural computation to predict TiO2 photocatalytic efficiency for nitrogen oxides removal. *Journal of Photochemistry and Photobiology A Chemistry*, *165*(1-3), 91–96. doi:10.1016/j.jphotochem.2004.03.004

Uyanık, G. K., & Güler, N. (2013). A study on multiple linear regression analysis. *Procedia: Social and Behavioral Sciences*, *106*, 234–240. doi:10.1016/j.sbspro.2013.12.027

Vapnik, V. N. (1995). *The nature of statistical learning*. Theory. doi:10.1007/978-1-4757-2440-0

Vlachas, P. R., Pathak, J., Hunt, B. R., Sapsis, T. P., Girvan, M., Ott, E., & Koumoutsakos, P. (2020). Backpropagation algorithms and reservoir computing in recurrent neural networks for the forecasting of complex spatiotemporal dynamics. *Neural Networks*, *126*, 191–217. doi:10.1016/j.neunet.2020.02.016 PMID:32248008

Vyas, M., Modhera, B., Vyas, V., & Sharma, A. (2011). Performance forecasting of common effluent treatment plant parameters by artificial neural network. *Journal of Engineering and Applied Sciences (Asian Research Publishing Network)*, *6*(1), 38–42.

Wilson, D. J. (2017). *Hazardous waste site soil remediation: theory and application of innovative technologies*. Routledge. doi:10.1201/9780203752258

Wu, G.-D., & Lo, S.-L. (2008). Predicting real-time coagulant dosage in water treatment by artificial neural networks and adaptive network-based fuzzy inference system. *Engineering Applications of Artificial Intelligence*, *21*(8), 1189–1195. doi:10.1016/j.engappai.2008.03.015

Xiang, H., Min, X., Tang, C.-J., Sillanpää, M., & Zhao, F. (2022). Recent advances in membrane filtration for heavy metal removal from wastewater: A mini review. *Journal of Water Process Engineering*, *49*, 103023. doi:10.1016/j.jwpe.2022.103023

Yang, Y., Ren, W., Hu, K., Zhang, P., Wang, Y., Duan, X., Sun, H., & Wang, S. (2022). Challenges in radical/nonradical-based advanced oxidation processes for carbon recycling. *Chem Catalysis*, *2*(8), 1858–1869. doi:10.1016/j.checat.2022.05.021

Yetilmezsoy, K. (2006). Determination of optimum body diameter of air cyclones using a new empirical model and a neural network approach. *Environmental Engineering Science*, *23*(4), 680–690. doi:10.1089/ees.2006.23.680

Yetilmezsoy, K., & Demirel, S. (2008). Artificial neural network (ANN) approach for modeling of Pb (II) adsorption from aqueous solution by Antep pistachio (Pistacia Vera L.) shells. *Journal of Hazardous Materials*, *153*(3), 1288–1300. doi:10.1016/j.jhazmat.2007.09.092 PMID:17980484

Yu, R.-F., Chen, H.-W., Cheng, W.-P., & Hsieh, P.-H. (2009). Dosage control of the Fenton process for color removal of textile wastewater applying ORP monitoring and artificial neural networks. *Journal of Environmental Engineering*, *135*(5), 325–332. doi:10.1061/(ASCE)EE.1943-7870.0000016

Zaghloul, M. S., Hamza, R. A., Iorhemen, O. T., & Tay, J. H. (2020). Comparison of adaptive neuro-fuzzy inference systems (ANFIS) and support vector regression (SVR) for data-driven modelling of aerobic granular sludge reactors. *Journal of Environmental Chemical Engineering*, *8*(3), 103742. Advance online publication. doi:10.1016/j.jece.2020.103742

Zaghloul, M. S., Hamza, R. A., Iorhemen, O. T., & Tay, J. H. (2020). Comparison of adaptive neuro-fuzzy inference systems (ANFIS) and support vector regression (SVR) for data-driven modelling of aerobic granular sludge reactors. *Journal of Environmental Chemical Engineering*, *8*(3), 103742. doi:10.1016/j.jece.2020.103742

Zhao, L., Dai, T., Qiao, Z., Sun, P., Hao, J., & Yang, Y. (2020). Application of artificial intelligence to wastewater treatment: A bibliometric analysis and systematic review of technology, economy, management, and wastewater reuse. *Process Safety and Environmental Protection*, *133*, 169–182. doi:10.1016/j.psep.2019.11.014

Compilation of References

Abad, M., Noguera, P., Puchades, R., Maquieira, A., & Noguera, V. (2002). Physico-chemical and chemical properties of some coconut coir dusts for use as a peat substitute for containerised ornamental plants. *Bioresource Technology*, *82*(3), 241–245. doi:10.1016/S0960-8524(01)00189-4 PMID:11991072

Afolalu, I., & Kareem, S. (2021). Enviable roles of manufacturing processes in sustainable fourth industrial revolution – A case study of mechatronics. *Materials Today: Proceedings*, *44*(Part 1), 2895–2901. doi:10.1016/j.matpr.2021.01.099

Ahangar-Asr, H., Mansouri, M., Teshnehlab, M., & Pazoki, A. R. (2011, September). A hybrid strategy for the control of Rotary Inverted Pendulum. In *2011 International Conference on Electrical and Control Engineering* (pp. 5656-5659). 10.1109/ICECENG.2011.6057971

Ahmed, M. E., Al-Dhafeeri, A., & Mydlarczyk, A. (2019). Predominance of attached versus suspended growth in a mixed-growth, continuous-flow biological reactor treating primary-treated petrochemical wastewater. *Arabian Journal for Science and Engineering*, *44*(5), 4111–4117. doi:10.100713369-018-3315-y

Akhtaruzzaman, M., & Shafie, A. A. (2010, August*). Modeling and control of a rotary inverted pendulum using various methods, comparative assessment and result analysis.* In *2010 IEEE International Conference on Mechatronics and Automation,* (pp. 1342-1347). IEEE. 10.1109/ICMA.2010.5589450

Albers, A., Brudniok, S., Ottnad, J., Sauter, C., & Sedchaicharn, K. (2006, December). Upper body of a new humanoid robot-the design of ARMAR III. In *6th IEEE-RAS International Conference on Humanoid Robots,* (pp. 308-313). IEEE.

Al-Betar, M. A., Alyasseri, Z. A. A., Awadallah, M. A., & Abu Doush, I. (2021). Coronavirus herd immunity optimizer (CHIO). *Neural Computing & Applications*, *33*(10), 5011–5042. doi:10.100700521-020-05296-6 PMID:32874019

Aleboyeh, A., Kasiri, M., Olya, M., & Aleboyeh, H. (2008). Prediction of azo dye decolorization by UV/H2O2 using artificial neural networks. *Dyes and Pigments*, *77*(2), 288–294. doi:10.1016/j.dyepig.2007.05.014

Alkalla, M. G., & Fanni, M. A. (2021). Integrated structure/control design of high-speed flexible robot arms using topology optimization. *Mechanics Based Design of Structures and Machines, 49*(3), 381–402. doi:10.1080/15397734.2019.1688170

Al-Mahturi, A., Santoso, F., Garratt, M. A., & Anavatti, S. G. (2019, June). An intelligent control of an inverted pendulum based on an adaptive interval type-2 fuzzy inference system. In *2019 IEEE International Conference on Fuzzy Systems (FUZZ-IEEE)* (pp. 1-6). IEEE. 10.1109/FUZZ-IEEE.2019.8858948

Al-Mutairi, N., Kartam, N., Koushki, P., & Al-Mutairi, M. (2004). Modeling and predicting biological performance of contact stabilization process using artificial neural networks. *Journal of Computing in Civil Engineering, 18*(4), 341–349. doi:10.1061/(ASCE)0887-3801(2004)18:4(341)

Ansari, M., Othman, F., Abunama, T., & El-Shafie, A. (2018). Analysing the accuracy of machine learning techniques to develop an integrated influent time series model: Case study of a sewage treatment plant, Malaysia. *Environmental Science and Pollution Research International, 25*(12), 12139–12149. doi:10.100711356-018-1438-z PMID:29455350

Anthopoulos, L., & Kazantzi, V. (2022). Urban energy efficiency assessment models from an AI and big data perspective: Tools for policy makers. *Sustainable Cities and Society, 76*, 103492. doi:10.1016/j.scs.2021.103492

Anvar, S. M. M., Hassanzadeh, I., & Alizadeh, G. (2010, October). Design and implementation of sliding mode-state feedback control for stabilization of Rotary Inverted Pendulum. In *ICCAS 2010* (pp. 1952–1957). IEEE. doi:10.1109/ICCAS.2010.5670108

An, W., Wang, Z., Jiang, T., Chen, P., Liang, X., Shao, J., Nie, J., Xu, M., & Wang, Z. L. (2020). Reliable mechatronic indicator for self-powered liquid sensing toward smart manufacture and safe transportation. *Materials Today, 41*, 10–20. doi:10.1016/j.mattod.2020.06.003

Archontoulis, S. V., & Miguez, F. E. (2015). Nonlinear regression models and applications in agricultural research. *Agronomy Journal, 107*(2), 786–798. doi:10.2134/agronj2012.0506

Areerachakul, S. (2012). Comparison of ANFIS and ANN for estimation of biochemical oxygen demand parameter in surface water. *International Journal of Chemical and Biological Engineering, 6*, 286–290.

Åström, K. J., Hägglund, T., Hang, C. C., & Ho, W. K. (1993). Automatic tuning and adaptation for PID controllers - a survey. *Control Engineering Practice, 1*(4), 699–714. doi:10.1016/0967-0661(93)91394-C

Austin, R. (2011). *Unmanned Aircraft Systems,* (1st ed.). Wiley. https://www.perlego.com/book/1009794/unmanned-aircraft-systems-pdf

Bahaa, M. K., Ayman, G. A., Hussein, K., & Ashraf, E.-S. (2012). Application of artificial neural networks for the prediction of water quality variables in the Nile Delta. *Journal of Water Resource and Protection.*

Baimukashev, D., Rakhim, B., Rubagotti, M., & Varol, H. A. (2022). End-to-end deep fault-tolerant control. *IEEE/ASME Transactions on Mechatronics*, *27*(4), 2224–2234. doi:10.1109/TMECH.2021.3100150

Banakh, L., & Kempner, M. (2010). *Vibrations of mechanical systems with regular structure*. Springer. doi:10.1007/978-3-642-03126-7

Baryshev, Y., & Biryokov, S. (2021). Application of the CDIO standards for cyber-physical education in mechatronics and robotics in a research university on the example of development of digital electronic skills. *Procedia Computer Science*, *190*, 45–50. doi:10.1016/j.procs.2021.06.006

Bates, D. M., & Chambers, J. M. (2017). *Nonlinear models Statistical models in S*. Routledge.

Bazavan, L. C., Roibu, H., & George Bizdoaca, N. (2022). Strategy control of drive mechatronic system for Virtusphere with two actuators. *2022 11th International Conference on Modern Circuits and Systems Technologies, MOCAST 2022*. 10.1109/MOCAST54814.2022.9837781

Bekiroglu, K., Duru, O., Gulay, E., Su, R., & Lagoa, C. (2018). Predictive analytics of crude oil prices by utilizing the intelligent model search engine. *Applied Energy*, *228*, 2387–2397. doi:10.1016/j.apenergy.2018.07.071

Bektaş, M., & Guler, M. A. (2018).Steady-state CFD Analysis of 3D Bio-inspired Flapping Wing Models.

Bektaş, M., Kurtuluş, D. F., & Güler, M. A. (2018). Steady-State CFD Analysis of 3D Bio-inspired Flapping Wing Models. *International Symposium on Sustainable Aviation*, Rome, Italy. https://hdl.handle.net/11511/70891

Belokobylsky, S. V., Eliseev, S. V., & Kashuba, V. B. (2013). *Prikladnye zadachi strukturnoj teorii vibrozashchitnyh sistem* [Applied problems of the structural theory of vibration protection systems]. Polytechnic.

Bendsoe, M. P., & Sigmund, O. (2003). *Topology Optimization*. Springer Science & Business Media. https://link.springer.com/book/10.1007/978-3-662-05086-6

Benedetti, L., Dirckx, G., Bixio, D., Thoeye, C., & Vanrolleghem, P. A. (2008). Environmental and economic performance assessment of the integrated urban wastewater system. *Journal of Environmental Management*, *88*(4), 1262–1272. doi:10.1016/j.jenvman.2007.06.020 PMID:17766032

Bhowmick, P., & Ganguly, A. (2022). Vibration control of a class of mechatronic systems using Negative Imaginary theory. *IFAC-PapersOnLine*, *55*(1), 14–19. doi:10.1016/j.ifacol.2022.04.003

Bhuyan, R. K., Routara, B. C., & Parida, A. K. (2015). An approach for optimization the process parameter by using TOPSIS Method of Al-24%SiC metal matrix composite during EDM. *Materials Today: Proceedings*, *2*(4–5), 3116–3124. doi:10.1016/j.matpr.2015.07.272

Bishop, R. H. (2005). *Mechatronics: An Introduction*. Taylor & Francis Group. doi:10.1201/9781420037241.ch1

Bongiorno, J. J., & Park, K. (2020). Design of linear multivariable feedback control systems. In Design of Linear Multivariable Feedback Control Systems. doi:10.1007/978-3-030-44356-6

Boopathi, S. (2013). Experimental Study and Multi- Objective Optimization of Near-Dry Wire-Cut Electrical Discharge Machining Process Faculty of Mechanical Engineering. In *Shodhganga@ INFLIBNET.* http://hdl.handle.net/10603/16933

Boopathi, S. (2019). Experimental investigation and parameter analysis of LPG refrigeration system using Taguchi method. *SN Applied Sciences*, *1*(8), 892. doi:10.100742452-019-0925-2

Boopathi, S. (2021a). An extensive review on sustainable developments of dry and near-dry electrical discharge machining processes. *Journal of Manufacturing Science and Engineering*, *144*(5), 1–37. doi:10.1115/1.4052527

Boopathi, S. (2021b). An investigation on gas emission concentration and relative emission rate of the near-dry wire-cut electrical discharge machining process. *Environmental Science and Pollution Research International.* doi:10.100711356-021-17658-1 PMID:34837614

Boopathi, S., Balasubramani, V., Kumar, R. S., & Singh, G. R. (2021). The influence of human hair on kenaf and Grewia fiber-based hybrid natural composite material: An experimental study. *Functional Composites and Structures*, *3*(4), 045011. doi:10.1088/2631-6331/ac3afc

Boopathi, S., & Sivakumar, K. (2012). Experimental comparative study of near-dry wire-cut electrical discharge machining (WEDM). *European Journal of Scientific Research*, *75*(4), 472–481.

Boopathi, S., & Sivakumar, K. (2013). Experimental investigation and parameter optimization of near-dry wire-cut electrical discharge machining using multi-objective evolutionary algorithm. *International Journal of Advanced Manufacturing Technology*, *67*(9–12), 2639–2655. doi:10.100700170-012-4680-4

Boopathi, S., & Sivakumar, K. (2016a). Optimal parameter prediction of oxygen-mist near-dry Wire-cut EDM. *International Journal of Manufacturing Technology and Management*, *30*(3–4), 164–178. doi:10.1504/IJMTM.2016.077812

Boopathi, S., Sivakumar, K., & Kalidas, R. (2012). Parametric Study of Dry WEDM Using Taguchi Method. *International Journal of Engineering Research and Development*, *2*(4), 1–6.

Boopathi, S., Thillaivanan, A., Pandian, M., Subbiah, R., & Shanmugam, P. (2021). Friction stir processing of boron carbide reinforced aluminium surface (Al-B4C) composite: Mechanical characteristics analysis. *Materials Today: Proceedings*, *50*, 2430–2435. doi:10.1016/j.matpr.2021.10.261

Boopathi, S., v, H. B., Mageswari, M., & Asif, M. M. (2022). Influences of Boron Carbide Particles on the Wear Rate and Tensile Strength of Aa2014 Surface Composite Fabricated By Friction-Stir Processing. *Materiali in Tehnologije*, *56*(3), 263–270. doi:10.17222/mit.2022.409

Borase, R. P., Maghade, D. K., Sondkar, S. Y., & Pawar, S. N. (2020). A review of PID control, tuning methods and applications. *International Journal of Dynamics and Control*, *9*(2), 818–827. doi:10.100740435-020-00665-4

Boser, B. E., Guyon, I. M., & Vapnik, V. N. (1992). Training algorithm for optimal margin classifiers. *Proceedings of the Fifth Annual ACM Workshop on Computational Learning Theory*. 10.1145/130385.130401

Braun, R., & Jazdi, S. (2022). A methodology for the detection of functional relations of mechatronic components and assemblies in brownfield systems. *Procedia CIRP*, *107*, 119–124. doi:10.1016/j.procir.2022.04.020

Bula, I., Bula, E., & Hajrizi, E. (2021). Cost Oriented Autonomous Window Cleaning Robot from Mechatronic Scrap – MechCleanBot. *IFAC-PapersOnLine*, *54*(13), 233–238. doi:10.1016/j.ifacol.2021.10.451

Burges, C. J. (1998). A tutorial on support vector machines for pattern recognition. *Data Mining and Knowledge Discovery*, *2*(2), 121–167. doi:10.1023/A:1009715923555

Bykhovsky, I. I. (1969). Osnovy teorii vibracionnoj tekhniki [Fundamentals of the theory of vibration technique]. *Mechanical Engineering (New York, N.Y.)*.

Cakmakci, M. (2007). Adaptive neuro-fuzzy modelling of anaerobic digestion of primary sedimentation sludge. *Bioprocess and Biosystems Engineering*, *30*(5), 349–357. doi:10.100700449-007-0131-2 PMID:17593401

Calone, R., & Orsini, F. (2022). Aquaponics: A Promising Tool for Environmentally Friendly Farming. *Front. Young Minds*, *10*, 707801. Advance online publication. doi:10.3389/frym.2022.707801

Carvalho, Hora, & Fernandes (2021). A process for designing innovative mechatronic products. *International Journal of Production Economics,* Volume 231, 107887. *Computer-aided design*, https://en.m.wikipedia.org/wiki/Computer-aided_design

Castillo-Botón, C., Casillas-Pérez, D., Casanova-Mateo, C., Ghimire, S., Cerro-Prada, E., Gutierrez, P. A., Deo, R. C., & Salcedo-Sanz, S. (2022). Machine learning regression and classification methods for fog events prediction. *Atmospheric Research*, *272*, 106157. doi:10.1016/j.atmosres.2022.106157

CDIO. (2010). The CDIO Standards v 2.0 . CDIO. http://www.cdio.org/files/standards/CDIOStds&Rubricsv2.0_2010Dec8.pdf

Chen, C.-h. (2013). *Pattern recognition and artificial intelligence*. Elsevier.

Chen, Y. F., & Huang, A. C. (2014). Adaptive control of rotary inverted pendulum system with time-varying uncertainties. *Nonlinear Dynamics*, *76*(1), 95–102. doi:10.100711071-013-1112-4

Chidambaram, M., & Saxena, N. (2018a). Auto-tuning of decentralized unstable system with refined ZN method. In Advances in Industrial Control, (pp. 167–182). doi:10.1007/978-981-10-7727-2_9

Chidambaram, M., & Saxena, N. (2018b). Tuning rules for PID controllers for unstable systems. In Advances in Industrial Control, (pp. 151–165). doi:10.1007/978-981-10-7727-2_8

Chowdhury, P., Viraraghavan, T., & Srinivasan, A. (2010). Biological treatment processes for fish processing wastewater – A review. *Bioresource Technology*, *101*(2), 439–449. doi:10.1016/j.biortech.2009.08.065 PMID:19751972

Chun, M.-G., Kwak, K.-C., & Ryu, J.-W. (1999). *Application of ANFIS for coagulant dosing process in a water purification plant.* Paper presented at the FUZZ-IEEE'99. 1999 IEEE International Fuzzy Systems. Conference Proceedings (Cat. No. 99CH36315).

Civelekoglu, G., Perendeci, A., Yigit, N. O., & Kitis, M. (2007). Modeling Carbon and Nitrogen Removal in an Industrial Wastewater Treatment Plant Using an Adaptive Network-Based Fuzzy Inference System. *CLEAN–Soil, Air, Water*, *35*(6), 617–625.

Coelho, F. F., Giasson, E., Campos, A. R., de Oliveira e Silva, R. G. P., & Costa, J. J. F. (2021). Geographic object-based image analysis and artificial neural networks for digital soil mapping. *Catena*, *206*, 105568. doi:10.1016/j.catena.2021.105568

Cohen, P. R., & Feigenbaum, E. A. (2014). *The Handbook of Artificial Intelligence* (Vol. 3). Butterworth-Heinemann.

Combes, S. A. (2010). Materials, Structure, and Dynamics of Insect Wings as Bioinspiration for MAVs. In R. Blockley & W. Shyy (Eds.), *Encyclopedia of Aerospace Engineering*. doi:10.1002/9780470686652.eae404

Cong, Q., Yu, W., & Chai, T. (2009). *Hierarchical neural network model for water quality prediction in wastewater treatment plants. In Advances in Computational Intelligence*. Springer.

Copot, C., Mac Thi, T., & Ionescu, C. (2018). PID based particle swarm optimization in offices light control. *IFAC-PapersOnLine*, *51*(4), 382–387. doi:10.1016/j.ifacol.2018.06.096

Cortes, C., & Vapnik, V. (1995). Support-Vector Networks. *Machine Learning*, *20*(3), 273–297. doi:10.1007/BF00994018

Cortez, P. (2012). *Data Mining with Multilayer Perceptrons and Support Vector Machines* (Vol. 24). Intelligent Systems Reference Library. doi:10.1007/978-3-642-23241-1_2

Cruz-López, J. A., Boyer, V., & El-Baz, D. (2017). *Training many neural networks in parallel via back-propagation.* Paper presented at the 2017 IEEE International Parallel and Distributed Processing Symposium Workshops (IPDPSW). 10.1109/IPDPSW.2017.72

Dargahi, J., Arbatani, S., Sokhanvar, S., Xie, W. F., & Ramezanifard, R. (2014). A novel tactile softness display for minimally invasive surgery. *Mechatronics*, *24*(8), 1144–1156. doi:10.1016/j.mechatronics.2014.09.002

David, G. C., Geoffrey, T. W., & Paula, M. C. (2018). *The Routledge Companion to International Human Resource Management.* Routledge Publisher.

Deaton, J. D., & Grandhi, R. V. (2014). A survey of structural and multidisciplinary continuum topology optimization: Post 2000. *Structural and Multidisciplinary Optimization*, *49*(1), 1–38. doi:10.100700158-013-0956-z

DeSilva, C. W. (2000). *Vibration. Fundamentals and practice*. CRC Press.

Ding, H., Gao, R. X., Isaksson, A. J., Landers, R. G., Parisini, T., & Yuan, Y. (2020). State of AI-based monitoring in smart manufacturing and introduction to focused section. *IEEE/ASME Transactions on Mechatronics*, *25*(5), 2143–2154. doi:10.1109/TMECH.2020.3022983

Di, X., & Shi, R. (2021). A survey on autonomous vehicle control in the era of mixed-autonomy: From physics-based to AI-guided driving policy learning. *Transportation Research Part C, Emerging Technologies*, *125*, 103008. doi:10.1016/j.trc.2021.103008

Djojodihardjo, H., Syamim, A., Ramli, S., Wiriadidjaja, S., & Rafie, A. M. (2016). Kinematic And Aerodynamic Modeling Of Bi- And Quad-wing Flapping Wing *Micro-air-vehicle, Advances in Aerospace Science and Technology* *1*(1),83-101.

Domingues, D. S., Takahashi, H. W., Camara, C. A. P., & Nixdorf, S. L. (2012). Automated system developed to control pH and concentration of nutrient solution evaluated in hydroponic lettuce production. *Computers and Electronics in Agriculture*, *84*, 53–61. doi:10.1016/j.compag.2012.02.006

Dorigo, M., Birattari, M., & Stutzle, T. (2006). Ant colony optimization. *IEEE Computational Intelligence Magazine*, *1*(4), 28–39. doi:10.1109/MCI.2006.329691

Duong, V. U., Dong, T. L., Tran., B., & Le, N. (2017). *Refocus of CDIO Standards to enhance students' entrepreneurship skills.* Proceedings of the 13th International CDIO Conference, University of Calgary, Calgary, Canada, June 18-22.

Duong, V., Mikhail, M., & Thu, Q. L. (2021). Study on the Stability of a Wheeled Climbing Robot, *Proceedings of the 2nd Annual International Conference on Materials, Machines and Methods for Sustainable Development (MMMS 2020)*, 638-645. doi:10. 1007/978-3-030-69610-885.

Dybov, A., & Stark, R. (2021). Analysis of common function models for mechatronic products. *Procedia CIRP*, *100*, 391–396. doi:10.1016/j.procir.2021.05.091

Ebert-Uphoff, I., Gardner, J. F., Murray, W. R., & Perez, R. (2000). Preparing for the next century: The state of mechatronics education. *IEEE/ASME Transactions on Mechatronics*, *5*(2), 226–227. doi:10.1109/3516.847096

Edukit. (2021). *Rotary Inverted Pendulum* https://sites.google.com/view/ucla-st-motor-control/home

Eliseev, S.V. (2018). *Prikladnoj sistemnyj analiz i strukturnoe matematicheskoe modelirovanie (dinamika transportnyh i tekhnologicheskih mashin: svyaznost' dvizhenij, vibracionnye vzaimodejstviya, rychazhnye svyazi)* [Applied system analysis and structural mathematical modeling (dynamics of transport and technological machines: connectivity of movements, vibration interactions, lever connections)]. Publishing house of Irkutsk State Transport University.

Eliseev, A. V., Kuznetsov, N. K., & Moskovskikh, A. O. (2019). *Sistemnye predstavleniya, strukturnye skhemy i svyazi elementov* [Dynamics of machinery. System representations, block diagrams and connections of elements]. Innovative Mechanical Engineering.

Eliseev, S. V., & Artyunin, A. I. (2016). Prikladnaya teoriya kolebanij v zadachah dinamiki linejnyh mekhanicheskih sistem [Applied theory of oscillations in problems of dynamics of linear mechanical systems]. *Science*.

Eliseev, S. V., & Eliseev, A. V. (2020). *Theory of oscillations. Structural mathematical modeling in problems of dynamics of technical objects*. Springer International Publishing. doi:10.1007/978-3-030-31295-4

Eliseev, S. V., Eliseev, A. V., Bolshakov, R. S., & Khomenko, A. P. (2021). Metodologiya sistemnogo analiza v zadachah ocenki, formirovaniya i upravleniya dinamicheskim sostoyaniem tekhnologicheskih i transportnyh mashin [Methodology of system analysis in the tasks of assessment, formation and management of the dynamic state of technological and transport machines]. *Science*.

Eliseev, S. V., Lukyanov, A. V., Reznik, Y. N., & Khomenko, A. P. (2006). *Dynamics of mechanical systems with additional ties*. Publishing House of Irkutsk State University.

Eliseev, S. V., Reznik, Y. N., & Khomenko, A. P. (2011). Mekhatronnye podhody v dinamike mekhanicheskih kolebatel'nyh sistem [Mechatronic approaches in the dynamics of mechanical oscillatory systems]. *Science*.

Eliseev, S. V., Reznik, Y. N., Khomenko, A. P., & Zasyadko, A. A. (2008). *Dinamicheskij sintez v obobshchennyh zadachah vibrozashchity i vibroizolyacii tekhnicheskih obektov* [Dynamic synthesis in generalized problems of vibration protection and vibration isolation of technical objects]. Publishing House of Irkutsk State University.

Elmolla, E. S., Chaudhuri, M., & Eltoukhy, M. M. (2010). The use of artificial neural network (ANN) for modeling of COD removal from antibiotic aqueous solution by the Fenton process. *Journal of Hazardous Materials*, *179*(1-3), 127–134. doi:10.1016/j.jhazmat.2010.02.068 PMID:20307930

Emaminejad, N., & Akhavian, R. (2022). Trustworthy AI and robotics: Implications for the AEC industry. *Automation in Construction*, *139*, 104298. doi:10.1016/j.autcon.2022.104298

Enelund, M. (2012). Integration of Education for Sustainable Development in a Mechanical Engineering Programme. *Proceeding of the 8th International CDIO Conference*, Queensland University of Technology, Brisbane, July 1-4, 2012.

EnergizedLab. (2021). *Rotary Inverted Pendulum* https://energizedlab.com/

Enjikalayin, A. R., Prabakaran, V., Htet, L. N., Selva, K., Raj, V. S., & Elara, M. R. (2020). A Magnetic Climbing Robot for Autonomous Thickness Measurement in Ship Hull Maintanance. *Journal of Marine Science and Engineering*, *469*. doi:10.3390/jmse8060469,8

Eric, R. (2011). The Learn Startups: How Today's Entrepreneurs Use Continuous Innovation to Create Radically Successful Businesses. Crown Business.

Espinal, C. A., & Matulić, D. (2019). Recirculating Aquaculture Technologies. *Aquaponics Food Prod. Syst.*, *01*, 35–76. doi:10.1007/978-3-030-15943-6_3

Fan, M., Hu, J., Cao, R., Ruan, W., & Wei, X. (2018). A review on experimental design for pollutants removal in water treatment with the aid of artificial intelligence. *Chemosphere*, *200*, 330–343. doi:10.1016/j.chemosphere.2018.02.111 PMID:29494914

Fantoni, I., Lozano, R., & Lozano, R. (2002). *Non-linear control for underactuated mechanical systems*. Springer Science & Business Media. doi:10.1007/978-1-4471-0177-2

Feng, L., Wang, Y., Zhang, Z., & Du, Q. (2021). Geographically and temporally weighted neural network for winter wheat yield prediction. *Remote Sensing of Environment*, *262*, 112514. doi:10.1016/j.rse.2021.112514

Ferreira, L., Figuera, Y., Iglesiaz, I., & Souto, M. (2017). Offline CAD-based robot programming and welding parametrization of a flexible and adoptive robotic cell using enriched CAD/CAM system for shipbuilding. *Procedia Manufacturing*, *11*, 215–223. doi:10.1016/j.promfg.2017.07.228

Fornes, F., Belda, R. M., Abad, M., Noguera, P., Puchades, R., Maquieira, A., & Noguera, V. (2003). The microstructure of coconut coir dusts for use as alternatives to peat in soilless growing media. *Australian Journal of Experimental Agriculture*, *43*(9), 1171–1179. doi:10.1071/EA02128

Freedman, D. A. (2009). *Statistical models: Theory and practice*. Cambridge University Press.

Galiev, I.I., Nekhaev, V.A., & Nikolaev, V.A. (2010). *Metody i sredstva vibrozashchity zheleznodorozhnyh ekipazhej* [Methods and means of vibration protection of railway railway train]. Publishing house of the Educational and Methodological Center for Education on railway transport.

Galluzzi, R., Circosta, S., Amati, N., & Tonoli, A. (2021). Rotary regenerative shock absorbers for automotive suspensions. *Mechatronics*, *77*, 102580. doi:10.1016/j.mechatronics.2021.102580

Gao, X., Cui, D., Gou, W., Mu, Y., & Li, B. (2017). Dynamic and stability analysis on stairs climbing of wheel-track mobile robot. *International Journal of Advanced Robotic Systems*, *14*(4), 1–13. doi:10.1177/1729881417720783

Garre, A., Ruiz, M. C., & Hontoria, E. (2020). Application of Machine Learning to support production planning of a food industry in the context of waste generation under uncertainty. *Operations Research Perspectives*, 100147.

George, T., & Ganesan, V. (2022). Optimal tuning of PID controller in time delay system: A review on various optimization techniques. *Chemical Product and Process Modeling*, *17*(1), 1–28. doi:10.1515/cppm-2020-2001

Geraldes, W., Witvoet, G., & Vermeulen, J. P. M. B. (2022). The mechatronic architecture and design of the High-Dynamic Double-Crystal Monochromator for Sirius light source. *Precision Engineering*, *77*, 110–126. doi:10.1016/j.precisioneng.2022.05.009

Gerdes, J., Gupta, S., & Wilkerson, S. (2012). A Review of Bird-inspired Flapping Wing Miniature Air Vehicle Designs, Khan, M. A., & Padhy, C. (2021). Influence of aerodynamic parameters on dragonfly inspired corrugated aerofoil. In K. B. Rushi, R. Sivaraj, & J. Prakash (Eds.), *Advances in Fluid Dynamics*. Lecture Notes in Mechanical EngineeringSpringer. doi:10.1007/978-981-15-4308-1_23

Ghaedi, A. M., & Vafaei, A. (2017). Applications of artificial neural networks for adsorption removal of dyes from aqueous solution: A review. *Advances in Colloid and Interface Science*, *245*, 20–39. doi:10.1016/j.cis.2017.04.015 PMID:28473053

Giroto, J. A., Guardani, R., Teixeira, A. C. S., & Nascimento, C. A. O. (2006). Study on the photo-Fenton degradation of polyvinyl alcohol in aqueous solution. *Chemical Engineering and Processing: Process Intensification*, *45*(7), 523–532. doi:10.1016/j.cep.2005.12.001

Goncharevich, I. F., & Frolov, K. V. (1981). Teoriya vibracionnoj tekhniki [Theory of vibration technique]. *Science*.

Goodall, D., Dixon, R., Hamadache, M., & Ward, C. (2020). Railways Discovering Mechatronics and Monitoring - An Overview. *IFAC-PapersOnLine*, *53*(2), 8488–8493. doi:10.1016/j.ifacol.2020.12.1427

Graham, J. A. (2016). Practical Arc Welding. Flairnet publisher, Australia.

Granata, F., Papirio, S., Esposito, G., Gargano, R., & de Marinis, G. (2017). Machine learning algorithms for the forecasting of wastewater quality indicators. *Water (Switzerland)*, *9*(2), 105. Advance online publication. doi:10.3390/w9020105

Grekousis, G. (2019). Artificial neural networks and deep learning in urban geography: A systematic review and meta-analysis. *Computers, Environment and Urban Systems*, *74*, 244–256. doi:10.1016/j.compenvurbsys.2018.10.008

Guimarães, O. L. C., dos Reis Chagas, M. H., Villela Filho, D. N., Siqueira, A. F., Izário Filho, H. J., de Aquino, H. O. Q., & Silva, M. B. (2008). Discoloration process modeling by neural network. *Chemical Engineering Journal*, *140*(1-3), 71–76. doi:10.1016/j.cej.2007.09.021

Guo, H., Jeong, K., Lim, J., Jo, J., Kim, Y. M., Park, J., Kim, J. H., & Cho, K. H. (2015). Prediction of effluent concentration in a wastewater treatment plant using machine learning models. *Journal of Environmental Sciences (China)*, *32*, 90–101. doi:10.1016/j.jes.2015.01.007 PMID:26040735

Guo, X., & Cheng, G.-D. (2010). Recent development in structural design and optimization. *Lixue Xuebao*, *26*(6), 807–823. doi:10.100710409-010-0395-7

Gurtner, Z., Heinz, A., & Ophey, T. (2022). Efficient oscillation detection for verification of mechatronic closed-loop systems using search-based testing. *Mechanical Systems and Signal Processing*, *163*, 108112. doi:10.1016/j.ymssp.2021.108112

Habib & Komoto. (2021). Comparative analysis of design concepts of mechatronics systems with a CAD tool for system architecting. *Mechatronics*, *24*(7), 788–804.

Hamadache, I., & Mellal, M. A. (2021). Design optimization of car side safety system by particle swarm optimization and grey wolf optimizer. In M. A. Mellal & G. M. Pecht (Eds.), *Nature-Inspired Computing Paradigms in Systems: Reliability, Availability, Maintainability, Safety and Cost (RAMS+C) and Prognostics and Health Management (PHM)*. Elsevier. doi:10.1016/B978-0-12-823749-6.00006-4

Hamed, M. M., Khalafallah, M. G., & Hassanien, E. A. (2004). Prediction of wastewater treatment plant performance using artificial neural networks. *Environmental Modelling & Software*, *19*(10), 919–928. doi:10.1016/j.envsoft.2003.10.005

Han, J. H., Lee, J. S., Kim, D. K. (2009). Bio-inspired Flapping UAV design: A University Perspective. *Proc. SPIE.7295,Health Monitoring of Structural and Biological Systems*. doi:. doi:10.1117/12.815337

Haq, R. H., Marwah, O. M., Rahman, M. N., Haw, H. F., Abdullah, H., & Ahmad, S. (2019). 3D Printer parameters analysis for PCL/PLA filament wire using Design of Experiment (DOE). *IOP Conference Series. Materials Science and Engineering*, *607*(1), 1–10. doi:10.1088/1757-899X/607/1/012001

Haribalaji, V., Boopathi, S., & Asif, M. M. (2021a). Optimization of friction stir welding process to join dissimilar AA2014 and AA7075 aluminum alloys. *Materials Today: Proceedings*. doi:10.1016/j.matpr.2021.09.499

Haribalaji, V., Boopathi, S., Asif, M. M., Jeyakumar, M., Subbiah, R., & Lewise, K. A. S. (2021). Influences of Friction stir tool parameters for joining two similar AZ61A alloy plates. *Materials Today: Proceedings*, *50*, 2547–2553. doi:10.1016/j.matpr.2021.12.074

Haribalaji, V., Venkatesan, G., Mohammed Asif, M., Pandian, M., Subbiah, R., & Boopathi, S. (2022). Investigation on corrosion and tensile Characteristics: Friction stir welding of AA7075 and AA2014. *Materials Today: Proceedings*. doi:10.1016/j.matpr.2022.04.037

Harris, S. M., & Cred, C. E. (2002). *Shock and vibration handbook*. McGraw – Hill Book Co.

Harrou, F., Dairi, A., Sun, Y., & Senouci, M. (2018). Statistical monitoring of a wastewater treatment plant: A case study. *Journal of Environmental Management*, *223*, 807–814. doi:10.1016/j.jenvman.2018.06.087 PMID:29986328

Hasheminasab, H., Gholipour, Y., Kharrazi, M., & Streimikiene, D. (2018). A novel metric of sustainability for petroleum refinery projects. *Journal of Cleaner Production*, *171*, 1215–1224. doi:10.1016/j.jclepro.2017.09.223

Heimicke, N., Ng, G.-L., Krüger, M., & Albers, A. (2021). A systematic for realizing agile principles in the process of mechatronic systems development through individual selection of suitable process models, methods and practices. *Procedia CIRP*, *100*, 619–624. doi:10.1016/j.procir.2021.05.133

Held, A., Knufer, S., & Seifried, R. (2015). *Topology Optimization of Members of Dynamically Loaded Flexible Multibody Systems using Integral Type Objective Functions and Exact Gradients*. 6.

Held, A., & Seifried, R. (2012). Topology Optimization of Members of Elastic Multibody Systems. *Proceedings in Applied Mathematics and Mechanics*, *12*(1), 67–68. doi:10.1002/pamm.201210025

Hennessey, M., Priebe, J., Huang, P., & Grommes, R. (1987, March). Design of a lightweight robotic arm and controller. In *Proceedings of the IEEE International Conference on Robotics and Automation,* (Vol. 4, pp. 779-785). IEEE. 10.1109/ROBOT.1987.1087976

Hong, J., Shin, J. H., La, J., Koh, H. S., Lee, C., & Kim, S. W. (2014). *Design optimization for high manufacturability of optical fiber endcap for range detection LADAR*. doi:10.1080/15599612.2014.915602

Howell, S. G. (1992). Risk Reduction Engineering Laboratory Cincinnati, A TenYear Review of the Plastics Recycling, Elsevier Science Publishers.

Huang, J., Zhang, T., Fan, Y., & Sun, J. Q. (2019). Control of rotary inverted pendulum using model-free backstepping technique. *IEEE Access: Practical Innovations, Open Solutions*, *7*, 96965–96973. doi:10.1109/ACCESS.2019.2930220

Hube, S., Eskafi, M., Hrafnkelsdóttir, K. F., Bjarnadóttir, B., Bjarnadóttir, M. Á., Axelsdóttir, S., & Wu, B. (2020). Direct membrane filtration for wastewater treatment and resource recovery: A review. *The Science of the Total Environment*, *710*, 136375. doi:10.1016/j.scitotenv.2019.136375 PMID:31923693

İlman, M. M., Yavuz, Ş., Karagülle, H., & Uysal, A. (2022). Hybrid vibration control of an industrial CFRP composite robot-manipulator system based on reduced order model. *Simulation Modelling Practice and Theory*, *115*, 102456. doi:10.1016/j.simpat.2021.102456

İlman, M. M., Yavuz, Ş., & Taser, P. Y. (2022). Generalized Input Preshaping Vibration Control Approach for Multi-Link Flexible Manipulators using Machine Intelligence. *Mechatronics*, *82*, 102735. doi:10.1016/j.mechatronics.2021.102735

Inci, C., & Bosmans, V. (2022). Offline adaptation of co-simulation time steps by leveraging past co-simulation runs in multi-level mechatronic systems. *Mechanism and Machine Theory*, *171*, 104740. doi:10.1016/j.mechmachtheory.2022.104740

ISO. (2017). Non-destructive testing of weld - Ultrasonic testing - Techniques, testing levels, and assessment. *ISO Standards*.

Iwnicki, S. D. (2006). *Handbook of railway vehicle dynamics,* (1st ed.). CRC Press. doi:10.1201/9781420004892

Jabbar, K. Q., Barzinjy, A. A., & Hamad, S. M. (2022). Iron oxide nanoparticles: Preparation methods, functions, adsorption and coagulation/flocculation in wastewater treatment. *Environmental Nanotechnology, Monitoring & Management*, *17*, 100661. doi:10.1016/j.enmm.2022.100661

Jaroszewicz, A. (2013). Modeling And Simulation of Flapping Wing Entomopter In Martian Atmosphere, Springer, GeoPlanet. *Earth and Planetary Sciences*, *8*, 143–162. doi:10.1007/978-3-642-34020-8_12

Jefferson, H., Robert, D., & Edward, K. (2009, July). Plastics recycling: Challenges and opportunities. *Philosophical Transactions of the Royal Society of London. Series B, Biological Sciences*, *364*(1526), 27. doi:10.1098/rstb.2008.0311 PMID:19528059

Jiang, L., Liu, S., Cui, Y., & Jiang, H. (2022). Path planning for robotic manipulator in complex multi-obstacle environment based on improved_RRT. *IEEE/ASME Transactions on Mechatronics*, 1–12. Advance online publication. doi:10.1109/TMECH.2022.3165845

Jiao, S., Zhang, X., Zhang, X., Jia, J., & Zhang, M. (2022). Magnetic, Circuit Analysis of Halbach, Array and Improvement of Permanent, Magnetic Adsorption, Device for Wall- Climbing Robot. *Symmetry*, *14*(2), 429. doi:10.3390ym14020429

Johnson, E. B. (2001). Contextual teaching and learning: What It Is and Why It's Here to Say. *Corwin Publisher.*, *ISBN-13*, 978–0761978657.

Johnson, K. W., Torres Soto, J., Glicksberg, B. S., Shameer, K., Miotto, R., Ali, M., Ashley, E., & Dudley, J. T. (2018). Artificial intelligence in cardiology. *Journal of the American College of Cardiology*, *71*(23), 2668–2679. doi:10.1016/j.jacc.2018.03.521 PMID:29880128

Jose, J., Dinakaran, D., Ramya, M. M., Samuel, D. G. H. (2018). A survey on magnetic wall-climbing robots for inspection, *International Journal of Mechanical and Production Engineering Research and Development, 8*(6), 59-68.

Joseph, S. B., Dada, E. G., Abidemi, A., Oyewola, D. O., & Khammas, B. M. (2022). Metaheuristic algorithms for PID controller parameters tuning: Review, approaches and open problems. *Heliyon*, *8*(5), e09399. doi:10.1016/j.heliyon.2022.e09399 PMID:35600459

Kannan, E., Trabelsi, Y., Boopathi, S., & Alagesan, S. (2022). Influences of cryogenically treated work material on near-dry wire-cut electrical discharge machining process. *Surface Topography: Metrology and Properties*, *10*(1), 015027. doi:10.1088/2051-672X/ac53e1

Karaboga, D., Gorkemli, B., Ozturk, C., & Karaboga, N. (2014). A comprehensive survey: Artificial bee colony (ABC) algorithm and applications. *Artificial Intelligence Review*, *42*(1), 21–57. doi:10.100710462-012-9328-0

Karagülle, H., Malgaca, L., Dirilmiş, M., Akdağ, M., & Yavuz, Ş. (2017). Vibration control of a two-link flexible manipulator. *Journal of Vibration and Control*, *23*(12), 2023–2034. doi:10.1177/1077546315607694

Karlaftis, M. G., & Vlahogianni, E. I. (2011). Statistical methods versus neural networks in transportation research: Differences, similarities and some insights. *Transportation Research Part C, Emerging Technologies*, *19*(3), 387–399. doi:10.1016/j.trc.2010.10.004

Karnovsky, I. A., & Lebed, E. (2016). *Theory of vibration protection.* Springer International Publishing. doi:10.1007/978-3-319-28020-2

Keck, A., Schwarz, C., Meurer, T., Heckmann, A., & Grether, G. (2021). Estimating the wheel lateral position of a mechatronic railway running gear with nonlinear wheel–rail geometry. *Mechatronics*, *73*, 102457. doi:10.1016/j.mechatronics.2020.102457

Kennedy, J., & Eberhart, R. (1995). Particle swarm optimization. *Proceedings of ICNN'95 - International Conference on Neural Networks, 4*, 1942–1948. 10.1109/ICNN.1995.488968

Khan, M. A., El Sayed, H., Malik, S., Zia, M. T., Alkaabi, N., & Khan, J. (2022). A journey towards fully autonomous driving - fueled by a smart communication system. *Vehicular Communications, 100476*, 100476. Advance online publication. doi:10.1016/j.vehcom.2022.100476

Khan, N. A., Khan, A. H., Tiwari, P., Zubair, M., & Naushad, M. (2021). New insights into the integrated application of Fenton-based oxidation processes for the treatment of pharmaceutical wastewater. *Journal of Water Process Engineering, 44*, 102440. doi:10.1016/j.jwpe.2021.102440

Kim, S., & Kwon, S. (2017). Nonlinear optimal control design for underactuated two-wheeled inverted pendulum mobile platform. *IEEE/ASME Transactions on Mechatronics, 22*(6), 2803–2808. doi:10.1109/TMECH.2017.2767085

Kisi, O., & Ay, M. (2013). *Modelling COD concentration by using three different ANFIS techniques*. Academic Press.

Kizir, S., Bingul, Z., & Oysu, C. (2010). *Fuzzy control of a real time inverted pendulum system. Journal of Intelligent & Fuzzy Systems, 21*(1, 2), 121-133.

Klein, T. P., & Reinhart, G. (2016). Towards Agile Engineering of Mechatronic Systems in Machinery and Plant Construction. *Procedia CIRP, 52*, 68–73. doi:10.1016/j.procir.2016.07.077

Komoto, H., & Tomiyama, T. (2012). A framework for computer-aided conceptual design and its application to system architecting of mechatronics products. *Computer Aided Design, 44*(10), 931–946. doi:10.1016/j.cad.2012.02.004

Kondratenko, Y. P., & Chikrii, A. A. (2019). Advanced control techniques in complex engineering systems: Theory and applications. In *Studies in Systems*. Decision and Control. doi:10.1007/978-3-030-21927-7

Kosmal, T., Beaumont, K., Link, E., Phillips, D., Pulling, C., Wotton, H., Kudrna, C., Kubalak, J., & Williams, C. (2022). Hybrid additive robotic workcell for autonomous fabrication of mechatronic systems - A case study of drone fabrication. *Additive Manufacturing Letters, 3*, 100100. doi:10.1016/j.addlet.2022.100100

Krishen, J., & Becerra, V. M. (2006, October). *Efficient fuzzy control of a rotary inverted pendulum based on LQR mapping*. In IEEE Conference on Computer Aided Control System Design, IEEE International Conference on Control Applications, IEEE International Symposium on Intelligent Control, (pp. 2701-2706). IEEE. 10.1109/ISIC.2006.285548

Kulunk, Z., Cakiroglu, S., Sahin, M., & Esen, Z. (2016). Designing spring load system for mechatronic actuator. *Proceedings - 2016 10th International Conference on Compatibility, Power Electronics and Power Engineering, CPE-POWERENG 2016*, 276–279. 10.1109/CPE.2016.7544199

Kumar, A., & Rai, R. N. (2020). Grey-Taguchi and TOPSIS-Taguchi-Based Optimisation of Performance Parameters of Spark EDM on Heat-Treated AA7050/5 B4C Composite. *Journal of The Institution of Engineers (India), 101*(1), 71–79. doi:10.1007/s40033-020-00211-8

Kuo, T. C., Huang, Y. J., & Hong, B. W. (2009, July). Adaptive PID with sliding mode control for the rotary inverted pendulum system. In *2009 IEEE/ASME International Conference on Advanced Intelligent Mechatronics* (pp. 1804-1809). IEEE. 10.1109/AIM.2009.5229784

Kurniawan, S. B., Imron, M. F., Chik, C. E. N. C. E., Owodunni, A. A., Ahmad, A., Alnawajha, M. M., Rahim, N. F. M., Said, N. S. M., Abdullah, S. R. S., Kasan, N. A., Ismail, S., Othman, A. R., & Hasan, H. A. (2022). What compound inside biocoagulants/bioflocculants is contributing the most to the coagulation and flocculation processes? *The Science of the Total Environment, 806*, 150902. doi:10.1016/j.scitotenv.2021.150902 PMID:34653447

Kwok, K. O., Lai, F., Wei, W. I., Wong, S. Y. S., & Tang, J. W. T. (2020). Herd immunity – estimating the level required to halt the COVID-19 epidemics in affected countries. *The Journal of Infection, 80*(6), e32–e33. doi:10.1016/j.jinf.2020.03.027 PMID:32209383

Kwon, T., & Hodgins, J. K. (2010, July). *Control Systems for Human Running using an Inverted Pendulum Model and a Reference Motion Capture Sequence.* In *Symposium on Computer Animation,* (pp. 129-138).

Lakhiar, I. A., Gao, J., Syed, T. N., Ali Chandio, F., Tunio, M. H., Ahmad, F., & Ali Solangi, K. (2020). Overview of the aeroponic agriculture – An emerging technology for global food security. *International Journal of Agricultural and Biological Engineering, 13*(1), 1–10. doi:10.25165/j.ijabe.20201301.5156

Lee, J., Yoo, C., Park, Y. S., Park, B., Lee, S. J., Gweon, D. G., & Chang, P. H. (2012). An experimental study on time delay control of actuation system of tilt rotor unmanned aerial vehicle. *Mechatronics, 22*(2), 184–194. doi:10.1016/j.mechatronics.2012.01.005

Leitão, R. C., van Haandel, A. C., Zeeman, G., & Lettinga, G. (2006). The effects of operational and environmental variations on anaerobic wastewater treatment systems: A review. *Bioresource Technology, 97*(9), 1105–1118. doi:10.1016/j.biortech.2004.12.007 PMID:16551532

Leskovec, J., Rajaraman, A., & Ullman, J. D. (2020). *Mining of massive data sets*. Cambridge University Press. doi:10.1017/9781108684163

Liang, J. B., Lin, W. P., Wu, R., & Chen, C. (2021). Correlation of accommodation and lens location with higher-order aberrations and axial length elongation during orthokeratology lens wear. *International Journal of Optomechatronics, 15*(1), 1–9. doi:10.1080/15599612.2020.1857889

Liao, H., Chan, H. H. T., Gao, F., Zhao, X., Liu, G., & Liao, W. H. (2022). Proxy-based torque control of motor-driven exoskeletons for safe and compliant human-exoskeleton interaction. *Mechatronics, 88*, 102906. doi:10.1016/j.mechatronics.2022.102906

Lidia-Cristina, B., Horatiu, R., & Nicu-George, B. (2021). Design of improved mechatronic system for Virtusphere. *2021 International Conference on Applied and Theoretical Electricity, ICATE 2021 - Proceedings*. 10.1109/ICATE49685.2021.9464938

Lin, R., Li, Y., Yong, T., Cao, W., Wu, J., & Shen, Y. (2022). Synergistic effects of oxidation, coagulation and adsorption in the integrated fenton-based process for wastewater treatment: A review. *Journal of Environmental Management*, *306*, 114460. doi:10.1016/j.jenvman.2022.114460 PMID:35026715

Liu, H., Yang, C., Huang, M., & Yoo, C. (2020). Soft sensor modeling of industrial process data using kernel latent variables-based relevance vector machine. *Applied Soft Computing*, *90*, 106149. Advance online publication. doi:10.1016/j.asoc.2020.106149

Liu, S., Cao, J., Wang, Y., Chen, W., & Liu, Y. (2021). Self-play reinforcement learning with comprehensive critic in computer games. *Neurocomputing*, *449*, 207–213. doi:10.1016/j.neucom.2021.04.006

Liu, T., Yao, B., Luo, Z., Li, W., Li, C., Ye, Z., Gong, X., Yang, J., & Zhou, Y. (2022). Applications and influencing factors of the biochar-persulfate based advanced oxidation processes for the remediation of groundwater and soil contaminated with organic compounds. *The Science of the Total Environment*, *836*, 155421. doi:10.1016/j.scitotenv.2022.155421 PMID:35472360

Liu, Y., Yang, C., Wei, P., Zhou, P., & Du, J. (2021). An ODE-driven level-set density method for topology optimization. *Computer Methods in Applied Mechanics and Engineering*, *387*, 114159. doi:10.1016/j.cma.2021.114159

Li, W., Ding, H., & Cheng, K. (2012, September). *An investigation on the design and performance assessment of double-PID and LQR controllers for the inverted pendulum*. In *Proceedings of 2012 UKACC international conference on control,* (pp. 190-196). IEEE. 10.1109/CONTROL.2012.6334628

Lloyd, J., Duvenaud, D., Grosse, R., Tenenbaum, J., & Ghahramani, Z. (2014). Automatic construction and natural-language description of nonparametric regression models. *Proceedings of the AAAI Conference on Artificial Intelligence*. 10.1609/aaai.v28i1.8904

Lohmeier, S., Buschmann, T., Schwienbacher, M., Ulbrich, H., & Pfeiffer, F. (2006, December). Leg design for a humanoid walking robot. In *6th IEEE-RAS International Conference on Humanoid Robots,* (pp. 536-541). IEEE.

Loram, I. D., & Lakie, M. (2002). Human balancing of an inverted pendulum: Position control by small, ballistic-like, throw and catch movements. *The Journal of Physiology*, *540*(3), 1111–1124. doi:10.1113/jphysiol.2001.013077 PMID:11986396

Mandal, U. K., & Sarkar, B. (2012). Selection of Best Intelligent Manufacturing System (IMS) Under Fuzzy MOORA Conflicting MCDM Environment. *(International Journal of Emerging Technology and Advanced Engineering)*, *2*(9), 301–310. http://www.ijetae.com/files/Volume2Issue9/IJETAE_0912_52.pdf

Mansi S. Chabukswar, R., Nanwatkar, K., & Aparna, M. B. (2019). Design and Experimental Analysis of Magnetic Climbing Robot, *International Journal of Research in Engineering, Science and Management*, *2*(12), 418-425.

Manzoor, M., Wu, Q., & Masood, R.. (2014). Qinghe Vu & Rana Javed Masood. Stability, Trajectory Following and Formation Control of Wheel Mobile Robot. *The Open Automation and Control Systems Journal*, *6*(1), 1798–1809. doi:10.2174/1874444301406011798

Mao, N., Song, M., Pan, D., & Deng, S. (2018). Comparative studies on using RSM and TOPSIS methods to optimize residential air conditioning systems. *Energy*, *144*, 98–109. doi:10.1016/j.energy.2017.11.160

Margetts, R., & Ngwompo, R. F. (2015). Mode switching in causally dynamic hybrid bond graphs. *Mechatronics*, *28*, 96–102. doi:10.1016/j.mechatronics.2015.04.013

Mark, D. C., & Amy, C. E. (2005). Failing to Learn and Learning to Fail (Intelligently): How Great Organizations Put Failure to Work to Innovate and Improve. *Long Range Planning*, *38*(3), 299–319. doi:10.1016/j.lrp.2005.04.005

Martin, V.B., (2014) -Mobile Robotics R&D student team at Universidade Federal de Santa Catarina.

Mata, J. (2011). Interpretation of concrete dam behaviour with artificial neural network and multiple linear regression models. *Engineering Structures*, *33*(3), 903–910. doi:10.1016/j.engstruct.2010.12.011

Matsuyama, T., & Hwang, V. S.-S. (2013). *SIGMA: A knowledge-based aerial image understanding system*. Springer Science & Business Media.

Maucieri, C., Nicoletto, C., van Os, E., Anseeuw, D., Van Havermaet, R., & Junge, R. (2019) Hydroponic Technologies. In Aquaponics Food Production Systems. doi:10.1007/978-3-030-15943-6_4

Maulud, D., & Abdulazeez, A. M. (2020). A review on linear regression comprehensive in machine learning. *Journal of Applied Science and Technology Trends*, *1*(4), 140–147. doi:10.38094/jastt1457

McCaffrey, P. (2020). Introduction to machine learning: Regression, classification, and important concepts. In P. McCaffrey (Ed.), *An Introduction to Healthcare Informatics* (pp. 191–210). Academic Press. doi:10.1016/B978-0-12-814915-7.00014-4

Mehdi & Boudi. (2021). Towards a sustainable conceptual design of mechatronic systems application to a regenerative braking system. *Materials Today: Proceedings*, *45*(Part 8), 7588–7596.

Mellal, M. A. (2020). *Soft Computing Methods for System Dependability*. IGI Global. doi:10.4018/978-1-7998-1718-5

Mellal, M. A. Frik, A., & Boutiche, R. (2021). Reliability optimization of power plant safety system using grey wolf optimizer and shuffled frog-leaping algorithm. In Mohamed Arezki Mellal & M. G. Pecht (Eds.), Nature-Inspired Computing Paradigms in Systems: Reliability, Availability, Maintainability, Safety and Cost (RAMS+C) and Prognostics and Health Management (PHM), (pp. 1–13). doi:10.1016/B978-0-12-823749-6.00008-8

Mellal, M. A., & Zio, E. (2022). Multi-objective availability and cost optimization by PSO and COA for series-parallel systems with subsystems failure dependencies. *Microprocessors and Microsystems, 89*, 104422.

Mellal, M. A. (2020). Obsolescence – A review of the literature. *Technology in Society, 101347*. Advance online publication. doi:10.1016/j.techsoc.2020.101347

Mellal, M. A., Adjerid, S., & Benazzouz, D. (2011). Modeling and Simulation of Mechatronic System to Integrated Design of Supervision: Using a Bond Graph Approach. *Applied Mechanics and Materials, 86*, 467–470. . doi:10.4028/www.scientific.net/AMM.86.467

Mellal, M. A., & Salhi, A. (2021). Multi-objective System Design Optimization via PPA and a Fuzzy Method. *International Journal of Fuzzy Systems, 23*(5), 1213–1221. doi:10.100740815-021-01068-z

Mellal, M. A., & Williams, E. J. (2018). A survey on ant colony optimization, particle swarm optimization, and cuckoo algorithms. In *Handbook of Research on Emergent Applications of Optimization Algorithms*. IGI Global. doi:10.4018/978-1-5225-2990-3.ch002

Meng, X., Rozycki, P., Qiao, J. F., & Wilamowski, B. M. (2018). Nonlinear System Modeling Using RBF Networks for Industrial Application. *IEEE Transactions on Industrial Informatics, 14*(3), 931–940. doi:10.1109/TII.2017.2734686

Michelson, R. C. (2011), Slow flight in the lower Mars Atmosphere in support of NASA science missions.

Michelson, R.C., & Reece, S. (1998). Update on flapping wing micro air vehicle research ongoing work to Develop a Flapping Wing, Crawling "Entomopter".

Michelson, R. C. (2004). Novel approaches to miniature flight platforms. *Proceedings of the Institution of Mechanical Engineers. Part G, Journal of Aerospace Engineering, 218*(6), 363–373. doi:10.1243/0954410042794911

Mirjalili, S., Mirjalili, S. M., & Lewis, A. (2014). Grey Wolf Optimizer. *Advances in Engineering Software, 69*, 46–61. doi:10.1016/j.advengsoft.2013.12.007

Mjalli, F. S., Al-Asheh, S., & Alfadala, H. (2007). Use of artificial neural network black-box modeling for the prediction of wastewater treatment plants performance. *Journal of Environmental Management, 83*(3), 329–338. doi:10.1016/j.jenvman.2006.03.004 PMID:16806660

MMR. (2020). Computer Aided Design (CAD) Market: Global Industry Analysis and Forecast (2021-2027). *Technology Trends*. https://www.maximizemarketresearch.com/market-report/global-computer-aided-design-cad-market/63443/

Mohamed, Z., Martins, J. M., Tokhi, M. O., Sá da Costa, J., & Botto, M. A. (2005). Vibration control of a very flexible manipulator system. *Control Engineering Practice, 13*(3), 267–277. doi:10.1016/j.conengprac.2003.11.014

Mohan, V. G., Ali, A. F. M., Vijayan, B. L., Azad, S., & Ameedeen, M. A. B. (2022). *A Supervised Neural Network-based predictive model for petrochemical wastewater treatment dataset.* Paper presented at the 2022 First International Conference on Electrical, Electronics, Information and Communication Technologies (ICEEICT). 10.1109/ICEEICT53079.2022.9768566

Mohanty, C. P., Satpathy, M. P., Mahapatra, S. S., & Singh, M. R. (2018). Optimization of cryo-treated EDM variables using TOPSIS-based TLBO algorithm. *Sadhana - Academy Proceedings in Engineering Sciences, 43*(4), 1–18. doi:10.1007/s12046-018-0829-7

Mohan, V. G., Ameedeen, M. A., & Azad, S. (2022). *A Supervised Learning Neural Network Approach for the Prediction of Supercapacitive Energy Storage Materials.* doi:10.1007/978-981-33-4597-3_76

Montgomery, D. C., Peck, E. A., & Vining, G. G. (2021). *Introduction to linear regression analysis.* John Wiley & Sons.

Mori, T. (1969). Mechatronics. Academic Press.

Muskinja, N., & Tovornik, B. (2006). Swinging up and stabilization of a real inverted pendulum. [IEEE.]. *IEEE Transactions on Industrial Electronics, 53*(2), 631–639. doi:10.1109/TIE.2006.870667

Mutlu, R., Alici, G., & Li, W. (2016). A soft mechatronic microstage mechanism based on electroactive polymer actuators. *IEEE/ASME Transactions on Mechatronics, 21*(3), 1467–1478. doi:10.1109/TMECH.2015.2502597

Myilsamy, S., & Sampath, B. (2021). Experimental comparison of near-dry and cryogenically cooled near-dry machining in wire-cut electrical discharge machining processes. *Surface Topography: Metrology and Properties, 9*(3), 035015. doi:10.1088/2051-672X/ac15e0

Naveen, G., Naidu, M. A., Rao, B. T., & Radha, K. (2019). *A Comparative Study on Artificial Intelligence and Expert Systems.* Academic Press.

Nielsen, C. J., Ferrin, D. M., & Stanghellini, M. E. (2006). Efficacy of biosurfactants in the management of Phytophthora capsici on pepper in recirculating hydroponic systems. *Canadian Journal of Plant Pathology, 28*(3), 450–460. doi:10.1080/07060660609507319

Nilesh, T. Mohite, P., Shinde, A. P. K., Amar, B., Shubhash, S. T., & Vhagade, R. C. K. (2020). Multi-Response Optimization of Process Parameters of WEDM using TOPSIS Approach. *International Journal of Engineering Research and Technology (IJERT), 9*(3), 182–185. www.ijert.org

Nise, N. S. (2020). *Control systems engineering.* John Wiley & Sons.

Numan, J., Jorge M, B., Rodrigo, A., & Esteban, C. (2016). A Computer-Simulated Environment for Modeling and Dynamic- Behavior-Analysis of Special Brushless Motors for Mechatronic Mobile Robotics Systems. *IFAC-PapersOnLine*, *49*(29), 12–17. doi:10.1016/j.ifacol.2016.11.060

Nwosu, U., Wang, A., Palma, B., Zhao, H., Khan, M. A., Kibria, M., & Hu, J. (2021). Selective biomass photoreforming for valuable chemicals and fuels: A critical review. *Renewable & Sustainable Energy Reviews*, *148*, 111266. doi:10.1016/j.rser.2021.111266

Ogata, K. (2010). *Modern control engineering* (Vol. 5). Prentice hall.

Oguz, E., Tortum, A., & Keskinler, B. (2008). Determination of the apparent rate constants of the degradation of humic substances by ozonation and modeling of the removal of humic substances from the aqueous solutions with neural network. *Journal of Hazardous Materials*, *157*(2), 455–463. doi:10.1016/j.jhazmat.2008.01.018 PMID:18289778

Öksüz, M., Önal, M. B., Halicioğlu, R., & Dülger, L. C. (2018). Alternative controller design for rotary inverted pendulum. *Tehnički glasnik, 12*(3), 139-145.

Olivier, K. A. (2018). A magnetic climbing robot to perform autonomous welding in the shipbuilding industry. *Robotics and Computer-integrated Manufacturing*, *53*, 178–186. doi:10.1016/j.rcim.2018.04.008

Oller, I., Malato, S., & Sánchez-Pérez, J. (2011). Combination of advanced oxidation processes and biological treatments for wastewater decontamination—A review. *The Science of the Total Environment*, *409*(20), 4141–4166. doi:10.1016/j.scitotenv.2010.08.061 PMID:20956012

Onkal-Engin, G., Demir, I., & Engin, S. N. (2005). Determination of the relationship between sewage odour and BOD by neural networks. *Environmental Modelling & Software*, *20*(7), 843–850. doi:10.1016/j.envsoft.2004.04.012

Oosting, K., & Dickerson, S. L. (1988, April). Simulation of a high-speed lightweight arm. In *Proceedings. 1988 IEEE International Conference on Robotics and Automation,* (pp. 494–496). IEEE. doi:10.1109/ROBOT.1988.12100

Pai, T.-Y., Wang, S., Chiang, C., Su, H., Yu, L., Sung, P., Lin, C. Y., & Hu, H. (2009). Improving neural network prediction of effluent from biological wastewater treatment plant of industrial park using fuzzy learning approach. *Bioprocess and Biosystems Engineering*, *32*(6), 781–790. doi:10.100700449-009-0304-2 PMID:19253022

Panovko, G. Y. (2006). Dinamika vibracionnyh tekhnologicheskih processov [Dynamics of vibration technological processes]. *Regular and Chaotic Dynamics*.

Park, M. S., & Chwa, D. (2009). Swing-up and stabilization control of inverted-pendulum systems via coupled sliding-mode control method. *IEEE Transactions on Industrial Electronics*, *56*(9), 3541–3555. doi:10.1109/TIE.2009.2012452

Park, Y., Kim, M., Pachepsky, Y., Choi, S. H., Cho, J. G., Jeon, J., & Cho, K. H. (2018). Development of a nowcasting system using machine learning approaches to predict fecal contamination levels at recreational beaches in Korea. *Journal of Environmental Quality*, *47*(5), 1094–1102. doi:10.2134/jeq2017.11.0425 PMID:30272778

Parthiban, R., & Parthiban, L. (2012). Back propagation Neural network modeling approach in the anaerobic digestion of wastewater treatment. *International Journal of Environmental Sciences*, *2*(4), 1944.

Patnarin, B., Pankai, K., Gregory, K., John, B., &Dharmendra, K. M. (2022). Polyethylene Terephthalate (PET) Bottle -to- Bottle Recycling for the Beverage Industry: A Review. *Polymers 14*(2366).

Perendeci, A., Arslan, S., Çelebi, S. S., & Tanyolaç, A. (2008). Prediction of effluent quality of an anaerobic treatment plant under unsteady state through ANFIS modeling with on-line input variables. *Chemical Engineering Journal*, *145*(1), 78–85. doi:10.1016/j.cej.2008.03.008

Perendeci, A., Arslan, S., Tanyolaç, A., & Çelebi, S. S. (2007). Evaluation of input variables in adaptive-network-based fuzzy inference system modeling for an anaerobic wastewater treatment plant under unsteady state. *Journal of Environmental Engineering*, *133*(7), 765–771. doi:10.1061/(ASCE)0733-9372(2007)133:7(765)

Perendeci, A., Arslan, S., Tanyolaç, A., & Çelebi, S. S. (2009). Effects of phase vector and history extension on prediction power of adaptive-network based fuzzy inference system (ANFIS) model for a real scale anaerobic wastewater treatment plant operating under unsteady state. *Bioresource Technology*, *100*(20), 4579–4587. doi:10.1016/j.biortech.2009.04.049 PMID:19525108

Perez, E., Landers, R., & Bristow, D. (2022). Optimal trajectory smoothing for direct energy deposition additive manufacturing processes. *Mechatronics*, *88*, 102910. doi:10.1016/j.mechatronics.2022.102910

Pickering, H. W., Menzies, N. W., & Hunter, M. N. (2002). Zeolite/rock phosphate - A novel slow release phosphorus fertiliser for potted plant production. *Scientia Horticulturae*, *94*(3–4), 333–343. doi:10.1016/S0304-4238(02)00006-7

Pivnenko, K., Jabobsen, L. G., Eriksen, M. K., Damgaard, A., & Astrup, T. F. (2015). Challenges in Plastics Recycling. *Proceedings Sardinia 15th International Waste Management and Landfill Symposium.*

Plants, M. C. (1988). *Media and mixes container-grown for plants*. British Library.

Prasad, L. B., Tyagi, B., & Gupta, H. O. (2012, May). *Modelling and simulation for optimal control of nonlinear inverted pendulum dynamical system using PID controller and LQR*. In *2012 Sixth Asia Modelling Symposium,* (pp. 138-143). IEEE. 10.1109/AMS.2012.21

Puga, A. V. (2016). Photocatalytic production of hydrogen from biomass-derived feedstocks. *Coordination Chemistry Reviews*, *315*, 1–66. doi:10.1016/j.ccr.2015.12.009

Quanser. (2021). *Rotary Inverted Pendulum* https://www.quanser.com/products/rotary-inverted-pendulum/

Raheem, S. A., & Ibrahim, A. A. (2022). Effects of CLEM cloud based E-learning management system for teaching and learning process in mechatronics. *Optik (Stuttgart)*, *271*, 170048. doi:10.1016/j.ijleo.2022.170048

Rahimpour, F., Shojaeimehr, T., & Khadivi, M. A. (2013). A modeling study by response surface methodology (RSM) and artificial neural network (ANN) on Cu2+ adsorption optimization using light expended clay aggregate (LECA). *Journal of Industrial and Engineering Chemistry*, *20*. Advance online publication. doi:10.1016/j.jiec.2013.06.017

Rahmad, I. F., Nababan, E. B., Tanti, L., Triandi, B., Ekadiansyah, E., & Fragastia, V. A. (2019). Application of the Alcohol Sensor MQ-303A to Detect Alcohol Levels on Car Driver. *2019 7th Int. Conf. Cyber IT Serv. Manag. CITSM 2019,* 247–255. 10.1109/CITSM47753.2019.8965395

Rajendran, S., Priya, A. K., Senthil Kumar, P., Hoang, T. K. A., Sekar, K., Chong, K. Y., Khoo, K. S., Ng, H. S., & Show, P. L. (2022). A critical and recent developments on adsorption technique for removal of heavy metals from wastewater-A review. *Chemosphere*, *303*, 135146. doi:10.1016/j.chemosphere.2022.135146 PMID:35636612

Rana, S. C., & Patel, J. N. (2020). Selection of best location for small hydro power project using AHP, WPM and TOPSIS methods. *ISH Journal of Hydraulic Engineering*, *26*(2), 173–178. doi:10.1080/09715010.2018.1468827

Randolph, H. E., & Barreiro, L. B. (2020). Herd Immunity: Understanding COVID-19. *Immunity*, *52*(5), 737–741. doi:10.1016/j.immuni.2020.04.012 PMID:32433946

Ravindra, S. B., Pushparaj, M. P., & Saroj, K. P. (2018). Experimental investigation on permanent magnet based wheel mechanism for safe navigation of climbing robot. *Procedia Computer Science*. doi:10.1016/j.procs.2018.07.046,133:377-384

Rayyan (2019). *Mechatronics are essential for engineering factories of the future*. Construction Week Online. https://www.constructionweekonline.com/training/257088-mechatronics-are-essential-for-engineering-factories-of-the-future

Razi, K., Mokogwu, C. N., & Hashtrudi-Zaad, K. (2022). Coupled stability analysis of cascaded leader–follower teleoperation networks using Möbius transformation. *Mechatronics*, *88*, 102897. doi:10.1016/j.mechatronics.2022.102897

Rea, P., & Ottaviano, E. (2020). Mechatronic design of low-cost control systems for rehabilitation and assisting devices. In Handbook of Research on Advanced Mechatronic Systems and Intelligent Robotics (pp. 82–97). doi:10.4018/978-1-7998-0137-5.ch004

Ribeiro, D., Sanfins, A., & Belo, O. (2013) Wastewater treatment plant performance prediction with support vector machines. Lecture Notes in Computer Science, 7987, 99-111.

Rocard, Y. (1949). *Dynamique generale des vibrations* [General dynamics of oscillations]. Masson.

Roman, M., Bobasu, E., & Sendrescu, D. (2008, May). *Modelling of the rotary inverted pendulum system.* In 2008 IEEE International Conference on Automation, Quality and Testing [). IEEE.]. *Robotics*, *2*, 141–146.

Rouphael, Y., & Colla, G. (2005). Growth, yield, fruit quality and nutrient uptake of hydroponically cultivated zucchini squash as affected by irrigation systems and growing seasons. *Scientia Horticulturae*, *105*(2), 177–195. doi:10.1016/j.scienta.2005.01.025

Rox, M. F., Ropella, D. S., Hendrick, R. J., Blum, E., Naftel, R. P., Bow, H. C., Herrell, S. D., Weaver, K. D., Chambless, L. B., & Webster, R. J. (2020). Mechatronic design of a two-arm concentric tube robot system for rigid neuroendoscopy. *IEEE/ASME Transactions on Mechatronics*, *25*(3), 1432–1443. doi:10.1109/TMECH.2020.2976897 PMID:33746503

Ryu, J., Sung-Soo, K., & Sang-Sup, K. (1992). An efficient computational method for dynamic stress analysis of flexible multibody systems. *Computers & Structures*, *42*(6), 969–977. doi:10.1016/0045-7949(92)90107-B

Sagir, M., Tahir, M. B., Waheed, U., Akram, J., Ullah, S., & Waseem, A. (2022). Photocatalysts for Hydrogen Production. In A.-G. Olabi (Ed.), *Encyclopedia of Smart Materials* (pp. 371–377). Elsevier. doi:10.1016/B978-0-12-815732-9.00039-5

Sahoo, S. K., Naik, S. S., & Rana, J. (2019). Optimisation of WEDM process parameters during machining of HCHCr steel using TOPSIS method. *International Journal of Process Management and Benchmarking*, *9*(2), 216–231. doi:10.1504/IJPMB.2019.099332

Saifan Rafiq. (2017). *Design, Fabrication and Control of 2-DOF Rotary Inverted Pendulum* [Bachelors' dissertation]. National Institute of Technology, Srinagar, Kashmir, India.

Salari, D., Daneshvar, N., Aghazadeh, F., & Khataee, A. (2005). Application of artificial neural networks for modeling of the treatment of wastewater contaminated with methyl tert-butyl ether (MTBE) by UV/H2O2 process. *Journal of Hazardous Materials*, *125*(1-3), 205–210. doi:10.1016/j.jhazmat.2005.05.030 PMID:15996818

Sampath, B., & Haribalaji, V. (2021). Influences of welding parameters on friction stir welding of aluminum and magnesium: A review. *Materials Research Proceedings*, *19*, 222–230. doi:10.21741/9781644901618-28

Sampath, B., & Myilsamy, S. (2021). Experimental investigation of a cryogenically cooled oxygen-mist near-dry wire-cut electrical discharge machining process. *Strojniski Vestnik. Jixie Gongcheng Xuebao*, *67*(6), 322–330. doi:10.5545v-jme.2021.7161

Sampath, B., Naveenkumar, N., Sampathkumar, P., Silambarasan, P., Venkadesh, A., & Sakthivel, M. (2021). Experimental comparative study of banana fiber composite with glass fiber composite material using Taguchi method. *Materials Today: Proceedings*, *49*, 1475–1480. doi:10.1016/j.matpr.2021.07.232

Sanchez, O., Ordaz, P., & Poznyak, A. (2022). Model based reduced-order observers for a class of mechatronic systems with mitigation of disturbances effects using the Attractive Ellipsoid Method. *Mechatronics, 84*, 102778. doi:10.1016/j.mechatronics.2022.102778

Sander, W., André, S., Martin, C. & Roy, T. (2010). The relationship between Entrepreneurship and economic development: is it U-shaped? *Foundations and Trends in Entrepreneurship*, *6*(3), pp. 167-237. doi:. doi:10.1561/0300000023

Sathyanarayana, S. R., Gangadhar, W. V., Badrinath, M. G., Ravindra, R. M., & Shriramrao, A. U. (2022). Hydroponics: An Intensified Agriculture Practice to Improve Food Production. In Reviews in Agricultural Science (Vol. 10, pp. 101–114). doi:10.7831/ras.10.0_101

Schmidhuber, J. (2015). Deep learning in neural networks: An overview. *Neural Networks, 61*, 85–117. doi:10.1016/j.neunet.2014.09.003 PMID:25462637

Serrano Cardona, L., & Muñoz Mata, E. (2013). Paraninfo Digital. *Early Human Development, 83*(1), 1–11. doi:10.1016/j.earlhumdev.2006.05.022

Sham Tickoo. (2016). *AutoCAD 2016: A Problem- Solving Approach, Basic and Intermediate,* (22nd ed.). Purdue University Calumet.

Sham Tickoo. (2016). *Autodesk – Inventor 2016 for Designers,* (16th ed.). Purdue University Calumet.

Shao, S., Chen, M., Hou, J., & Zhao, Q. (2021). Event-triggered-based discrete-time neural control for a quadrotor UAV using disturbance observer. *IEEE/ASME Transactions on Mechatronics, 26*(2), 689–699. doi:10.1109/TMECH.2021.3051835

Shen, Y., Hua, J., Fan, W., Liu, Y., Yang, X., & Chen, L. (2023). Optimal design and dynamic performance analysis of a fractional-order electrical network-based vehicle mechatronic ISD suspension. *Mechanical Systems and Signal Processing, 184*, 109718. doi:10.1016/j.ymssp.2022.109718

Shi, S., & Xu, G. (2018). Novel performance prediction model of a biofilm system treating domestic wastewater based on stacked denoising auto-encoders deep learning network. *Chemical Engineering Journal, 347*, 280–290. doi:10.1016/j.cej.2018.04.087

Shi, Y., & Zhang, K. (2021). Advanced model predictive control framework for autonomous intelligent mechatronic systems: A tutorial overview and perspectives. *Annual Reviews in Control, 52*, 170–196. doi:10.1016/j.arcontrol.2021.10.008

Siew, H. L., & Nordin, M. J. (2012). *Regression techniques for the prediction of stock price trend.* Paper presented at the 2012 International Conference on Statistics in Science, Business and Engineering (ICSSBE). 10.1109/ICSSBE.2012.6396535

Singhal, T. (2020). A Review of Coronavirus Disease-2019 (COVID-19). *Indian Journal of Pediatrics, 87*(4), 281–286. doi:10.100712098-020-03263-6 PMID:32166607

Sivaiah, P., & Chakradhar, D. (2017). Multi-objective optimisation of cryogenic turning process using Taguchi-based grey relational analysis. *International Journal of Machining and Machinability of Materials*, *19*(4), 297–312. doi:10.1504/IJMMM.2017.086161

Smalheiser, N. R. (2000). Walter pitts. *Perspectives in Biology and Medicine*, *43*(2), 217–226. doi:10.1353/pbm.2000.0009 PMID:10804586

Smith, B. (2019). *Automation and Robotics for Construction*. AZO Robotics. https://www.google.com/amp/s/www.azorobotics.com/amp/article.aspx%3fArticleID=315

Solihin, M. I., & Akmeliawati, R. (2010). Particle swam optimization for stabilizing controller of a self-erecting linear inverted pendulum. *International Journal of Electrical and Electronic Systems Research*, *3*, 410–415.

Song, Z., & Shen, S. (2019). The application of intelligent and modular development of mechatronics: A case of design of smart home system. *Proceedings - 2019 11th International Conference on Measuring Technology and Mechatronics Automation, ICMTMA 2019*, 216–219. 10.1109/ICMTMA.2019.00055

Srinivas, G. L., & Javed, A. (2020). Optimization approaches of industrial serial manipulators to improve energy efficiency: A review. *IOP Conference Series. Materials Science and Engineering*, *912*(3), 032058. doi:10.1088/1757-899X/912/3/032058

Statista. (2016). *Computer aided design (CAD) market worldwide 2016-2028, by technology*. Statista. https://www.statista.com/statistics/790083/worldwide-computer-aided-design-market-by-technology/

Sun, J., Tian, Q., & Hu, H. (2017). Topology optimization based on level set for a flexible multibody system modeled via ANCF. *Structural and Multidisciplinary Optimization*, *55*(4), 1159–1177. doi:10.100700158-016-1558-3

Szeląg, B., Barbusiński, K., & Studziński, J. (2018). Activated sludge process modelling using selected machine learning techniques. *Desalination and Water Treatment*, *117*, 78–87. doi:10.5004/dwt.2018.22095

Tang, T. F., Chong, S. H., & Pang, K. K. (2020). *Stabilisation of a rotary inverted pendulum system with double-PID and LQR control:* Experimental verification. *International Journal of Automation and Control*, *14*(1), 18–33. doi:10.1504/IJAAC.2020.103799

Tansel, Y. I., & Yldrm, S. (2013). MOORA-based Taguchi optimisation for improving product or process quality. *International Journal of Production Research*, *51*(11), 3321–3341. doi:10.1080/00207543.2013.774471

Tezel, G., Yel, E., & Sinan, R. K. (2010). *Artificial neural network (Ann) model for domestic wastewater treatment plant control*. Paper presented at the The fourth scientific conference on water observation and information system for decision support (BALWOIS), Ohrid, Macedonia.

Thomas, E., & Steven, S. (2006). Integrating the Study of Technology into the Curriculum: A Consulting Teacher Model. *Journal of Technology Education*, *18*(1), 26–37.

Throat, S. (2022). *Advantages and Disadvantages of Mechatronics System*. Learnmech. https://learnmech.com/advantages-and-disadvantages-of-mechatronics-system/

Tokarev, B., & Klimova, T. (2021). Application of VR instruments in preprofessional education in the area of mechatronics and robotics in a nuclear research university. *Procedia Computer Science, 190*, 745–749. doi:10.1016/j.procs.2021.06.087

Toma, F.-L., Guessasma, S., Klein, D., Montavon, G., Bertrand, G., & Coddet, C. (2004). Neural computation to predict TiO2 photocatalytic efficiency for nitrogen oxides removal. *Journal of Photochemistry and Photobiology A Chemistry, 165*(1-3), 91–96. doi:10.1016/j.jphotochem.2004.03.004

Tran, T. L., Tung., D. M., & Nguyen, T. C. (2014). Improvement of Individual Learning and Instructor's Role for Better Team Learning. *Proceedings of the 10th Annual International CDIO Conference*, Barcelona, Spain, June 16-19.

Turmanidze, R., Dašić, P. V., & Popkhadze, G. (2020). Investigation of operational characteristics of mechatronic systems in Industry 4.0. In Handbook of Research on Integrating Industry 4.0 in Business and Manufacturing (pp. 324–343). doi:10.4018/978-1-7998-2725-2.CH015

Uthayakumar, M., Babu, K. V., Kumaran, S. T., Kumar, S. S., Jappes, J. T. W., & Rajan, T. P. D. (2019). Study on the machining of Al–SiC functionally graded metal matrix composite using die-sinking EDM. *Particulate Science and Technology, 37*(1), 103–109. doi:10.1080/02726351.2017.1346020

Uyanık, G. K., & Güler, N. (2013). A study on multiple linear regression analysis. *Procedia: Social and Behavioral Sciences, 106*, 234–240. doi:10.1016/j.sbspro.2013.12.027

Vapnik, V. N. (1995). *The nature of statistical learning*. Theory. doi:10.1007/978-1-4757-2440-0

Verdonck, O., Penninck, R., & De Boodt, M. (1984) The Physical Properties of Different Horticultural Substrates. In Acta Horticulturae (Issue 150, pp. 155–160). doi:10.17660/ActaHortic.1984.150.16

Vlachas, P. R., Pathak, J., Hunt, B. R., Sapsis, T. P., Girvan, M., Ott, E., & Koumoutsakos, P. (2020). Backpropagation algorithms and reservoir computing in recurrent neural networks for the forecasting of complex spatiotemporal dynamics. *Neural Networks, 126*, 191–217. doi:10.1016/j.neunet.2020.02.016 PMID:32248008

Vogel-Heuser, B., & Bi, F. (2021). Interdisciplinary effects of technical debt in companies with mechatronic products— A qualitative study. *Journal of Systems and Software, 171*, 110809. https://en.m.wikipedia.org/wiki/Home_automation. doi:10.1016/j.jss.2020.110809

Vyas, M., Modhera, B., Vyas, V., & Sharma, A. (2011). Performance forecasting of common effluent treatment plant parameters by artificial neural network. *Journal of Engineering and Applied Sciences (Asian Research Publishing Network), 6*(1), 38–42.

Wang, R., & Yang, L. & Zhang, C. (2021). Primary failure mode and effects analysis with function-motion-action hierarchy structure for mechatronic systems. *Engineering Failure Analysis, 130*, 105775. https://esub.com/blog/what-is-cad-in-construction/

Wang, C. C., Lin, T. W., & Hu, S. S. (2007). Optimizing the rapid prototyping process by integrating the Taguchi method with the Gray relational analysis. *Rapid Prototyping Journal*, *13*(5), 304–315. doi:10.1108/13552540710824814

Wang, D., Wang, L., & Wu, J. (2021). Physics-based mechatronics modeling and application of an industrial-grade parallel tool head. *Mechanical Systems and Signal Processing*, *148*, 107158. doi:10.1016/j.ymssp.2020.107158

Wang, P., Li, Y., Wang, Y. H., & Zhu, Z. Q. (2015). A new method based on TOPSIS and response surface method for MCDM problems with interval numbers. *Mathematical Problems in Engineering*, *2015*, 1–11. doi:10.1155/2015/938535

Welch, H., & Mondal, S. (2019). Analysis of Magnetic Wheel Adhesion Force for Climbing Robot. *Journal of Mechatronics and Robot*, *3*(1), 534–541. doi:10.3844/jmrsp.2019.534.541

Wilson, D. J. (2017). *Hazardous waste site soil remediation: theory and application of innovative technologies*. Routledge. doi:10.1201/9780203752258

Wu, G.-D., & Lo, S.-L. (2008). Predicting real-time coagulant dosage in water treatment by artificial neural networks and adaptive network-based fuzzy inference system. *Engineering Applications of Artificial Intelligence*, *21*(8), 1189–1195. doi:10.1016/j.engappai.2008.03.015

Xiang, H., Min, X., Tang, C.-J., Sillanpää, M., & Zhao, F. (2022). Recent advances in membrane filtration for heavy metal removal from wastewater: A mini review. *Journal of Water Process Engineering*, *49*, 103023. doi:10.1016/j.jwpe.2022.103023

Xia, Q., & Zhang, Y. (2021). A modular low-cost atomic force microscope for precision mechatronics education. *Mechatronics*, *76*, 102550. doi:10.1016/j.mechatronics.2021.102550

Yakimov, P. I., & Iovev, A. N. (2019). Development of a Modern Learning Environment for Education in Mechatronics and Industrial Automation. *IFAC-PapersOnLine*, *52*(25), 441–444. doi:10.1016/j.ifacol.2019.12.578

Yang, Y., Ren, W., Hu, K., Zhang, P., Wang, Y., Duan, X., Sun, H., & Wang, S. (2022). Challenges in radical/nonradical-based advanced oxidation processes for carbon recycling. *Chem Catalysis*, *2*(8), 1858–1869. doi:10.1016/j.checat.2022.05.021

Yang, Z., Zhang, T., Hong, J., Zhang, H., Zhao, Q., & Meng, Z. (2021). Research on driving control strategy and Fuzzy logic optimization of a novel mechatronics-electro-hydraulic power coupling electric vehicle. *Energy*, *233*, 121221. doi:10.1016/j.energy.2021.121221

Yatim, H., & Mat Darus, I. (2014). Self-tuning active vibration controller using particle swarm optimization for flexible manipulator system. *WSEAS Transactions on Systems and Control*, *9*, 55–66.

Yavuz, Ş., İlman, M. M., & Binici, B. (2020). Modeling approach and analysis time comparison of single-link flexible steel- and epoxy-glass/carbon-fiber composite manipulators. *Structures, 26*, 396–405.

Yavuz, Ş., Malgaca, L., & Karagülle, H. (2016). Vibration control of a single-link flexible composite manipulator. *Composite Structures, 140*, 684–691. doi:10.1016/j.compstruct.2016.01.037

Yetilmezsoy, K. (2006). Determination of optimum body diameter of air cyclones using a new empirical model and a neural network approach. *Environmental Engineering Science, 23*(4), 680–690. doi:10.1089/ees.2006.23.680

Yetilmezsoy, K., & Demirel, S. (2008). Artificial neural network (ANN) approach for modeling of Pb (II) adsorption from aqueous solution by Antep pistachio (Pistacia Vera L.) shells. *Journal of Hazardous Materials, 153*(3), 1288–1300. doi:10.1016/j.jhazmat.2007.09.092 PMID:17980484

Yuan, H., & Yuan, L. (2022). A design methodology for passive mechatronic vibration absorbers. *Mechanism and Machine Theory, 167*, 104523. doi:10.1016/j.mechmachtheory.2021.104523

Yudin, B., & Moshev, O. (2021). Application of the CDIO standards for cyber-physical education in mechatronics and robotics in a research university on the example of development of 3D-modeling and design skills. *Procedia Computer Science, 190*, 812–816. doi:10.1016/j.procs.2021.06.095

Yue, M., An, C., Du, Y., & Sun, J. (2016). Indirect adaptive fuzzy control for a nonholonomic/underactuated wheeled inverted pendulum vehicle based on a data-driven trajectory planner. *Fuzzy Sets and Systems, 290*, 158–177. doi:10.1016/j.fss.2015.08.013

Yu, M., Cheng, C., Evangelou, S., & Dini, D. (2022). Series active variable geometry suspension: Full-car prototyping and road testing. *IEEE/ASME Transactions on Mechatronics, 27*(3), 1332–1344. doi:10.1109/TMECH.2021.3097153

Yupapin, P., Trabelsi, Y., Nattappan, A., & Boopathi, S. (2022). Performance Improvement of Wire - Cut Electrical Discharge Machining Process Using Cryogenically Treated Super - Conductive State of Monel - K500 Alloy. *Iranian Journal of Science and Technology. Transaction of Mechanical Engineering, 0123456789*. doi:10.100740997-022-00513-0

Yu, R.-F., Chen, H.-W., Cheng, W.-P., & Hsieh, P.-H. (2009). Dosage control of the Fenton process for color removal of textile wastewater applying ORP monitoring and artificial neural networks. *Journal of Environmental Engineering, 135*(5), 325–332. doi:10.1061/(ASCE)EE.1943-7870.0000016

Yury, K., & Abramov, I. V. (2011). Building a knowledge base for intelligent control system of mechatronic machining center. *Proceedings of 14th International Conference on Mechatronics, MECHATRONIKA 2011*, 93–94. 10.1109/MECHATRON.2011.5961073

Zaghloul, M. S., Hamza, R. A., Iorhemen, O. T., & Tay, J. H. (2020). Comparison of adaptive neuro-fuzzy inference systems (ANFIS) and support vector regression (SVR) for data-driven modelling of aerobic granular sludge reactors. *Journal of Environmental Chemical Engineering, 8*(3), 103742. Advance online publication. doi:10.1016/j.jece.2020.103742

Zain, B. A. M., Tokhi, M. O., & Toha, S. F. (2009). PID-Based Control of a Single-Link Flexible Manipulator in Vertical Motion with Genetic Optimisation. *Third UKSim European Symposium on Computer Modeling and Simulation*, 355–360. doi:10.1109/EMS.2009.86

Zain, M. Z. M., Tokhi, M. O., & Mohamed, Z. (2006). Hybrid learning control schemes with input shaping of a flexible manipulator system. *Mechatronics*, *16*(3), 209–219. doi:10.1016/j.mechatronics.2005.11.004

Zebin, T., & Alam, M. S. (2012). Modeling and Control of a Two-link Flexible Manipulator using Fuzzy Logic and Genetic Optimization Techniques. *Journal of Computers, 7*, 578–585. doi:10.4304/jcp.7.3.578-585

Zeitler, R., & Franke, L. (2016). Computer-Aided Design and Simulation of Spatial Opto-Mechatronic Interconnect Devices. *Procedia CIRP*, *50*, 727–732. doi:10.1016/j.procir.2016.04.172

Zhang, R., Meng, L., Yu, Z., Chen, X., Liu, H., & Huang, Q. (2022). Stride length and stepping duration adjustments based on center of mass stabilization control. *IEEE/ASME Transactions on Mechatronics*, 1–11. Advance online publication. doi:10.1109/TMECH.2022.3159321

Zhao, L., Dai, T., Qiao, Z., Sun, P., Hao, J., & Yang, Y. (2020). Application of artificial intelligence to wastewater treatment: A bibliometric analysis and systematic review of technology, economy, management, and wastewater reuse. *Process Safety and Environmental Protection*, *133*, 169–182. doi:10.1016/j.psep.2019.11.014

Zhong, Z., Xu, M., Xiao, J., & Lu, H. (2021). Xu Minh, Xiao Junhao & Lu Huimin. Design and Control of an Omnidirectional Mobile Wall-Climbing Robot. *Applied Sciences (Basel, Switzerland)*, *11*(22), 11065. doi:10.3390/app112211065

Zhou, B., Liu, Y., Xiao, Y., Zhou, R., Gan, Y., & Fang, F. (2021). . Intelligent Guidance Programming of Welding Robot for 3D Curved Welding Seam, *Digital Object Identifier*.

Ziegler, J. G., & Nichols, N. B. (1993). Optimum settings for automatic controllers. *Journal of Dynamic Systems, Measurement, and Control*, *115*(2B), 220–222. doi:10.1115/1.2899060

Zippia. (2022). *Best States for a Mechatronics Engineer*. Zippia. https://www.zippia.com/mechatronics-engineer-jobs/best-states/

Zippia. (2022). *Mechatronics Engineer Demographics and Statistics in the US*. Zippia. https://www.zippia.com/mechatronics-engineer-jobs/demographics/

About the Contributors

Mohamed Arezki Mellal has a Ph.D. in Mechatronics. He is a Professor at the Department of Mechanical Engineering, Faculty of Technology, M'Hamed Bougara University, Algeria. He was a research scholar at various universities. He has edited seven books. He has authored and co-authored ten book chapters. He has written over fifty-four journal and conference papers. He was also a committee member of over sixty international conferences. He serves as a regular reviewer for several SCIE journals. His research interests include mechatronics and soft computing methods for engineering problems.

* * *

Doğukan Akgöl is a research assistant at Dokuz Eylul University working on automatic control and robotics field.

Al-Fahim Ali is a Senior Lecturer in Universiti Malaysia Pahang.

Mohamed Ameedeen is an Associate Professor and Chief Information Officer (CIO) in Universiti Malaysia Pahang.

Yılmaz Arikuşu received his B.Sc degree in Electrical and Electronic Engineering from Isık University in 2015. He completed his M.Sc. degree in Electrical and Electronics Engineering from Istanbul Bilgi University in 2018. Currently, he is a Ph.D. candidate in Electrical and Electronics Engineering at Istanbul University-Cerrahpasa and he is a research assistant at the Department of Electrical Engineering at Manisa Celal Bayar University. His research interests are automatic control systems, control systems design, robust control, and fuzzy logic control.

Tufail Ahmad Bhat graduated from the Islamic University of Science and Technology, Awantipora in 2021 with a Bachelor's degree in Mechanical Engineering. His Bachelor's thesis was about Underactuated Robotics. He is now a Research and Development intern at the IHFC, IIT Delhi's Technology Innovation Hub. His main areas of interest in research include underactuated systems, robotics, and control theory.

Sampath Boopathi completed his undergraduate in Mechanical engineering and postgraduate in the field of Computer-Aided Design. He completed his Ph.D. from Anna University and his field of research includes Manufacturing and optimization. He published 45 research articles in Internationally Peer-reviewed journals, one Patent grant, and Three published patents.

Sasikymar C. completed his undergraduate in Mechanical Engineering and postgraduate in the field of Engineering Design. He completed his Ph.D. from Anna University, Chennai, Tamil Nādu, India.

Mehmet Mert Ilman is a research assistant in the Department of Mechanical Engineering and the program of Machine Theory & Dynamics (MTD) at the M. Celal Bayar University since 2017. He finished his Ph.D. in the MTD program at Dokuz Eylül University. His main research interests include industrial flexible or soft robotic systems, active vibration control or control via machine intelligence, modeling, and simulation. He is also a reviewer for scientific journals and a member of the Turkish Machine Theory Association.

Swarnashree K. is working as Junior Research Fellow in DRDL, Hyderabad. My area of interest is Aerodynamics, Missiles dynamics, CFD and propulsion system.

Majid Koul is currently Assistant Professor at the Department of Mechanical Engineering, Islamic University of Science and Technology, J&K. He graduated in Mechanical Engineering from the National Institute of Technology, Srinagar, J&K, India in 2006. He completed his PhD from IIT Delhi in 2015 after working on a joint project between Mechatronics Lab, IIT Delhi and Touch Lab, IIT Madras, sponsored by the Department of Science and Technology, Government of India. His research interests include mechatronics, design dynamics and control of robots and haptic interfaces.

Nikolay Kuznetsov is Head of the Department of "Design and Standardization in Mechanical Engineering" ISTU. Doctor of Technical Sciences, Professor. Graduated from Irkutsk Polytechnic Institute with a degree in Aircraft Engineering. After working at the Irkutsk branch of the NIAT of the Ministry of Aviation Industry, he studied full-time postgraduate studies at the Irkutsk Polytechnic Institute, majoring in theoretical mechanics. In 1981 he defended his PhD thesis on the dynamics of industrial robots and until 1984 worked at the Polytechnic Institute as first an assistant and then an associate professor of the Department of Theoretical Mechanics. In 1984-1996 he worked at the Irkutsk Scientific Center of the Siberian Branch of the USSR Academy of Sciences (RAS), having worked his way up from senior researcher to head of the mechatronics laboratory. In 1996, he moved to work at Irkutsk State Technical University as an associate professor of the Department of Equipment and Automation of Mechanical Engineering. In 2006 he defended his doctoral dissertation on the topic "Methods of reducing dynamic errors of controlled machines with elastic links based on the concept of additional links". In May 2008 he was elected to the position of head of the Department of "Design and Standardization in Mechanical Engineering". N.K. Kuznetsov's scientific and pedagogical activity is connected with the solution of a number of modern scientific and technical problems of machine dynamics, robotics and automation of machine-building production. He developed theoretical foundations and practical recommendations for the creation of systems for active damping of elastic vibrations of robot actuators, developed a method of dynamic synthesis of controlled machines with elastic links based on the application of the ideas of purposeful mechanics in combination with the potential capabilities of software motion control systems, proposed a number of original technical solutions for automation of technological processes. Under his leadership and with his direct participation, the first robotic complexes were created and put into production in the city of Irkutsk, which are also one of the first in the country.

Artem Mironov is a specialist in the field of modern applied mechanics, actively working in the direction of development methods of structural mathematical modeling based on the use of the methodology of system analysis for solving problems of dynamics of machines for various purposes. Author of more than 70 scientific papers and 20 inventions.

Varun Mohan is a PhD student in Universiti Malaysia Pahang.

Sureshkumar Myilsamy completed his undergraduate in Mechanical Engineering and postgraduate in the field of Engineering Design. He completed his Ph.D. from Anna University, Chennai, Tamil Nādu, India.

Ahalya N. is working as an associate professor in biotechnology department of MSRIT. She is interested in subjects related to environmental biotechnology, microbiology, and nanotechnology.

Sanjeev Kumar R. is presently working as a Professor in the Department of Mechanical Engineering, Swarnandhra College of Engineering & Technology, Andhra Pradesh. He received Bachelor of Engineering in Mechanical Engineering from Anna University, Chennai in 2008. He secured M.E degree in CAD/CAM from Anna University in 2010. He received Ph.D. degree in Mechanical Engineering from Anna University, Chennai in 2018. He is having 12 years of professional teaching experience. He has received a recognized Research Supervisor order for guiding Ph.D scholar in Anna University, Chennai. He has published 17 research papers in the International Journals, 3 book chapters in Elsevier and 2 patents. He is a Reviewer in Journal of Industrial & Production Engineering & other reputed Journals. He is a life member of ISTE and IAENGG. His area of interest is Production Scheduling, Optimization Techniques.

Aditya Singh is a Civil Engineer, Independent Researcher, Ad Hoc Reviewer in 2 American International Journals (IJSSMET and IJRLEDM) and Alumni Mentor(LPU). He had completed his regular Bachelor of Technology degree in Civil Engineering from Lovely Professional University, India, in 2020. He had 8 publications under his name, in which 5 of them are published in national and international journals in the years 2018 and 2019, one as a book chapter in an international book for unpaid publication in USA in November 2021 and two international conference papers in (Springer Nature) LNNS Vol 371 and Vol 569 conference proceedings in 2022. Out of his 8 publications, he is a single author in 5 of them, whereas in one of them he is the main author and in other two he is the second author. He had also worked as a Peer Reviewer or Sub Reviewer for 37 times so far in international conferences and international books.

Duong Vu received PhD degree in Mechanics from Saint Petersburg University, Russia in 1993. He is currently a Vice Dean of the Technology Engineering School, Duy Tan University, a Prof on Faculty of Mechanical Engineering. He was also a Dean of Vocational School of Duy Tan University. His interests are in welding technology, Mechatronics and entrepreneurship, professional starter education, Manufacturing and Mechanical Engineering.

Şahin Yavuz is an experienced Associate Professor with a demonstrated history of working in the higher education industry. Skilled in Mathematical Modeling, Matlab, Programmable Logic Controller (PLC), Data Analysis, and Programming. Strong research professional with a Doctor of Philosophy (PhD) focused in Mechanical Engineering from Dokuz Eylul University.

Index

I

L

M

N

O

P

R

S

T

U

V

W

www.ingramcontent.com/pod-product-compliance
Lightning Source LLC
LaVergne TN
LVHW080859300926
849162LV00003B/478

* 9 7 8 1 6 6 8 4 5 8 8 7 7 *